Synthetic and Biological-Derived Hydroxyapatite Implant Coatings

Synthetic and Biological-Derived Hydroxyapatite Implant Coatings

Editors

Liviu Duta
Faik Nüzhet Oktar

MDPI

Basel • Beijing • Wuhan • Barcelona • Belgrade • Novi Sad • Cluj • Manchester

Editors
Liviu Duta
National Institute for Lasers,
Plasma and Radiation Physics (NILPRP)
Magurele
Romania

Faik Nüzhet Oktar
Marmara University
Istanbul
Turkey

Editorial Office
MDPI
St. Alban-Anlage 66
4052 Basel, Switzerland

This is a reprint of articles from the Special Issue published online in the open access journal *Coatings* (ISSN 2079-6412) (available at: https://www.mdpi.com/journal/coatings/special_issues/hydroxyapatite).

For citation purposes, cite each article independently as indicated on the article page online and as indicated below:

Lastname, A.A.; Lastname, B.B. Article Title. *Journal Name* **Year**, *Volume Number*, Page Range.

ISBN 978-3-7258-0457-3 (Hbk)
ISBN 978-3-7258-0458-0 (PDF)
doi.org/10.3390/books978-3-7258-0458-0

Cover image courtesy of Liviu Duta

Contents

About the Editors

Liviu Duta

Liviu Duta successfully completed his master's studies in Biomaterials in 2008 and his Ph.D. in Solid State Physics in 2013. Since 2024, he has held the position of Senior Scientific Researcher I-st degree in the "Laser-Surface-Plasma Interactions" (LSPI) laboratory, Lasers Department, National Institute for Lasers, Plasma and Radiation Physics, Magurele, Romania. He has over 17 years of experience in the field of laser processing and the characterization of various types of materials. Liviu Duta is recognized as one of the pioneers of synthesis via the pulsed laser deposition (PLD) technique of simple and doped hydroxyapatite coatings derived from biological resources. He has demonstrated his leadership and involvement in the academic community by organizing several international conferences. Additionally, he actively contributes as a reviewer for reputable ISI-ranked journals and is a member of a number of editorial boards. He has served as a guest editor for seven Special Issues and is a member of the Scientific Committee of the Doctorate School in Physics. Since 2009, Liviu Duta has taken on the responsibility of coordinating more than 10 international specialization stages for students from French universities. His current research interests include thin-film deposition (via PLD and matrix-assisted pulsed laser evaporation techniques), biomaterials, hard and protective coatings, advanced characterization methods, and the exploration of natural-origin calcium phosphates as sustainable biofunctional coatings for medical applications. Additionally, he is involved in research related to biomimetic metallic implants. Liviu Duta has made significant contributions to the field with more than 60 published scientific papers, 3 book chapters, 4 editorials, 1 eBook, 2 awarded invention Patents, and 6 international prizes. Moreover, he has over 1400 citations, corresponding to an H-index of 21 (according to Google Scholar).

Faik Nüzhet Oktar

Faik Oktar serves as a dedicated educator within the Department of Bioengineering at Marmara University in Turkey. Furthermore, he remains actively engaged as an educator at the Center for Nanotechnology and Biomaterials Research. Faik Oktar is a distinguished professor with an H-index of 37, showcasing his expertise in nanotechnology and biomaterials. His significant contributions are evident through numerous impactful multidisciplinary publications in his respective field. Having earned his Ph.D. from Bogazici University, Turkey, and possessing extensive research experience, he has made significant contributions to the field, particularly through groundbreaking work on biomaterials. He has authored more than 110 research articles published in SCI-Exp journals, demonstrating a high number of citations for his works, and has contributed to numerous book chapters as well. He has actively participated in over 100 national and international congresses, showcasing his distinguished work in the field. Additionally, he has provided supervision for numerous students' theses. To summarize, his interdisciplinary approach and commitment to fostering innovative solutions establish him as a respected leader in the academic world.

coatings

Editorial

Synthetic and Biological-Derived Hydroxyapatite Implant Coatings

Liviu Duta [1,*] and Faik Nüzhet Oktar [2,3]

1 National Institute for Lasers, Plasma and Radiation Physics, 077125 Magurele, Romania
2 Department of Bioengineering, Faculty of Engineering, University of Marmara, Istanbul 34722, Turkey; foktar@marmara.edu.tr
3 Advanced Nanomaterials Research Laboratory (ANRL), University of Marmara, Istanbul 34722, Turkey
* Correspondence: liviu.duta@inflpr.ro; Tel.: +40-(0)-214574450 (ext. 2023)

Bone and joint defects or diseases, coupled with edentation, pose challenging and commonly encountered clinical issues with a significant incidence in the medical domain. These conditions are primarily attributed to factors such as aging, trauma, infections, tumor resection, or congenital/hereditary disorders. It is important to acknowledge that the prevalence of these medical issues is anticipated to rise further due to the global increase in life expectancy [1]. Currently, orthopedic and dental implants predominantly utilize materials crafted from titanium (Ti) and its medical-grade alloys, owing to their exceptional performance attributes [2]. In recent years, there has been a growing interest among researchers in biodegradable materials, specifically magnesium and its alloys [3]. The need for a secondary surgical intervention to remove the implant, once the tissue has adequately healed, is eliminated when employing biodegradable implants. This stands as one of the most notable advantages associated with this type of devices.

A significant challenge in the biomedical field is the development of implants that can seamlessly integrate into the living body [4]. Despite ongoing efforts to enhance the corrosion resistance (an electrochemical process involving both reduction and oxidation reactions) of metallic materials, the underlying issue continues to be persistent. A viable approach to regulating the corrosion process and inhibiting direct interaction between body fluids and implants involves the use of coatings, which function as protective layers at the human body ("host")–implant interface [5]. To optimize this interface, the surface of the metallic substrate is typically coated with a bioceramic material. The objective of these coatings is to enhance fixation and expedite the osseointegration rate at the living tissue–implant interface, thereby improving its long-term functionality. Currently, the "gold" standard material utilized for coating implants is hydroxyapatite (HA). HA stands as one of the most extensively researched calcium phosphates (CaPs) for osseointegrative applications, owing to its exceptional biocompatibility [6], high biomineralization capacity, thermodynamic stability in body fluid environments [7], good osteoconductivity [8], and chemical–structural similarity to the inorganic component of human bone tissues [9]. HA can be either synthetically produced or derived from natural resources. It is important to highlight that synthetic HA is typically produced using intricate and chemically demanding protocols [1]. Therefore, an alternative and sustainable approach for HA production involves extracting it from abundant biological resources, such as animal bones resulting from the food-processing industry, fish discards, eggshells, sea shells, corals, and so on) [10,11]. Studies have reported that biological-derived HA (BHA), when compared to synthetic HA, comprises a variety of trace elements (e.g., Na, Mg, Sr, and K) that play specific bio-functional roles [12]. Additionally, BHA exhibits enhanced mechanical performance [13], a more dynamic response to the environment, tends to elicit less intense inflammatory reactions, and has demonstrated excellent biocompatibility, bioactivity, and osseoconduction characteristics [14].

Citation: Duta, L.; Oktar, F.N. Synthetic and Biological-Derived Hydroxyapatite Implant Coatings. *Coatings* **2024**, *14*, 39. https://doi.org/10.3390/coatings14010039

Received: 17 December 2023
Revised: 21 December 2023
Accepted: 24 December 2023
Published: 27 December 2023

It is worth noting that HA constitutes the primary inorganic component of bone tissue, featuring a structure with a unique affinity that permits a broad range of substitutions and ion doping. This resemblance to the composition found in natural, healthy bones includes elements such as Mg, Si, Zn, Sr, Li, and CO_3^{2-}. The objective is to foster bone remodeling, exhibit antibacterial activity, and promote enhanced bio-integration.

Among the various surface modification methods, the plasma spray technique is commonly employed in commercial applications to coat implants with bioactive ceramics, particularly HA coatings. Nevertheless, issues have arisen concerning the poor clinical performance of traditional plasma-sprayed coatings. Hence, within this Special Issue, various techniques have been introduced as alternative methods to plasma spray for coating metallic implants across diverse medical applications. These methods encompass, but are not limited to, pulsed laser deposition (PLD), magnetron sputtering (MS), electrophoretic deposition, dip-coating, hydrothermal treatment, and biomimetic coating [1,2,13,15–24].

The topic of interest for this Special Issue is, therefore, dedicated to the synthesis and characterization techniques of HA-based coatings. Both synthetic and naturally derived HA materials were considered. One should emphasize that a special focus was dedicated to the preparation of bioactive and biodegradable HA-based biomaterials with tunable properties. These biomaterials are intended for applications in bone repair and regeneration, tissue engineering, orthopedics and biosensing, dental implants, and in vivo applications.

Author Contributions: Conceptualization, L.D.; writing—original draft preparation, L.D.; writing—review and editing, L.D. and F.N.O. All authors have read and agreed to the published version of the manuscript.

Funding: L.D. acknowledges the support from the Romanian Ministry of Research, Innovation and Digitization under the Romanian National Core Program LAPLAS VII—contract no. 30N/2023.

Institutional Review Board Statement: Not applicable.

Informed Consent Statement: Not applicable.

Acknowledgments: L.D. and F.N.O. thank all the authors for their contributions to this Special Issue, entitled "Synthetic and Biological-Derived Hydroxyapatite Implant Coatings", and the editorial staff of the journal *Coatings*.

Conflicts of Interest: The authors declare no conflicts of interest.

References

1. Duta, L.; Stan, G.E.; Popescu-Pelin, G.; Zgura, I.; Anastasescu, M.; Oktar, F.N. Influence of Post-Deposition Thermal Treatments on the Morpho-Structural, and Bonding Strength Characteristics of Lithium-Doped Biological-Derived Hydroxyapatite Coatings. *Coatings* **2022**, *12*, 1883. [CrossRef]
2. Kylychbekov, S.; Allamyradov, Y.; Khuzhakulov, Z.; Majidov, I.; Banga, S.; ben Yosef, J.; Duta, L.; Er, A.O. Bioactivity and Mechanical Properties of Hydroxyapatite on Ti6Al4V and Si(100) Surfaces by Pulsed Laser Deposition. *Coatings* **2023**, *13*, 1681. [CrossRef]
3. Vranceanu, D.M.; Ionescu, I.C.; Ungureanu, E.; Cojocaru, M.O.; Vladescu, A.; Cotrut, C.M. Magnesium Doped Hydroxyapatite-Based Coatings Obtained by Pulsed Galvanostatic Electrochemical Deposition with Adjustable Electrochemical Behavior. *Coatings* **2020**, *10*, 727. [CrossRef]
4. Mariani, E.; Lisignoli, G.; Borzì, R.M.; Pulsatelli, L. Biomaterials: Foreign Bodies or Tuners for the Immune Response? *Int. J. Mol. Sci.* **2019**, *20*, 636. [CrossRef] [PubMed]
5. Dong, J.; Lin, T.; Shao, H.; Wang, H.; Wang, X.; Song, K.; Li, Q. Advances in degradation behavior of biomedical magnesium alloys: A review. *J. Alloys Compd.* **2022**, *908*, 164600. [CrossRef]
6. Yang, H.; Xia, K.; Wang, T.; Niu, J.; Song, Y.; Xiong, Z.; Zheng, K.; Wei, S.; Lu, W. Growth, in vitro biodegradation and cytocompatibility properties of nano-hydroxyapatite coatings on biodegradable magnesium alloys. *J. Alloys Compd.* **2016**, *672*, 366–373. [CrossRef]
7. Kim, H.; Camata, R.P.; Vohra, Y.K.; Lacefield, W.R. Control of phase composition in hydroxyapatite/tetracalcium phosphate biphasic thin coatings for biomedical applications. *J. Mater. Sci. Mater. Med.* **2005**, *16*, 961–966. [CrossRef]
8. Vladescu, A.; Pârâu, A.; Pana, I.; Cotrut, C.M.; Constantin, L.R.; Braic, V.; Vrânceanu, D.M. In vitro activity assays of sputtered HAp coatings with SiC addition in various simulated biological fluids. *Coatings* **2019**, *9*, 389. [CrossRef]
9. Graziani, G.; Boi, M.; Bianchi, M. A Review on ionic substitutions in hydroxyapatite thin films: Towards complete biomimetism. *Coatings* **2018**, *8*, 269. [CrossRef]

10. Oktar, F.N.; Unal, S.; Gunduz, O.; Nissan, B.B.; Macha, I.J.; Akyol, S.; Duta, L.; Ekren, N.; Altan, E.; Yetmez, M. Marine-derived bioceramics for orthopedic, reconstructive and dental surgery applications. *J. Aust. Ceram. Soc.* **2023**, *59*, 57–81. [CrossRef]
11. Dorcioman, G.; Grumezescu, V.; Stan, G.E.; Chifiriuc, M.C.; Gradisteanu, G.P.; Miculescu, F.; Matei, E.; Popescu-Pelin, G.; Zgura, I.; Craciun, V.; et al. Hydroxyapatite Thin Films of Marine Origin as Sustainable Candidates for Dental Implants. *Pharmaceutics* **2023**, *15*, 1294. [CrossRef] [PubMed]
12. Wang, B.; Feng, C.; Liu, Y.; Mi, F.; Dong, J. Recent advances in biofunctional guided bone regeneration materials for repairing defective alveolar and maxillofacial bone: A review. *Jpn. Dent. Sci. Rev.* **2022**, *58*, 233–248. [CrossRef] [PubMed]
13. Duta, L.; Oktar, F.; Stan, G.; Popescu-Pelin, G.; Serban, N.; Luculescu, C.; Mihailescu, I. Novel doped hydroxyapatite thin films obtained by pulsed laser deposition. *Appl. Surf. Sci.* **2013**, *265*, 41–49. [CrossRef]
14. Duta, L.; Neamtu, J.; Melinte, R.P.; Zureigat, O.A.; Popescu-Pelin, G.; Chioibasu, D.; Oktar, F.N.; Popescu, A.C. In Vivo Assessment of Bone Enhancement in the Case of 3D-Printed Implants Functionalized with Lithium-Doped Biological-Derived Hydroxyapatite Coatings: A Preliminary Study on Rabbits. *Coatings* **2020**, *10*, 992. [CrossRef]
15. Aravindakshan, R.; Saju, K.K.; Aruvathottil Rajan, R. Investigation into Effect of Natural Shellac on the Bonding Strength of Magnesium Substituted Hydroxyapatite Coatings Developed on Ti6Al4V Substrates. *Coatings* **2021**, *11*, 933. [CrossRef]
16. Al-Arjan, W.S.; Aslam Khan, M.U.; Nazir, S.; Abd Razak, S.I.; Abdul Kadir, M.R. Development of Arabinoxylan-Reinforced Apple Pectin/Graphene Oxide/Nano-Hydroxyapatite Based Nanocomposite Scaffolds with Controlled Release of Drug for Bone Tissue Engineering: In-Vitro Evaluation of Biocompatibility and Cytotoxicity against MC3T3-E1. *Coatings* **2020**, *10*, 1120. [CrossRef]
17. Hsu, H.-C.; Wu, S.-C.; Lin, C.-Y.; Ho, W.-F. Characterization of Hydroxyapatite/Chitosan Composite Coating Obtained from Crab Shells on Low-Modulus Ti–25Nb–8Sn Alloy through Hydrothermal Treatment. *Coatings* **2023**, *13*, 228. [CrossRef]
18. Ishizaki, C.; Yabutsuka, T.; Takai, S. Development of Apatite Nuclei Precipitated Carbon Nanotube-Polyether Ether Ketone Composite with Biological and Electrical Properties. *Coatings* **2020**, *10*, 191. [CrossRef]
19. Florian, P.E.; Duta, L.; Grumezescu, V.; Popescu-Pelin, G.; Popescu, A.C.; Oktar, F.N.; Evans, R.W.; Roseanu Constantinescu, A. Lithium-Doped Biological-Derived Hydroxyapatite Coatings Sustain In Vitro Differentiation of Human Primary Mesenchymal Stem Cells to Osteoblasts. *Coatings* **2019**, *9*, 781. [CrossRef]
20. Parau, A.C.; Dinu, M.; Cotrut, C.M.; Pana, I.; Vranceanu, D.M.; Constantin, L.R.; Serratore, G.; Marinescu, I.M.; Vitelaru, C.; Ambrogio, G.; et al. Effect of Deposition Temperature on the Structure, Mechanical, Electrochemical Evaluation, Degradation Rate and Peptides Adhesion of Mg and Si-Doped Hydroxyapatite Deposited on AZ31B Alloy. *Coatings* **2023**, *13*, 591. [CrossRef]
21. Predoi, D.; Iconaru, S.L.; Predoi, M.V. Fabrication of Silver- and Zinc-Doped Hydroxyapatite Coatings for Enhancing Antimicrobial Effect. *Coatings* **2020**, *10*, 905. [CrossRef]
22. Yu, H.-N.; Hsu, H.-C.; Wu, S.-C.; Hsu, C.-W.; Hsu, S.-K.; Ho, W.-F. Characterization of Nano-Scale Hydroxyapatite Coating Synthesized from Eggshells Through Hydrothermal Reaction on Commercially Pure Titanium. *Coatings* **2020**, *10*, 112. [CrossRef]
23. Duta, L.; Popescu, A.C. Current Status on Pulsed Laser Deposition of Coatings from Animal-Origin Calcium Phosphate Sources. *Coatings* **2019**, *9*, 335. [CrossRef]
24. Duta, L. In Vivo Assessment of Synthetic and Biological-Derived Calcium Phosphate-Based Coatings Fabricated by Pulsed Laser Deposition: A Review. *Coatings* **2021**, *11*, 99. [CrossRef]

coatings

Article

Influence of Post-Deposition Thermal Treatments on the Morpho-Structural, and Bonding Strength Characteristics of Lithium-Doped Biological-Derived Hydroxyapatite Coatings

L. Duta [1,*], G. E. Stan [2], G. Popescu-Pelin [1], I. Zgura [2], M. Anastasescu [3] and F. N. Oktar [4,5]

1 National Institute for Lasers, Plasma and Radiation Physics, 077125 Magurele, Romania
2 National Institute of Materials Physics, 077125 Magurele, Romania
3 Institute of Physical Chemistry "Ilie Murgulescu", Romanian Academy, 060021 Bucharest, Romania
4 Department of Bioengineering, Faculty of Engineering, Goztepe Campus, University of Marmara, Istanbul 34722, Turkey
5 Center for Nanotechnology & Biomaterials Research, Goztepe Campus, University of Marmara, Istanbul 34722, Turkey
* Correspondence: liviu.duta@inflpr.ro; Tel.: +40-(0)-214574450 (ext. 2023)

Abstract: We report on hydroxyapatite (HA) of biological-origin doped with lithium carbonate (LiC) and lithium phosphate (LiP) coatings synthesized by Pulsed laser deposition onto Ti6Al4V substrates fabricated by the Additive manufacturing technique. A detailed comparison from the structural, morphological, chemical composition, wetting behavior and bonding strength standpoints of as-deposited (NTT) and post-deposition thermal-treated (TT) coatings at temperatures ranging from 400 to 700 °C (i.e., TT400–TT700), was performed. Structural investigations indicated a complete crystallization of the initially amorphous HA-based layers at temperatures in excess of 500 °C. The morphological analyses emphasized the rough appearance of the film surfaces, consisting of particulates whose dimensions increased at higher temperatures, with an emphasis on LiC coatings. AFM investigations evidenced rough surfaces, with a clear tendency to increase in corrugation with the applied temperature, in the case of LiC coatings. A hydrophobic behavior was observed for control, NTT and TT400 samples, whilst a radical shift towards hydrophilicity was demonstrated for both types of structures at higher temperatures. In the case of TT500–TT700 coatings, the pull-out adherence values increased considerably compared to control ones. Taking into consideration the obtained results, the positive influence of post-deposition thermal treatments (performed at higher temperatures) on the physical–chemical and mechanical properties of LiC and LiP coatings was indicated. Alongside these improved characteristics observed at elevated temperatures, the sustainable nature of the used BioHA materials should recommend them as viable alternatives to synthetic HA ones for bone implant applications.

Keywords: biogenic hydroxyapatite; lithium doping; thermal treatment; implant coating; pull-out bonding strength; PLD

Citation: Duta, L.; Stan, G.E.; Popescu-Pelin, G.; Zgura, I.; Anastasescu, M.; Oktar, F.N. Influence of Post-Deposition Thermal Treatments on the Morpho-Structural, and Bonding Strength Characteristics of Lithium-Doped Biological-Derived Hydroxyapatite Coatings. *Coatings* **2022**, *12*, 1883. https://doi.org/10.3390/coatings12121883

Academic Editor: James Kit-Hon Tsoi

Received: 4 November 2022
Accepted: 29 November 2022
Published: 4 December 2022

1. Introduction

Bone joint defects/diseases, along with edentation, represent difficult and frequent clinical problems with high incidence in the medical field and are mainly caused by age, trauma, infections, tumor resection or congenital/hereditary disorders [1]. One should note that the frequency of these medical problems is foreseen to expand even more as a result of the increase in life expectancy worldwide. A testimony for an increased demand for advanced bone implants stand in global market research studies. For instance, Allied Market Research projected that the global market of orthopaedic and dental implants will experience a steep growth in the following next years, reaching a value of $12,743 million USD by 2023 [2], and of $66,636 million USD by 2025 [3].

Nowadays, the materials of choice for orthopedic and dental implants are generally fabricated from titanium (Ti) and its medical-grade alloys (e.g., Ti6Al4V, Ti6Al7Nb) [4,5]. Ti6Al4V was shown to possess important characteristics, such as high strength and fracture toughness, superior corrosion resistance, low density, excellent biocompatibility, and favorable osseointegration characteristics [6,7]. One should note here that, in the case of senior patients, the osseointegration of such implants is difficult to achieve in a short period of time and that is why bone cements are preferred [8]. Unfortunately, such an implant has a limited life-time and, considering the increase in life expectancy mentioned above, it can become a major deficit in the case of substitution surgeries. Despite the fact that, Ti6Al4V is considered highly biocompatible, there are studies from the literature which reported on both inflammatory reactions occurring around the implantation site [9,10], and the release of various chemical elements from the implants (i.e., vanadium and aluminum ions [4,9]), which can accumulate in the organs, with hard-to-predict long-term effects. Therefore, to surpass these inconveniences and to optimize the interface between the human body ("host") and the implants, a bioceramic material is usually deposited on the surface of the metallic substrate in the form of a thin film. The aim of this coating is both to accelerate the osseointegration rate, and to ensure the biomimetism of the implant, by enhancing its long-term functionality [11]. Moreover, the use of this coating as a buffer layer offers protection against the release of unwanted metallic ions from the implants. In the latter case, crystalline coatings, with reduced corrosion, are preferred. Nowadays, the "gold" standard material used to coat implants is hydroxyapatite (HA). HA is one of the most investigated calcium phosphates (CaPs) for osseointegrative applications due to its excellent biocompatibility, high biomineralization capacity, controlled degradation speed, and good osteoconductivity [12]. Synthetic HA is generally produced by chemical laborious protocols. Thus, one alternative, sustainable solution to produce natural HA is to extract it from abundant biological resources [13,14] (e.g., bones resulted from the food-processing industry). It was reported that animal bone-derived HA (further denoted as BHA), in comparison to synthetic HA, contains an array of trace-elements (e.g., Na, Mg, Sr, and K), with definite bio-functional roles [15], as well as an improved mechanical performance [16]. In addition, BHA demonstrated a high metabolic activity, a much more dynamic response to the environment [17], and less intense inflammatory reactions [18].

Pulsed laser deposition (PLD) was delineated as a promising technology for the synthesis of bioceramic layers [19–21], due to good control over film stoichiometry and thickness, as well as to its ability to yield rough and well-adhered films [22]. Post-deposition thermal treatments performed on the synthesized layers are also to be mentioned here. Their role is both to restore the stoichiometry and to improve the overall crystalline status of the fabricated coatings [23].

Nowadays, different solutions to further improve the mechanical and biological properties of the fabricated HA films are being investigated. One such strategy is doping with different concentrations of reagents [24], among which Lithium (Li) plays a key-role. Thus, Li is a metal with important therapeutic properties, which range from the treatment of bipolar disorders to antibacterial, anticancer, antiviral, and pro-regenerative effects [25,26]. Its mechanism of action was shown to be extremely complex [25,27]. It was also reported that, when Li is substituted into the HA structure, it plays a double role: (i) it could induce a decrease in the solubility of HA [28,29], without interfering on biocompatibility [30], and (ii) an enhancement of mechanical properties, without altering the HA structure [23].

In this respect, our previous PLD studies suggested that the Li addition to BHA can further improve the adhesion and differentiation of human mesenchymal stem cells to osteoblasts lineage, as compared to the undoped BHA coatings [31,32]. In the framework of the current study, we therefore aimed to survey if a fine tuning of the physical–chemical and surface energy characteristics (with direct consequences on the bio-functional response), as well as of the bonding strength of Li-doped BHA coatings could be further made possible via post-deposition annealing treatments performed at different temperatures (in the range of 400–700 °C) and ambient (i.e., air and water-vapors enriched atmosphere). To the best of our knowledge, this is the first report on the influence of post-deposition thermal treatments on the structural, morphological, compositional, wetting behavior and bonding strength characteristics of bone-derived HA coatings synthesized by the PLD technique.

2. Materials and Methods

2.1. Powder Preparation

The BHA powders were obtained from bovine femoral bones, which were subjected to a veterinary control before usage. The extremities of bones were cut off and only the femoral shafts were further used. The bone marrow along with all the other soft tissue residues were removed from the bones. The femoral shafts were cut into small pieces and washed thoroughly using distilled water. They were then introduced for two weeks into an alkaline sodium hypochlorite solution (1%), for deproteinization. After washing and drying, all the bone pieces were calcined at 850 °C for 4 h in air (using an electric laboratory oven, Protherm model, Alserteknik Inc., Istanbul, Turkey), for the complete elimination of all organic bone components. The as-obtained bone pieces were first crushed by grinding and then introduced into a ball mill to obtain fine powders, with particles of submicronic dimensions. Next, all powders were calcined again at 850 °C for 4 h in air, with a heating rate of 4 °C/min. The combined chemical inactivation treatment and thermal processing at a high temperature eliminates all risks of disease transmission [33].

The BHA powders were prepared in accordance with the EU settlement No. 722/2012 and ISO No. 22442-1:2015, which regulate the management of health security issues when working with products derived from tissues of animal origin.

Li (1 wt.%) was incorporated into BHA by using two source powders: Li_2CO_3 and Li_3PO_4 (Sigma Aldrich, St. Louis, MI, USA). Batches of Li-BHA of ~40 g were prepared by powder mechanical mixing using a high energy Retsch S 100 ball mill (200 rot/min, for ~15 min). To avoid any possible mixing with reused powders, virgin powders were always used.

2.2. Target Preparation

Amounts of 3 g of Li_2CO_3/BHA (further denoted as LiC) and Li_3PO_4/BHA (further denoted as LiP) powder mixtures were introduced into a stainless-steel mold (Specac Ltd., Orpington, UK), with a diameter of 2 cm. The LiC and LiP mixes were next compacted by pressing at ~6 MPa. Pellets with a thickness of ~0.5 cm were thus obtained. Further, the powder pellets were sintered into well-hardened, crack-free, PLD targets at 700 °C/4 h (heating rate 5 °C/min, natural cooling down). One should note here that sintering at temperatures higher than 700 °C was not possible owing to the low decomposition temperatures of Li_2CO_3.

2.3. Substrates Preparation

The Ti6Al4V substrates were fabricated by additive manufacturing, starting from a micro-sized powder (LPW Technology, Widness, UK), and using the experimental assembly presented in Ref. [34]. For the powder delivery, an automated powder feeder (Trumpf, Ditzingen, Germany) was used, connected through hoses (6 mm in diameter) to a nozzle with three flow channels, mounted on a robotic system TruLaser Robot 5020 (Trumph, Ditzingen, Germany), made of a Kr30HA robot (Kuka Robotics, Angsburg, Germany) with six movement axes. The translation speed was set at 1 m/min. The powder was transported to the robot using a He–Ar mixture (1:3 ratio) gas, with a flow rate of 3 g/min. The Gaussian laser beam used to melt the powder had a selected peak power of 1 kW and a diameter of the surface focused spot size of 600 μm. The beam was supplied through an optical fiber and generated by a Yb:YAG source (λ = 1030 nm, continuous wave), model TruDisk 3001 (Trumpf, Ditzingen, Germany). The technical drawing of the Ti implant substrates was designed in the graphical engineering software SolidWorks® (Dassault Systems, Vélizy-Villacoublay, France) and imported in TruTops Cell® (Trumpf, Ditzingen, Germany), which is a movement code generator for the robotic arm. Using these optimized parameters, a bulk shape was printed and further sliced using a disk cutting machine, model Brillant 200 (ATM, Mammelzen, Germany). Individual Ti6Al4V substrates, with thicknesses of ~3 mm, were therefore obtained.

2.4. PLD Coatings

All PLD experiments were performed inside a stainless-steel deposition chamber, using a pulsed UV KrF* COMPexPro 205F excimer laser source (λ = 248 nm, $\tau_{FWHM} \leq 25$ ns) (Coherent, Santa Clara, CA, USA), operated at a frequency repetition rate of 10 Hz.

The ablated material was transferred and collected both onto Ti6Al4V substrates (0.6×0.6) cm^2 and double-side polished Si (100) wafers of (1×1) cm^2 (Medapteh Plus Cert SRL, Magurele, Romania). They were placed parallel to the target surface, at a separation distance of 5 cm. Prior to introduction inside the deposition chamber, all substrates were cleaned in an ultrasonic bath Elmasonic X-tra 30H (Elma Schmidbauer GmbH, Singen, Germany), according to a protocol described in Ref. [32]. The Si wafers (which are easy to cleave) were used both for the cross-sectional scanning electron microscopy (SEM) analyses to infer the PLD films thickness, and AFM investigations. To ensure the best possible adherence of the coatings [35], the Ti and Si substrates were plasma cleaned using a Diener system (Diener Electronic, Ebhausen, Germany), prior to the introduction into the reaction chamber.

The laser beam was incident at 45° onto the target's surface and the fluence was set at 4 J/cm^2, with a corresponding laser energy of 360 mJ. During deposition, all substrates were heated and maintained at a constant temperature of 500 °C, by a PID-Excel temperature controller (Excel Instruments, Mumbai, India). The heating and cooling ramps were 20 and 10 °C/min, respectively. To avoid piercing and possible modification of the surface morphology induced by the laser radiation, the targets were continuously rotated during the multi-pulse deposition process with a frequency of 0.83 Hz (50 rot/min). For the deposition of one coating, 15,000 consecutive laser pulses were applied. All depositions were performed in water vapor-enriched ambient, at a residual pressure of 3.75×10^{-1} Torr (~50 Pa).

2.5. Post-Deposition Thermal Treatments

The post-deposition thermal processing of implant-type coatings could be of particularly great importance for the engineering of their physical–chemical (e.g., crystalline quality, morphology, stoichiometry, surface energy) and mechanical properties. Consequently, in the framework of this study, the as-deposited PLD coatings were subjected to post-deposition thermal treatments at different temperatures (i.e., 400, 500 and 700 °C, respectively), in both air and water-vapor ambient, for 6 h (further denoted as TT400A–TT700A and TT400W–TT700W, respectively). For comparison reasons, batches of samples not subjected to thermal treatments (further denoted as NTT) were also used.

2.6. Physical-Chemical and Mechanical Characterization Methods

(a). The crystalline status and phase composition of LiC and LiP coatings were investigated by X-ray diffraction (XRD), in grazing incidence (GI) geometry, using a Rigaku SmartLab (3 kW) equipment, with parallel beam setting and CuK$_\alpha$ radiation (λ = 1.5406 Å). The angle of incidence (α) was 2°. The patterns were recorded in the 2θ = (10–60)° range, at an acquisition time of 1°/min.

(b). The chemical structure of LiC and LiP coatings was investigated by Fourier transform infrared (FT-IR) spectroscopy, using a Perkin Elmer BX Spectrum equipment, operated in an attenuated total reflectance (ATR) mode. A Pike-MIRacle diamond head (0.18 cm diameter) was used. All measurements were performed on Ti substrates, in the range (1600–500) cm^{-1}, with a resolution of 4 cm^{-1} and a total number of 64 scans/experiment.

(c). The surface and bottom-to-top morphologies of LiC and LiP coatings were investigated using SEM, in top- and cross-view modes, respectively. An FEI Inspect S50 electron microscope (Field Electron and Ion Company, Hillsboro, OR, USA), used in the secondary electron mode, at an acceleration voltage of 20 kV, was used. Before examination, to prevent electrical charging, all samples were coated with a thin layer (10–12) nm of Au.

(d). To better emphasize the topological surface peculiarities of LiC and LiP coatings, high resolution Atomic Force Microscopy (AFM), performed in non-contact mode on a XE-100 apparatus from Park Systems, was used. Sharp tips (PPP-NCHR type from Nanosensors™, Neuchâtel, Switzerland), having a length of 125 μm, a 30 μm width and a radius of curvature of less than 8 nm, were used for surface scanning over areas of (20 × 20) μm^2. All AFM measurements were performed at room temperature (RT). The textural (amplitude) parameters, namely average roughness (R_q), surface skewness (R_{sk}) and surface kurtosis (R_{ku}), were determined.

(e). The chemical composition of LiC and LiP samples was assessed by energy dispersive X-ray spectroscopy (EDS). For these measurements, a SiLi EDAX Inc. detector (Mahwah, NJ, USA), operated at 20 kV, was used. The elemental analyses were performed in triplicate, on different areas ((250 × 250) μm^2) of the deposited layers.

(f). The wetting properties of the LiC and LiP coatings were investigated by static (equilibrium) contact angle (CA) measurements, using a Drop Shape Analysis System, model DSA100 (Krüss GmbH, Hamburg, Germany). The samples were positioned on a flat platform, under the tip of a blunt-end, stainless steel needle, with an outer diameter of 0.5 mm. The needle was attached to a syringe pump controlled by the DSA3® PC software and used for dripping liquids with a controlled volume on the test surface and for the CA evaluation. The volume of one liquid droplet was of ~1 μL. For each sample, the measurements were performed on two different regions of the surface. The CA measurements were performed by fitting the experimental profile of the sessile drop with a second-degree polynomial or with the equation of the circle, after which the slope of the tangent to the drop at the point of intersection with the line separating the liquid–solid–vapor interface was calculated. The video camera was tilted at an angle of ~2° to the sample plane. All measurements were performed at RT, in duplicate (on two identical samples). Based on the measured CA values, the solid Surface Free Energy (SFE) was determined. For the CA and SFE measurements, two standard liquids (water and diiodomethane) were used, for which one knows: the surface tension (γ_{water} = 72.8 mN/m; $\gamma_{diiodmethane}$ = 50.8 mN/m), and the dispersion and polar components (${\gamma_{water}}^{d}$ = 21.8 mN/m, ${\gamma_{water}}^{p}$ = 51 mN/m, and ${\gamma_{diiodmethane}}^{d}$ = 48.5 mN/m, ${\gamma_{diiodmethane}}^{p}$ = 2.3 mN/m). Using the CA values obtained for the two testing liquids, both the polar and dispersion parts of the synthesized coatings, and the total SFE, was estimated by applying the Owens–Wendt method [36].

(g). The adherence of the LiC and LiP coatings to the Ti substrate was estimated by the pull-out method. A standardized instrument, PAThandy AT101 (maximum pulling force = 1 kN) from DFD Instruments® (Kristiansand, Norway), equipped with stainless steel test elements with a diameter of Φ = 0.28 cm, was used. The complete procedure was detailed elsewhere [37], and it complies with the ASTM D4541 and ISO 4624 standards. The measurements were performed on quadruplicates.

3. Results and Discussion

3.1. X-ray Diffraction

The GIXRD patterns of the LiC- and LiP-based PLD coatings, thermally processed under different conditions (air and water vapor ambient), are comparatively presented in Figure 1a,b. As expected, due to the low thickness of the deposited layers, the diffractograms were dominated by the high intensity peaks (i.e., 1 0 0, 0 0 2, and 1 0 1) of the Ti substrate (ICDD: 00-044-1294). While initially amorphous (i.e., NTT—as-deposited)—as indicated by the shallow amorphous halo situated in the angular region 2θ ≈ (23–33)°—the films became progressively more crystalline with the increase of the post-deposition temperature. Consequently, the diffraction maxima of HA (ICDD: 00-009-0432), as the sole or dominant phase of the PLD films, became incrementally more salient.

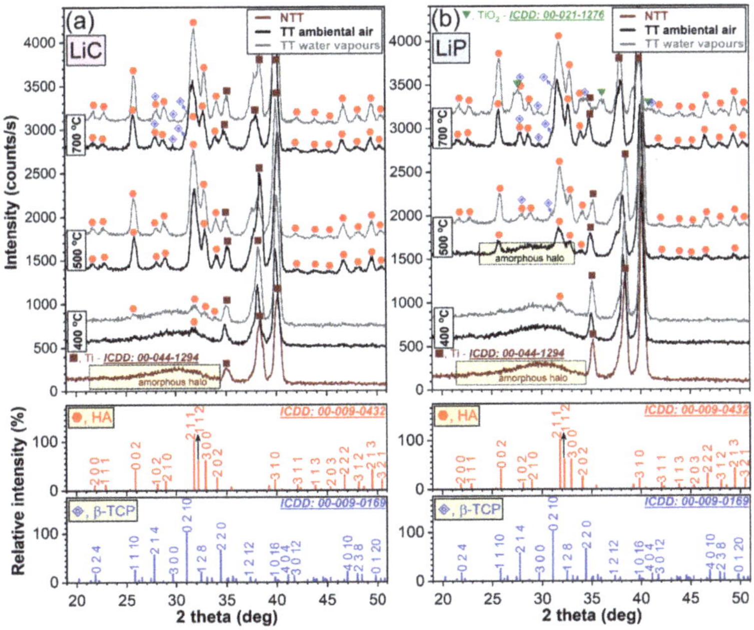

Figure 1. The GIXRD patterns of the NTT and TT (**a**) LiC- and (**b**) LiP-type coatings, in air and water vapors ambient. The reference ICDD-PDF4 files of hydroxyapatite (No. 00-009-0432) and β-tricalcium phosphate (No. 00-009-0169) are presented at the bottom of the graphs for comparison.

The LiP coatings heat-treated in air (Figure 1b) elicited a latency of crystallization with respect to the LiC ones (Figure 1a). This was suggested by the persistence of the amorphous halo (while reduced in intensity, and accompanied by defined peaks of nano-crystalline HA) up to a temperature of 500 °C. More advanced structuring processes (i.e., HA crystallization followed by its decomposition) were generally noticed for the films annealed in water-vapor ambient. The formation of the β-TCP phase (ICDD: 00-009-0169) was hinted by (i) the left-hand side asymmetry of the 2 1 1 peak of HA, followed the advent of a definite shoulder appertaining to the maximum intensity 0 2 10 peak of β-TCP, and (ii) the increase in intensity of the 1 0 2 peak of HA, as consequence of the superimposition in the same angular region of the prominent 2 1 4 maximum of β-TCP. If in the case of LiC samples (Figure 1a), the partial decomposition of HA was disclosed only after the TT performed at 700 °C in both air and water-vapor ambient, in the case of LiP samples, the first signs of HA decomposition were evidenced at a lower temperature (i.e., TT performed at 500 °C in water-vapor ambient). Furthermore, while after the TT at 700 °C in air, the HA → β-TCP transformation of the LiP coatings was still marginal, in the case of the ones heat-treated in water-vapor ambient, the weight of the β-TCP was visibly increased (Figure 1b). This was supported by both the increase of the 0 2 10 β-TCP/2 1 1 HA peaks ratio, and the supplemental easier discrimination of diffraction maxima related to the 2 1 4 and 2 2 0 lines of β-TCP. Interestingly, in the case of LiP coatings annealed at 700 °C in water-vapor ambient, the additional presence of a well-defined TiO_2-rutile phase (ICDD: 00-021-1276) was revealed. This is not unprecedented, since the formation of rutile was previously reported in the case of magnetron-sputtered HA films subjected to post-deposition heat-treatments at 750 °C [38]. It is most likely a consequence of the Ti substrate oxidation at the interface with the deposited film during the re-structuring and decomposition stages of HA (with oxygen diffusing from the film to the substrate surface).

The progressive incorporation of hydroxyl and carbonate functional groups into the HA lattice is known as a driving force for its crystallization [39,40]. Conversely, the gradual de-hydroxylation and de-carbonation of HA at higher temperatures, represents the starting point of HA restructuring, since its crystalline lattice will become faulty and unstable, and thus, susceptible to chemical reactions and decomposition [41,42].

The stronger HA → β-TCP decomposition, signaled in the case of LiP coatings annealed in water-vapor ambient at 700 °C, could be linked to a decrease in the heat-treatment efficacy (with the water vapors being more readily evaporated from the HA-based film surface at this high temperature). Furthermore, the excess of orthophosphate groups supplied by the Li source material (i.e., Li_3PO_4) used in the case of LiP samples, could act as a supplemental chemical destabilizer and decomposition adjuvant.

The crystallite size (crystalline coherence length) along the *a*- and *c*-axis of HA, was determined from the integral breaths of the 0 0 2 and 3 0 0 diffraction maxima by applying the Scherrer equation [43], after the subtraction of the instrumental breadth (using a NIST 1976b corundum standard reference material). The estimated *a*- and *c*-axis crystallite size values are plotted comparatively in Figure 2a,b.

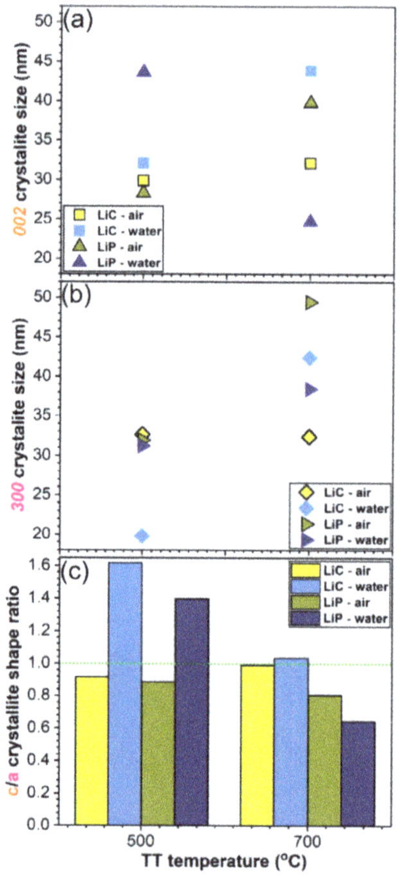

Figure 2. Average crystallite sizes determined by Scherrer equation along the (**a**) *c*-axis and (**b**) *a*-axis for the HA constituent phase of the LiC and LiP samples heat-treated in air and water-vapor ambient at 500 and 700 °C, respectively. (**c**) Crystallite shape anisotropy factors.

The *c*/*a* crystallite shape anisotropy factors are presented in Figure 2c. As anticipated, the (*a*- and *c*-axis, respectively) crystallite sizes experienced a moderate (i.e., LiC in air) or markedly increase (i.e., LiP in air, LiC in water-vapor ambient) with the heat-treatment temperature, with a single exception: LiP coatings annealed in water-vapor ambient at 700 °C. In this particular case, a strong diminution of the crystalline coherence length along the *c*-axis was recorded. This should be considered a direct consequence of the more prominent structural changes experienced by this type of coating (namely, increase of β-TCP share, and emergence of the rutile phase at the film–substrate interface). Interestingly, the *c/a* shape factors (Figure 2c) indicated that both LiC and LiP samples heat-treated in air were quasi-isotropic (ratios close to unity) irrespective of temperature, whilst the ones annealed in water-vapor ambient at 500 °C were markedly oblong. When the heat-treatment temperature is increased (to 700 °C) for the latter samples, their crystallite shape became either isotropic (for LiC) or flattened (for LiP).

3.2. Fourier Transformed Infra-Red Spectroscopy

The FTIR-ATR spectra of the NTT and TT LiC and LiP coatings are presented comparatively in Figure 3a,b. All spectra featured the characteristic vibrational modes of the orthophosphate structural units in an apatitic material [44,45]: ν_4 triply degenerated bending (~564–569 and ~590–600 cm^{-1}, respectively); ν_1 non-degenerated symmetric stretching (~941–962 cm^{-1}); and ν_3 triply degenerated asymmetric stretching (~993–1018 and ~1084–1087 cm^{-1}, respectively).

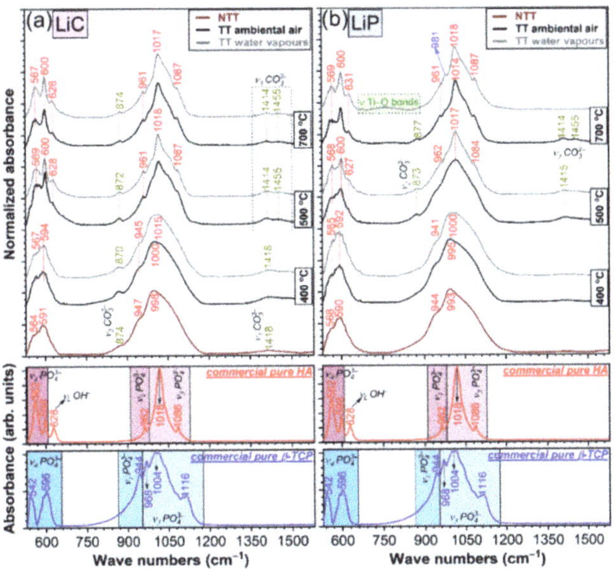

Figure 3. The FTIR-ATR spectra of the NTT and TT (**a**) LiC- and (**b**) LiP-type coatings, in air and water-vapor ambient. The reference spectra of pure (commercial) hydroxyapatite and β-tricalcium phosphate compounds are presented at the bottom of the graphs for comparison.

The broad envelopes of FTIR-ATR spectra of both LiC and LiP NTT samples suggested their decreased degree of structural order, in good agreement with the afore-presented GIXRD results. As frequently reported in literature for amorphous calcium phosphates [38,46,47], the stretching modes of the $(PO_4)^{3-}$ groups were found to be shifted to lower wave numbers with respect to the stoichiometric crystalline HA positions. Furthermore, the absence of the libration mode (ν_L) of structural OH^- groups at ~628 cm^{-1} indicated the strong de-hydroxylation of NTT LiC and LiP coatings, which is a typical characteristic for the films prepared by physical vapors deposition methods [23,48,49]. To the difference of LiP NTT coatings, in the case of LiC NTT ones, the emergence of new low intensity IR absorption bands at ~874 and ~1418 cm^{-1}, respectively, was revealed. They are associated with the ν_2 symmetric and ν_3 asymmetric stretching modes of carbonate groups [44,45], which suggests that they originated (and are incorporated) from the Li_2CO_3 additive, used as Li source material in this specific case.

Subsequent to the heat-treatment applied at 400 °C, marginal structural modifications in terms of short-range order were recorded. However, the increase of the annealing temperature up to 500 °C produced noteworthy changes. Both types of coatings elicited narrower, well-defined orthophosphate IR bands, peaking at positions close to the ones of the stoichiometric (commercial) HA. The only exception was constituted by the LiP samples heat-treated in air (thereby, in good agreement with the GIXRD data which indicated the incomplete crystallization of this specimen) (Figure 1b vs. Figure 3b). The LiC and LiP samples annealed at 500 °C featured both the libration mode of OH^- and stretching modes of $(CO_3)^{2-}$, incorporated from the TT ambient, further stressing the importance of these functional groups for the crystallization of HA. The carbonation of HA should be emphasized, as it is also characteristic for the main mineral component of bone, and furthermore, it is known to positively influence the osseointegration ability [48,50,51].

The osteoclasts dissolve more readily this type of bioceramics, leading to faster bone regeneration and remodeling [48,50].

The 700 °C heat treatment seemingly produced the complete crystallization of all type of samples, annealed in both air and water-vapor ambient. Important structural alterations were noticed in the case of LiP coatings annealed at 700 °C in water-vapor ambient: (i) two new bands arose at ~750 and ~981 cm^{-1}, respectively, associated with the vibrations of Ti–O bonds in rutile [38,52], and stretching modes of orthophosphate units in β-TCP; and (ii) complete elimination of the carbonate groups. The latter event might be linked with the HA re-structuring process and its incipient decomposition. The Ca–CO_3 bonds in HA are known to be weaker with respect to the Ca–PO_4 ones [53], and thereby, when the decomposition process starts, they will be supposedly the first ones to break apart.

3.3. Scanning Electron Microscopy

SEM micrographies corresponding to the control Ti, NTT LiC and LiP coatings and to those subjected to post-deposition thermal treatments (at different temperatures), in air and water-vapor ambient, are comparatively presented in Figures 3 and 4. SEM images were recorded at two different magnifications of 5000 and 200,000×, respectively. At smaller magnifications, for all synthesized coatings, surfaces with rough and irregular morphologies could be observed. They generally consisted of spherical, ovoidal or even elongated formations, but rounded at their extremities (Figures 4 and 5), which are known in the literature as "particulates". Their dimensions ranged from a few hundred nanometers to a few microns (Figure 6). In the case of LiC TT400A and LiP TT400W, some bigger, isolated particles were also detected. One can note that the presence of these particulates of various dimensions on the surface of the synthesized coatings and their dependence on the composition of the target are well-known characteristics of the PLD process [19]. These particulates are mainly resulting by the expulsion of the liquid phase from the surface of the target during the irradiation process, as an effect of expansive waves, hydrodynamic instabilities, or melting in the sub-surface area [54]. It can be emphasized that both the particulate sizes and their random distribution were defining factors in obtaining rough morphologies. Such type of surface is acquainted for its positive impact in medical applications (e.g., implant-type coatings), thus promoting the colonization of osteogenic cells in their microcavities, followed by their intense proliferation [19,55,56].

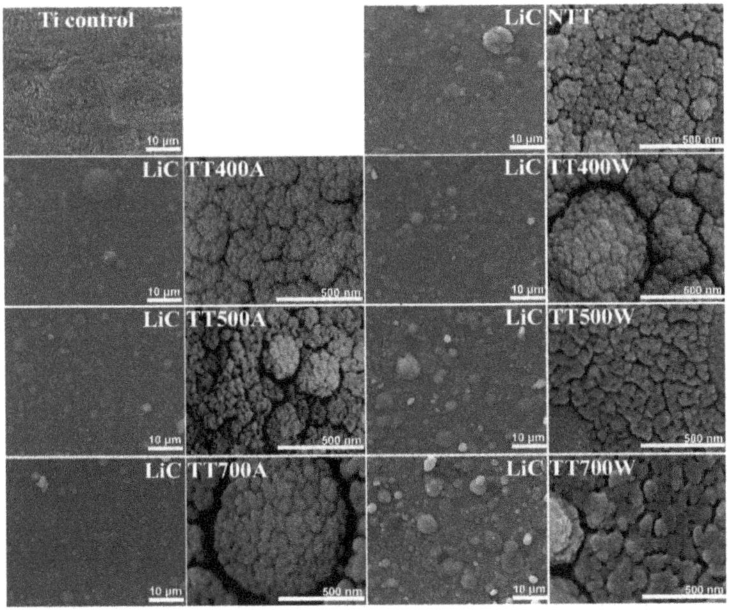

Figure 4. Top-view general and detailed SEM micrographs of control Ti, NTT LiC samples and of those subjected to post-deposition thermal treatments (at different temperatures), in air and water-vapor ambient, synthesized onto Ti substrates.

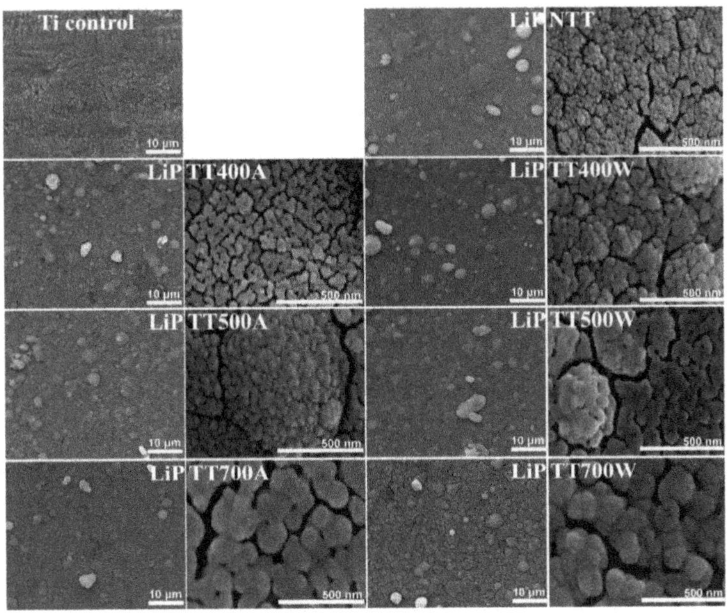

Figure 5. Top-view general and detailed SEM micrographs of control Ti, NTT LiP samples and of those subjected to post-deposition thermal treatments (at different temperatures), in air and water-vapor ambient, synthesized onto Ti substrates.

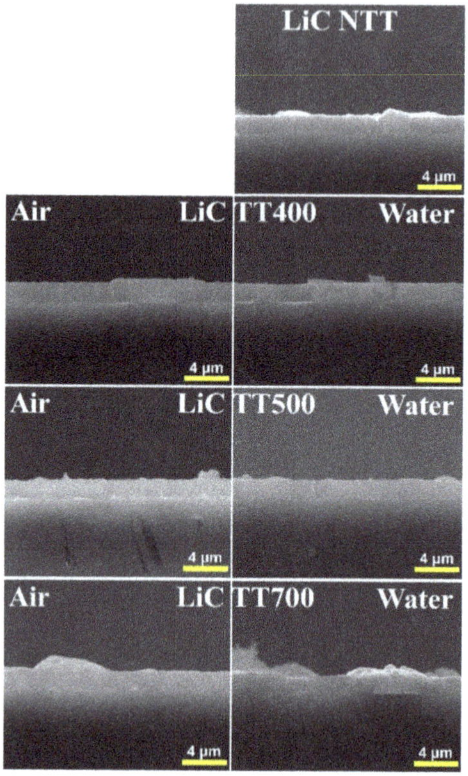

Figure 6. Cross-view SEM micrographs of control Ti, NTT LiC samples and of those subjected to post-deposition thermal treatments (at different temperatures), in air and water-vapor ambient, synthesized onto Si substrates.

Further, higher magnification SEM images revealed the appearance of some minor cracks on the surface of the coatings. They were visible in the case of NTT samples and seem to develop with the increase of the post-deposition temperatures. Their formation can be explained in terms of the release of thermal stress developed at the substrate–coating interface, because of the differences in the thermal expansion coefficients between the substrate and the ceramic films [57].

The slight widening of the cracks, which are more noticeable in the case of LiC-TT700A samples, might be attributed to the crystallization and hydroxylation of amorphous phases. The constriction of BHA coatings thus followed which widened the cracks [58].

It is important to point out that the structural integrity of the synthesized coatings was well maintained even after the thermal treatment of the coatings. Thus, no delamination phenomena, immediately after deposition or after a longer period of time (i.e., months), was observed.

Cross-sectional SEM images characteristic to NTT LiC and LiP coatings and to those subjected to post-deposition thermal treatments (at different temperatures), in air and water-vapor ambient, are comparatively presented in Figures 6 and 7. In all studied cases, generally uniform coatings with a compact microstructure were highlighted.

The average thickness values of the NTT LiC and LiP coatings and of those subjected to post-deposition thermal treatments (at different temperatures), in air and water-vapor ambient, were estimated from the cross-sectional SEM images. A slightly higher average thickness (~1.4–1.5 µm) was evidenced in the case of LiP coatings, in comparison to LiC ones (~1.1–1.2 μm). This corresponds to an average deposition rate of ~0.1 nm/pulse. Thus, in the case of LiC samples subjected to thermal treatments in air and water-vapor ambient, the highest thickness values were inferred for TT400A and TT500W coatings, whilst for NTT, TT500A/W and TT700A/W ones, comparable thickness values were deduced. For LiP samples subjected to thermal treatments in air, the highest thickness values were inferred for NTT samples, followed by a decrease for TT400 and TT500 ones, and a slight increase towards higher temperatures (i.e., TT700A coatings). In the case of LiP samples submitted to thermal treatments in water-vapor ambient, a decrease for NTT–TT400 coatings, followed by a slight increase for TT500 ones and a decrease towards higher temperatures (quite abrupt for TT700W coatings), was noticed. This trend was similar to the one observed for LiC samples subjected to the same post-deposition thermal treatments (i.e., water-vapor ambient).

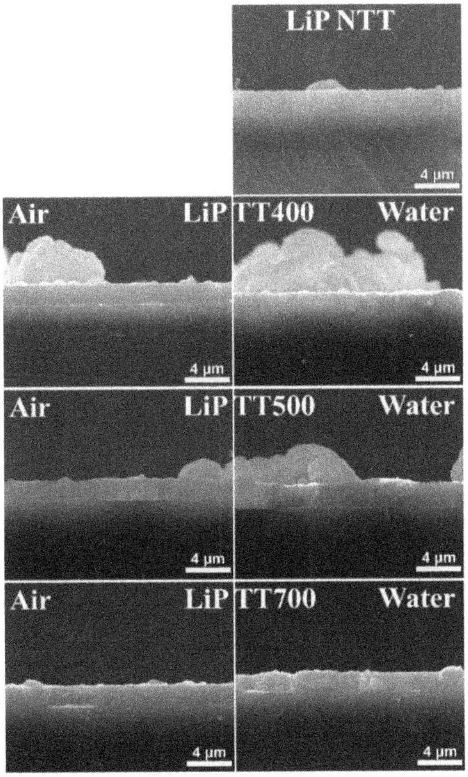

Figure 7. Cross-view SEM micrographs of control Ti, NTT LiP samples and of those subjected to post-deposition thermal treatments (at different temperatures), in air and water-vapor ambient, synthesized onto Si substrates.

3.4. Atomic Force Microscopy

Typical bi-dimensional (2D) topographic images of the control Si, NTT LiC and LiP coatings and of those subjected to post-deposition thermal treatments (at different temperatures), in air and water-vapor ambient, collected at a scanning area of (20 × 20) μm^2, are comparatively presented in Figures 10 and 11. The three-dimensional (3D) AFM images, which are topographic maps of the scanned areas, are presented in the "Enhanced ColorTM" view mode (Park Systems) to better emphasize the morphological surface peculiarities. The 2D-AFM images are presented in the "classical" view mode, with one color gradient (brown in this case) for the *z*-scale depth.

From Figures 8 and 9, morphological differences between the investigated surfaces could be observed, depending on the temperature and ambient atmosphere during thermal treatment on one hand, and on the chemical nature of the coatings, on the other hand. A common morphological characteristic for all imaged surfaces was the presence of surface small particles together with large clusters or material, in the form of protruding hills ("bumps"), with various forms. Random small pits or even surface cavities could be also observed (see the dark regions in the 2D AFM images or holes/valleys in the 3D images (more intuitive as aspect)). From a quantitative point of view, all these differences lead to the variation of the values corresponding to the amplitude parameters given in Table 1, particularly on the roughness behavior.

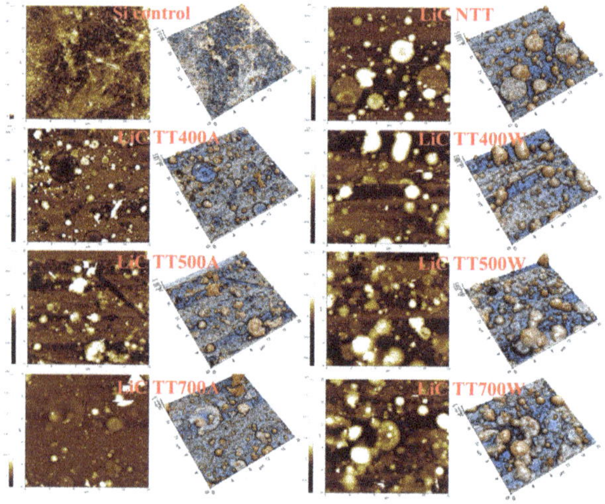

Figure 8. The 2D (classical view mode) and 3D (enhanced color view mode) AFM images of control Si, NTT LiC coatings and of those subjected to post-deposition thermal treatments (at different temperatures), in air (A–left columns) and water-vapor (W–right columns) ambient.

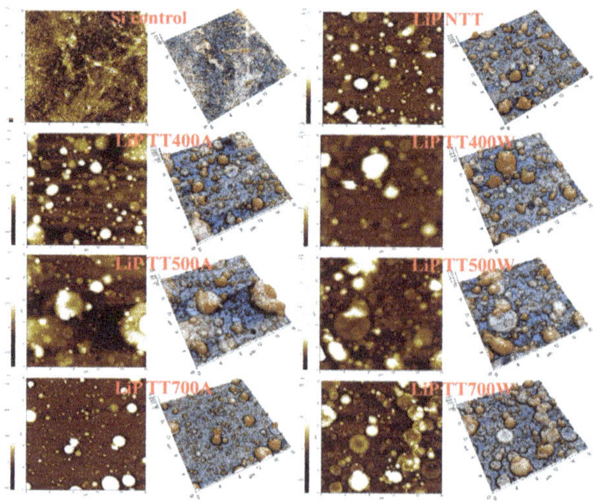

Figure 9. The 2D (classical view mode) and 3D (enhanced color view mode) AFM images of control Si, NTT LiP coatings and of those subjected to post-deposition thermal treatments (at different temperatures), in air (A–left columns) and water-vapor (W–right columns) ambient.

Table 1. Surface amplitude parameters (R_q, R_{sk} and R_{ku}) for a (20 × 20 μm^2) scanned surface area of control Si, NTT LiC and LiP coatings and of those subjected to post-deposition thermal treatments (at different temperatures), in air and water-vapor ambient.

Amplitude Parameter			**Si Control**	
R_q [nm]			0.31	
R_{sk}			−40.59	
R_{ku}			3606.86	
			LiC	
			Air	
	NTT	**TT400**	**TT500**	**TT700**
R_{sk}	−2.11	−1.56	−2.27	−6.49
R_{ku}	10.76	13.26	13.80	53.05
			Water	
	NTT	**TT400**	**TT500**	**TT700**
R_{sk}	−2.11	−1.38	−0.99	−1.16
R_{ku}	10.76	5.09	6.16	6.50
			LiP	
			Air	
	NTT	**TT400**	**TT500**	**TT700**
R_{sk}	−3.61	−1.54	−1.21	−3.38
R_{ku}	21.59	9.02	6.15	17.04
			Water	
	NTT	**TT400**	**TT500**	**TT700**
R_{sk}	−3.61	−3.98	−3.15	−1.28
R_{ku}	21.59	20.80	21.28	5.65

The peak-to-valley parameter (R_{pv}), which is the height difference between the lowest and the highest points, and the root mean square (RMS) roughness (R_q) quantitatively describe the corrugation of the scanned areas (a measure of the roughness degree), while the skewness (R_{sk}) and kurtosis (R_{ku}) are linked to the surface morphological features of the samples. Related to the distribution of the height histograms, the R_{ku} parameter mathematically describes the randomness of heights profile while R_{sk} is connected to the asymmetry of the height distribution [59].

For scientific consistency, all parameters are provided in Table 1. Apart from Table 1, the RMS roughness (R_q) of the investigated samples was plotted for different thermal treatment temperatures and ambient, together with the peak-to-valley parameter (R_{pv}) (Figure 10). It should be mentioned here that roughness is one of the important parameters that should be considered when investigating the wetting behavior of a certain solid surface [60]. Moreover, it was demonstrated that the roughness influence can prove significant for dynamic or static wetting [61].

As expected, the RMS roughness of the bare Si sample is very low ~0.3 nm. The R_{pv} of the same scanned area, of (20 × 20) μm^2, is ~58 nm, most probably due to some scratches appearing during manipulation. On the other hand, the high value of the kurtosis parameter can be associated with a Cauchy distribution of the surface events, being characteristic for a symmetric distribution with heavy tails and a single peak at the center of the distribution [62], while the negative value of the skewness parameter can be related, in our case, to the large number of superficial pores naturally formed in the SiO_2 layer on top of Si. Further on, as can be observed from Figure 10a, there is a clear trend of increases in corrugation with the temperature of the thermal treatment in the case of LiC structures. While at 400 and 500 °C, the TT under water-vapor ambient seems to be slightly higher, at 700 °C, the R_q parameters had almost equal values (~160 nm) for both air and water-vapor ambient. An abrupt increase was observed for the R_{pv} which exceeded 2100 nm, slightly higher for air (2193 nm) in comparison with water-vapor ambient (2127 nm). For the LiP structures, R_q decreases for water-vapor ambient (Inset to Figure 10b), while for the samples thermally treated in air ambient, there was first a small increase followed by a decrease. This is in good agreement with the results obtained from SEM and cross-SEM investigations.

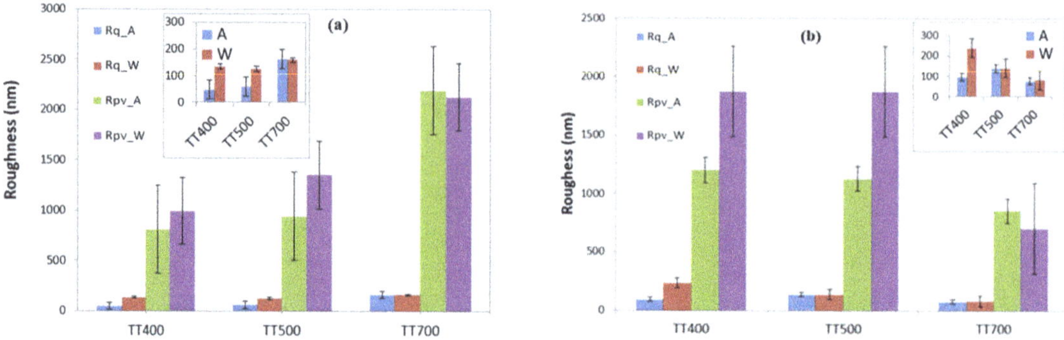

Figure 10. Roughness values inferred in the case of LiC (**a**) and LiP (**b**) coatings subjected to post-deposition thermal treatments (at different temperatures), in air and water-vapor ambient. Inset: detailed roughness values.

From the data presented in Table 1, one can observe that the R_{sk} and R_{ku} values were also very different. According to the ISO 4287/1997, the R_{sk} parameter represents the asymmetry of the profile in relation to the midline. The negative values of this parameter suggest that sharp "valleys" are predominant and that there are only a few peaks above the value that deviate from the asymmetry around the mean. It was reported in the literature that surfaces with negative R_{sk} values generally cause lower friction coefficients [63].

One should also note that the R_{ku} parameter is a measure of the disorder of heights ("hills") on the surface. For typical Gaussian height distributions, the R_{ku} value is close to three, smaller values corresponding to a wide height distribution, with low "hills" and "valleys", whilst values greater than three indicate restricted height distributions [64,65]. The relatively large values (over three) observed for kurtosis parameters (so called "Leptokurtic distribution") are related to the protruding particles formed on the surface of LiC and LiP series which prevails over the small pores and pits as well as over the small superficial particles (which determines the formation of long tails in the height histograms associated with the AFM images).

3.5. Energy Dispersive X-ray Spectroscopy

Besides the main elements which can be found in the chemical composition of HA materials (e.g., Ca, P, and O), the performed quantitative compositional analyses of NTT LiC and LiP coatings, and of those subjected to post-deposition thermal treatments (400, 500, and 700 °C, respectively), in air and water-vapor ambient, indicated also the presence of trace-elements characteristic to the bone mineral phase (e.g., Na, Mg, and Si). Such elements play a key-role in HA's functionality. Because of the low energy of characteristic radiation, located below the detection limit of EDS, the presence of light elements (i.e., Li) could not be emphasized. One should mention that Carbon (C) was present in the composition of the synthesized coatings. Its origin could be partly due to ambient pollution during various sample manipulations after deposition. However, HA containing C was demonstrated to be more osteoconductive and more resorbable in comparison to stoichiometric HA. This allows the osteoclasts to dissolve the ceramic rapidly and determines a faster osteosynthesis and bone remodeling [48,50]. In the case of animal origin materials, where the mineral component of bone is a non-stoichiometric carbonated HA material, C is a ubiquitous element. Further, no contaminants in the chemical composition of all investigated samples were detected.

It was demonstrated that the Ca/P atomic ratio plays a key-role for the cytocompatibility of HA-based coatings [66]. One should note that, in the case of LiC and LiP fabricated targets, the Ca/P molar ratio was of 1.72 ± 0.1 and 1.84 ± 0.1, respectively.

In the case of LiC and LiP coatings (Table 2), the Ca/P molar ratios showed a non-stoichiometry located in the range (1.61–1.85), which corresponds to values generally found in the case of apatites of biological origin [19]. In addition, for implant-type coatings, it was shown that the Ca/P molar ratio depends very much on the type of deposition technique [67,68].

Table 2. Ca/P atomic ratio corresponding to NTT LiC and LiP coatings, and of those subjected to post-deposition thermal treatments (at different temperatures), in air and water-vapor ambient.

Element	Composition [Mean ± SD (at.%)]			
	LiC			
	Air			
	NTT	**TT400**	**TT500**	**TT700**
P	6.7 ± 0.9	7.2 ± 0.6	6.3 ± 0.6	7.1 ± 0.3
Ca	11.9 ± 3.5	12.6 ± 1.05	10.8 ± 1.1	11.6 ± 0.7
Ca/P ratio	1.76 ± 0.3	1.73 ± 0.1	1.71 ± 0.1	1.64 ± 0.1
	Water			
	NTT	**TT400**	**TT500**	**TT700**
P	6.7 ± 0.9	6.4 ± 0.4	7.2 ± 0.8	7 ± 0.09
Ca	11.9 ± 3.5	10.7 ± 0.8	13.3 ± 3.4	11.4 ± 0.3
Ca/P ratio	1.76 ± 0.3	1.68 ± 0.1	1.85 ± 0.3	1.63 ± 0.1
	LiP			
	Air			
	NTT	**TT400**	**TT500**	**TT700**
P	7.6 ± 0.07	6.9 ± 0.6	7.2 ± 0.5	7.1 ± 0.3
Ca	12.8 ± 0.3	11.2 ± 1.1	11.9 ± 0.8	11.6 ± 0.4
Ca/P ratio	1.68 ± 0.1	1.60 ± 0.1	1.65 ± 0.1	1.62 ± 0.1
	Water			
	NTT	**TT400**	**TT500**	**TT700**
P	7.6 ± 0.07	6.5 ± 0.3	7.2 ± 0.5	7.1 ± 0.5
Ca	12.8 ± 0.3	10.5 ± 0.4	11.9 ± 0.8	12.2 ± 1.01
Ca/P ratio	1.68 ± 0.1	1.61 ± 0.1	1.66 ± 0.1	1.71 ± 0.1

A decreasing tendency of the Ca/P ratio with the applied temperature was observed (Table 2). This trend was more visible in the case of LiC samples subjected to post-deposition thermal treatments, in ambient air. In the case of LiP-based coatings thermally treated in water-vapor ambient, a Ca/P increase tendency with the temperature was noticed (1.61 < 1.66 < 1.71). Moreover, the LiP-TT400W, TT500W and TT700W coatings elicited lower Ca/P values as compared to the TT400A, TT500A and TT700A ones.

3.6. Contact Angle Measurements

The CA characteristic values, recorded both in the case of water and diiodomethane (used as testing liquids), for control (bare Ti), NTT LiC and LiP coatings, and for those subjected to post-deposition thermal treatments (at different temperatures), in air and water-vapor ambient, are comparatively presented in Figures 11 and 12. The corresponding representative images of water droplets on sample surfaces are also given in Figures 13 and 14.

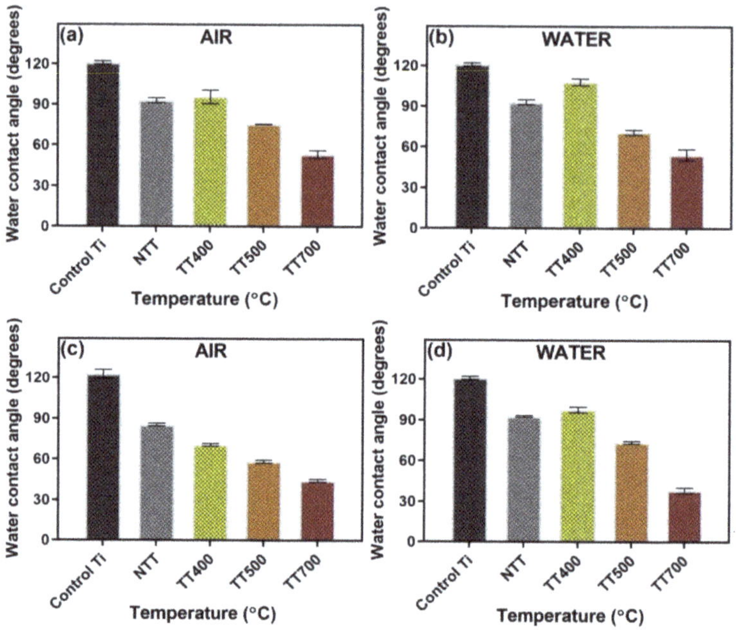

Figure 11. The values of the contact angle obtained in the case of control (Ti), NTT LiC (**a**,**b**) and LiP (**c**,**d**) coatings, and of those subjected to post-deposition thermal treatments (at different temperatures), in air (**a**,**c**) and water-vapor (**b**,**d**) ambient. Test liquid: water.

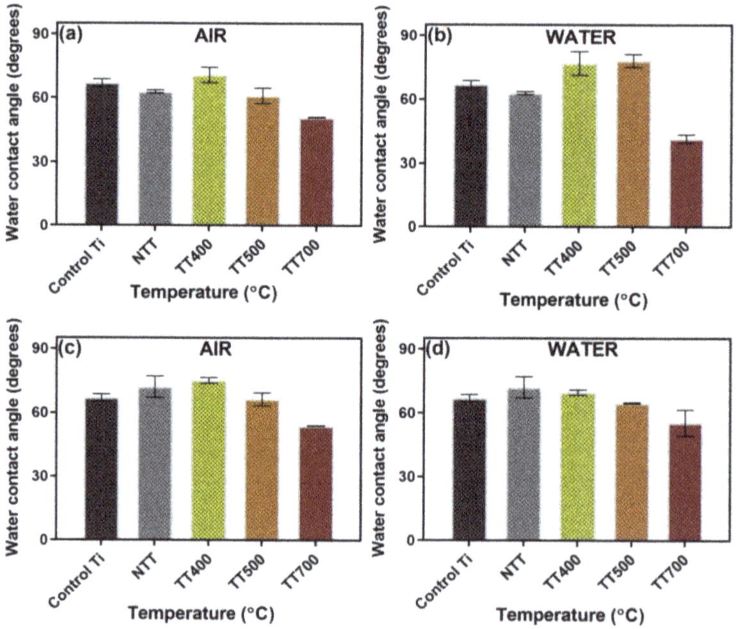

Figure 12. The values of the contact angle obtained in the case of control (Ti), NTT LiC (**a**,**b**) and LiP (**c**,**d**) coatings, and of those subjected to post-deposition thermal treatments (at different temperatures), in air (**a**,**c**) and water-vapor (**b**,**d**) ambient. Test liquid: diiodomethane.

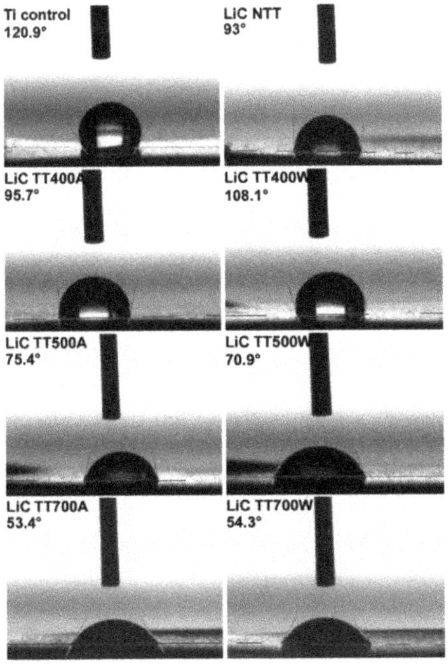

Figure 13. Representative images of the water droplets on control (Ti), NTT LiC and of those subjected to post-deposition thermal treatments (at different temperatures), in air and water-vapor ambient.

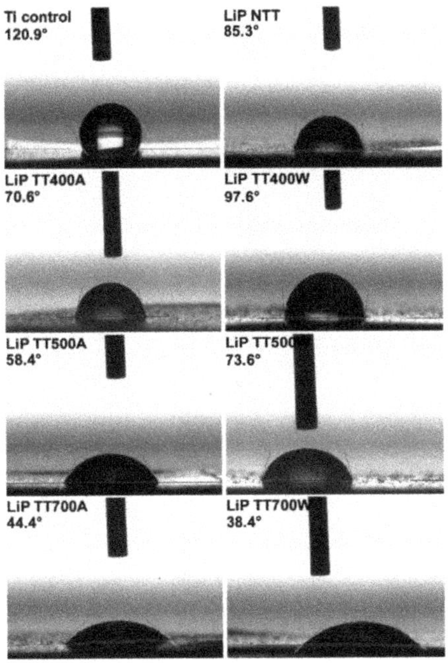

Figure 14. Representative images of the water droplets on control (Ti), NTT LiP and of those subjected to post-deposition thermal treatments (at different temperatures), in air and water-vapor ambient.

The obtained CA values for water indicated surfaces with a hydrophobic behavior (CA > 90°) in the case of control and NTT samples (Figure 11a–d). It should be emphasized that a radical change of the coatings' behavior towards hydrophilicity with the applied temperature was indicated for all TT samples. Thus, in the case of LiP coatings, the CA values dropped down to ~38° (i.e., TT700W samples, Figure 11d).

In the case of diiodomethane (Figure 12), the control and all synthesized coatings presented hydrophilic behaviors. The same trend met in the case of water used as testing liquid was observed also here, with CA values decreasing with the applied temperature. In this case, the smallest values (i.e., ~41°) were obtained in the case of LiC coatings (i.e., TT700W samples, Figure 12b).

It should be emphasized that, highly similar CA values were revealed by two consecutive measurements, performed on different areas of the films surface, which point toward the homogeneity of the synthesized coatings, which is in good agreement with the results of the cross-sectional SEM analysis.

Immediately after an implant is introduced inside the human body, the first events which occur imply its "wetting" by the physiological fluids, followed by cells' attachment onto the surface [69]. It was demonstrated that a hydrophilic surface favors adhesion, migration, and cellular proliferation, and can therefore benefit of a rapid osseous regeneration [70,71]. The low CA values obtained in the case of TT structures (especially TT 700 ones) should be therefore emphasized.

It is important to mention that the SFE characteristics of an implant coating have an important role on the biological response. In general, SFE is influenced by the polar (γ_p) and dispersive (γ_d) components. There is a great dependence of these components not only on the roughness of the coatings but also on their nature, degree of atomic packing, electronic density, or satisfied/unsatisfied atomic bond ratio. The evolution of SFE, in the case of control (Ti), NTT LiC and LiP coatings and for those subjected to post-deposition thermal treatments (at different temperatures), in air and water-vapor ambient, are shown in Figures 15 and 16. It is known that γ_p is generated by the chemical bonds/interactions (e.g., dipole–dipole interactions) within a material, whilst γ_d is related to the movement of electrons around atoms/molecules and temporary variation in the electron density with associated temporary dipoles [31]. It was demonstrated that SFE leads to the arrangement of functional groups and electrical charges on the surface of the biomaterials in contact with the living environment, and thus, governs the initial interactions with the intercellular fluid and the adherence of cells [69,72,73]. One should note here that a higher value of the SFE is indicative for an increased hydrophilicity (the contact angle decreases). Thus, the ~(50–80)° range was suggested to be optimal for proper wettability and cell survival [31,74], which consequently leads to an improved biological response [75].

In the case of control (Ti), and LiC and LiP (NTT–TT500A&W) coatings, the dispersive component (γ_d) was the prominent one (Figures 15 and 16, cross-hatched areas).

In comparison, in the case of both LiC and LiP coatings (i.e., TT700A&W), the polar component (γ_p) was significantly increased (Figures 15 and 16, solid gray regions). This is indicative for a noteworthy modification of the surface chemistry and electron density and might be probably due to the abundance of polar groups present on the surface. This consequently determines an increased wettability due to the polar nature of the water droplet itself [31].

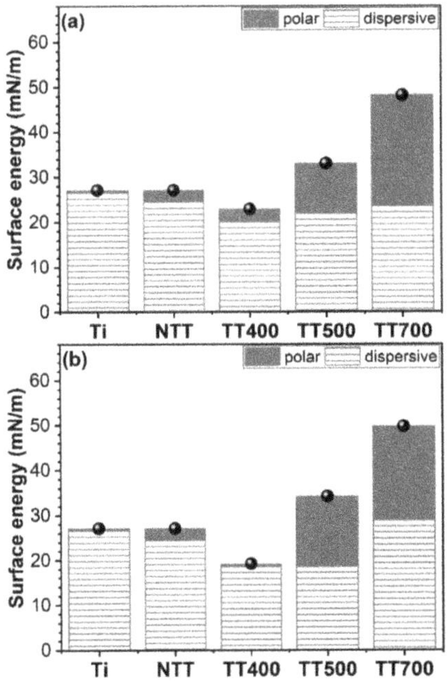

Figure 15. The surface free energy values recorded for the control (Ti), NTT LiC and for those subjected to post-deposition thermal treatments (at different temperatures), in air (**a**) and water-vapor (**b**) ambient.

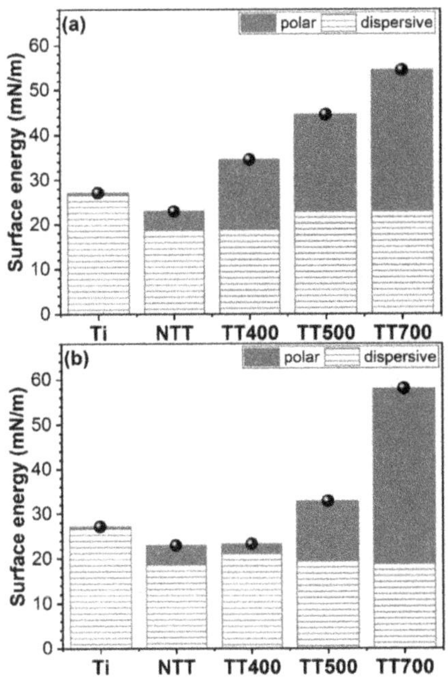

Figure 16. The surface free energy values recorded for the control (Ti), NTT LiP and for those subjected to post-deposition thermal treatments (at different temperatures), in air (**a**) and water-vapor (**b**) ambient.

3.7. Pull-Out Bonding Strength Tests

The bonding force at the film–substrate interface is recognized as a key-parameter in the manufacturing process of high-quality implants [76], as it regulates both their initial stability and long-term functionality [48,77,78].

Prior to starting the measurements, four tests to control the quality of the bonding adhesive using bare Ti substrates (control) were performed under identical conditions. The results of all measurements performed at the stainless-steel test element–control Ti interface were in accordance with the specifications provided by the manufacturer.

In Figure 17, the results of the adherence tests (calculated as mean $\pm$ SD) in the case of NTT LiC and LiP coatings, and of those subjected to post-deposition thermal treatments (at different temperatures), in air and water-vapor ambient are presented. It should be mentioned that, for these measurements, there were considered the events when the fracturing was of adhesive type only, i.e., it occurred at the coating–Ti substrate interface.

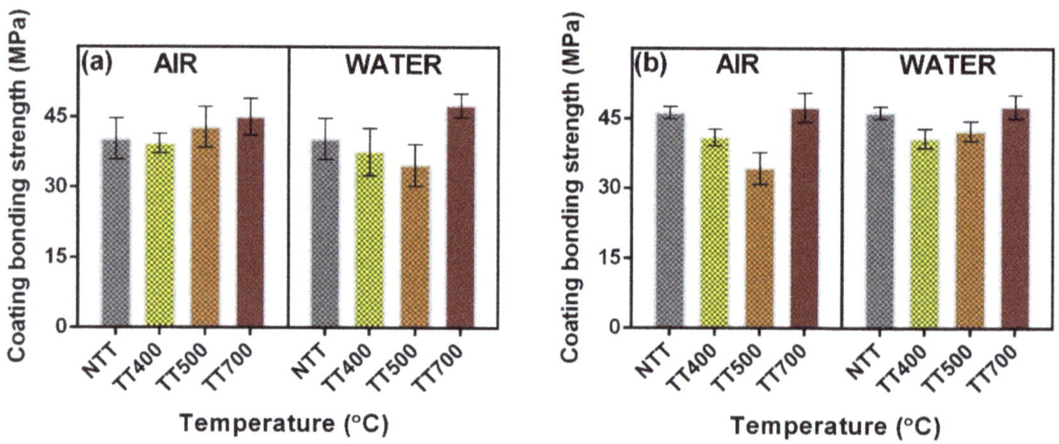

Figure 17. Mean bonding strength values recorded in the case of NTT LiC (**a**) and LiP (**b**) coatings, and of those subjected to post-deposition thermal treatments (at different temperatures), in air and water-vapor ambient.

It is interesting to observe from the data presented in Figure 17 that there exist two trends: (i) an increase of the adherence values with the applied temperature, in the case of LiC coatings subjected to post-deposition thermal treatments in air (Figure 17a), and LiP ones subjected to post-deposition thermal treatments in water-vapor ambient (Figure 15b), and (ii) a decrease of the adherence values for NTT–TT500 coatings, followed by an increase towards higher temperatures, both in the case of LiC coatings subjected to post-deposition thermal treatments in water-vapor ambient (Figure 17a) and LiP ones subjected to post-deposition thermal treatments in air (Figure 17b). A possible explanation for the increase of the adherence values with the applied temperature can be attributed to the processes of atomic inter-diffusion at the coating–Ti substrate interface, which occur during the post-deposition thermal treatments. Therefore, metallic ions can decrease the porosity of the samples, which can eventually lead to much more compact and hard structures [79].

It is important to mention that according to Ref. [76], adherence values exceeding 15 MPa are considered acceptable for the use of these structures as coatings for metallic implants. According to this criterion, the measured adherence values in the case of LiC and LiP coatings (more than three times higher than the imposed standard) should be therefore considered remarkable.

4. Conclusions

Hydroxyapatite (HA) of biological-origin doped with lithium carbonate (LiC) and lithium phosphate (LiP) coatings were synthesized by Pulsed laser deposition technique onto Ti6Al4V substrates, fabricated by additive manufacturing technique. The role of the post-deposition thermal treatments, performed in the range 400–700 °C (TT400–TT700) on the physical–chemical and mechanical characteristics of the synthesized coatings was thoroughly assessed and compared with the case of non-thermally treated (NTT) coatings.

From the structural, morphological, compositional, wetting behavior and mechanical points of view, the best overall properties were obtained in the case of both LiC and LiP synthesized coatings, submitted to the highest post-deposition thermal treatment temperature (i.e., 700 °C), in a water-vapor ambient. In addition, one should emphasize that the surface coating of orthopedic and dental implants, fabricated from medical-grade Ti alloys, with adherent bioceramic layers obtained from sustainable resources doped with Lithium is expected to foster the development of a new generation of osseous implants, which will (i) harmoniously couple the excellent mechanical properties of the metallic substrate with the proven capacity of the Lithium-doped bioceramic to both improve cell adhesion and differentiation, and induce fast osseointegration rates, and (ii) reduce the risk of bio-functional coating mechanical failure reported for the commercially available thick layers (i.e., hundreds of micrometers to millimeters range coatings fabricated by plasma spray). By combining Li-doped bioceramics with various polymers, one can obtain novel biomaterials with proper flexibility (i.e., bio-ink used in 3D printing for clinical applications of tissue engineering). Such types of Li-doped biomaterials might have a significant clinical potential in the predicted future.

Author Contributions: Conceptualization, L.D.; methodology, L.D. and G.E.S.; software, L.D. and G.E.S.; validation, L.D., G.E.S., G.P.-P., I.Z., M.A. and F.N.O.; formal analysis, L.D.; investigation, L.D., G.E.S., G.P.-P., I.Z. and M.A.; resources, L.D. and F.N.O.; data curation, L.D., G.E.S., G.P.-P., I.Z. and M.A.; writing—original draft preparation, L.D.; writing—review and editing, L.D., G.E.S., G.P.-P., I.Z., M.A. and F.N.O.; visualization, L.D., G.E.S., G.P.-P., I.Z., M.A. and F.N.O.; supervision, L.D.; project administration, L.D.; funding acquisition, L.D. All authors have read and agreed to the published version of the manuscript.

Funding: This research received no external funding.

Institutional Review Board Statement: Not applicable.

Informed Consent Statement: Not applicable.

Data Availability Statement: Not applicable.

Acknowledgments: L.D. acknowledges the support from two grants of the Romanian Ministry of Education and Research, CNCS-UEFISCDI, project numbers PN-III-P1-1.1-PD-2016-1568 (PD 6/2018) and PN-III-P1-1.1-TE2019-1449 (TE 189/2021), within PNCDI III. L.D and G.P.P. acknowledge the partial support from the Romanian Ministry of Education and Research, under Romanian National Core Program LAPLAS VI—Contract 16N/2019. G.E.S. and I.Z. acknowledge the partial support from the Romanian Ministry of Education and Research, under Romanian National Core Programme 21N. All authors thank A.C. Popescu and D. Chioibasu for the fabrication of the Ti6Al4V substrates.

Conflicts of Interest: The authors declare no conflict of interest.

References

1. Denry, I.; Kuhn, L.T. Design and characterization of calcium phosphate ceramic scaffolds for bone tissue engineering. *Dent. Mater.* **2016**, *32*, 43–53. [CrossRef]
2. Allied Market Research. Dental Implants and Prosthetics Market by Products (Dental Implants, Dental Prosthetics), by Materials (Metals, Polymers, Ceramics, Biomaterial): Global Opportunity Analysis and Industry Forecast, 2020–2030. Available online: https://www.alliedmarketresearch.com/dental-implants-and-prosthetics-market (accessed on 3 November 2022).
3. Allied Market Research. Orthopedic Implants Market Expected to Reach $68.80 Billion by 2030. Available online: https://www.alliedmarketresearch.com/press-release/orthopedic-implants-market.html (accessed on 3 November 2022).

4. Medany, S.S.; Elkamel, R.S.; Abdel-Gawad, S.A.; Fekry, A.M. A Novel Nano-Composite CSNPs/PVP/CoONPs Coating for Improving Corrosion Resistance of Ti-6Al-4V Alloy as a Dental Implant. *Metals* **2022**, *12*, 1784. [CrossRef]
5. Hulka, I.; Florido-Suarez, N.R.; Mirza-Rosca, J.C.; Saceleanu, A. Mechanical Properties and Corrosion Behavior of Thermally Treated Ti-6Al-7Nb Dental Alloy. *Materials* **2022**, *15*, 3813. [CrossRef] [PubMed]
6. Banerjee, D.; Williams, J.C. Perspectives on titanium science and technology. *Acta Mater.* **2013**, *61*, 844–879. [CrossRef]
7. Agius, D.; Kourousis, K.I.; Wallbrink, C. A review of the as-built SLM Ti-6Al-4V mechanical properties towards achieving fatigue resistant designs. *Metals* **2018**, *8*, 75. [CrossRef]
8. Rau, J.V.; Fadeeva, I.V.; Forysenkova, A.A.; Davydova, G.A.; Fosca, M.; Filippov, Y.Y.; Antoniac, I.V.; Antoniac, A.; D'Arco, A.; Di, M.; et al. Strontium Substituted Tricalcium Phosphate Bone Cement: Short and Long-Term Time-Resolved Studies and In Vitro Properties. *Adv. Mater. Interfaces* **2022**, *9*, 2200803. [CrossRef]
9. Sargeant, A.; Goswami, T. Hip implants-Paper VI-Ion concentrations. *Mater. Des.* **2007**, *28*, 155–171. [CrossRef]
10. Koch, C.; Johnson, S.; Kumar, D.; Jelinek, M.; Chrisey, D.; Doraiswamy, A.; Jin, C.; Narayan, R.J.; Mihailescu, I.N. Pulsed laser deposition of hydroxyapatite thin films. *Mater. Sci. Eng. C* **2007**, *27*, 484–494. [CrossRef]
11. Ananth, H.; Kundapur, V.; Mohammed, H.S.; Anand, M.; Amarnath, G.S.; Mankar, S. A review on biomaterials in dental implantology. *Int. J. Biomed. Sci.* **2015**, *11*, 113–120.
12. Li, J.; Yang, L.; Guo, X.; Cui, W.; Yang, S.; Wang, J.; Qu, Y.; Shao, Z.; Xu, S. Osteogenesis effects of strontium-substituted hydroxyapatite coatings on true bone ceramic surfaces in vitro and in vivo. *Biomed. Mater.* **2018**, *13*, 015018. [CrossRef]
13. Oktar, F.N.; Unal, S.; Gunduz, O.; Nissan, B.B.; Macha, I.J.; Akyol, S.; Duta, L.; Ekren, N.; Altan, E.; Yetmez, M. Marine-derived bioceramics for orthopedic, reconstructive and dental surgery applications. *J. Aust. Ceram. Soc.* **2022**. [CrossRef]
14. Akram, M.; Ahmed, R.; Shakir, I.; Ibrahim, W.A.W.; Hussain, R. Extracting hydroxyapatite and its precursors from natural resources. *J. Mater. Sci.* **2014**, *49*, 1461–1475. [CrossRef]
15. Wang, B.; Feng, C.; Liu, Y.; Mi, F.; Dong, J. Recent advances in biofunctional guided bone regeneration materials for repairing defective alveolar and maxillofacial bone: A review. *Jpn. Dent. Sci. Rev.* **2022**, *58*, 233–248. [CrossRef]
16. Qiao, W.; Liu, Q.; Li, Z.; Zhang, H.; Chen, Z. Changes in physicochemical and biological properties of porcine bone derived hydroxyapatite induced by the incorporation of fluoride. *Sci. Technol. Adv. Mater.* **2017**, *18*, 110–121. [CrossRef] [PubMed]
17. Zhou, H.; Lee, J. Nanoscale hydroxyapatite particles for bone tissue engineering. *Acta Biomater.* **2011**, *7*, 2769–2781. [CrossRef] [PubMed]
18. Inoue, M.; Rodriguez, A.P.; Nagai, N.; Nagatsuka, H.; LeGeros, R.Z.; Tsujigiwa, H.; Inoue, M.; Kishimoto, E.; Takagi, S. Effect of Fluoride-substituted Apatite on *In Vivo* Bone Formation. *J. Biomater. Appl.* **2011**, *25*, 811–824. [CrossRef] [PubMed]
19. Eason, R. *Pulsed Laser Deposition of Thin Films: Applications-Led Growth of Functional Materials*, 1st ed.; Wiley & Sons Interscience: Hoboken, NJ, USA, 2007; pp. 1–705.
20. Montazerian, M.; Hosseinzadeh, F.; Migneco, C.; Fook, M.V.L.; Baino, F. Bioceramic coatings on metallic implants: An overview. *Ceram. Int.* **2022**, *48*, 8987–9005. [CrossRef]
21. Duta, L.; Popescu, A.C. Current Status on Pulsed Laser Deposition of Coatings from Animal-Origin Calcium Phosphate Sources. *Coatings* **2019**, *9*, 335. [CrossRef]
22. Duta, L. In Vivo Assessment of Synthetic and Biological-Derived Calcium Phosphate-Based Coatings Fabricated by Pulsed Laser Deposition: A Review. *Coatings* **2021**, *11*, 99. [CrossRef]
23. Duta, L.; Oktar, F.N.; Stan, G.E.; Popescu-Pelin, G.; Serban, N.; Luculescu, C.; Mihailescu, I.N. Novel doped hydroxyapatite thin films obtained by pulsed laser deposition. *Appl. Surf. Sci.* **2013**, *265*, 41–49. [CrossRef]
24. Graziani, G.; Boi, M.; Bianchi, M. A Review on Ionic Substitutions in Hydroxyapatite Thin Films: Towards Complete Biomimetism. *Coatings* **2018**, *8*, 269. [CrossRef]
25. Farmani, A.R.; Salmeh, M.A.; Golkar, Z.; Moeinzadeh, A.; Ghiasi, F.F.; Amirabad, S.Z.; Shoormeij, M.H.; Mahdavinezhad, F.; Momeni, S.; Moradbeygi, F.; et al. Li-Doped Bioactive Ceramics: Promising Biomaterials for Tissue Engineering and Regenerative Medicine. *J. Funct. Biomater.* **2022**, *13*, 162. [CrossRef] [PubMed]
26. Duta, L.; Chifiriuc, M.C.; Popescu-Pelin, G.; Bleotu, C.; Gradisteanu, G.(P.); Anastasescu, M.; Achim, A.; Popescu, A. Pulsed Laser Deposited Biocompatible Lithium-Doped Hydroxyapatite Coatings with Antimicrobial Activity. *Coatings* **2019**, *9*, 54. [CrossRef]
27. Glenske, K.; Donkiewicz, P.; Köwitsch, A.; Milosevic-Oljaca, N.; Rider, P.; Rofall, S.; Franke, J.; Jung, O.; Smeets, R.; Schnettler, R.; et al. Applications of Metals for Bone Regeneration. *Int. J. Mol. Sci.* **2018**, *19*, 826. [CrossRef] [PubMed]
28. Mayer, I.; Berger, U.; Markitziu, A.; Gidalia, I. The uptake of lithium ions by synthetic carbonated hydroxyapatite. *Calcif. Tissue Int.* **1986**, *38*, 293–295. [CrossRef]
29. Koutsoukos, P.G.; Nancollas, G.H. The effect of lithium on the precipitation of hydroxyapatite from aqueous solutions. *Colloids Surf.* **1986**, *17*, 361–370. [CrossRef]
30. Shainberg, A.P.M.; Valério, P.; Zonari, A.; Oktar, F.N.; Ozyegin, L.S.; Graça, M.P.F.; Leite, M.F.; Goes, A.M. Attachment and proliferation of osteoblasts on lithium-hydroxyapatite composites. *Adv. Mater. Sci. Eng.* **2012**, *2012*, 650574. [CrossRef]
31. Popescu, A.C.; Florian, P.E.; Stan, G.E.; Popescu-Pelin, G.; Zgura, I.; Enculescu, M.; Oktar, F.N.; Trusca, R.; Sima, L.E.; Roseanu, A.; et al. Physical-chemical characterization and biological assessment of simple and lithium-doped biological-derived hydroxyapatite thin films for a new generation of metallic implants. *Appl. Surf. Sci.* **2018**, *439*, 724–735. [CrossRef]

32. Florian, P.E.; Duta, L.; Grumezescu, V.; Popescu-Pelin, G.; Popescu, A.C.; Oktar, F.N.; Evans, R.W.; Constantinescu, A.R. Lithium-Doped Biological-Derived Hydroxyapatite Coatings Sustain *In Vitro* Differentiation of Human Primary Mesenchymal Stem Cells to Osteoblasts. *Coatings* **2019**, *9*, 781. [CrossRef]
33. Sakudo, A. Inactivation Methods for Prions. *Curr. Issues Mol. Biol.* **2020**, *36*, 23–32. [CrossRef]
34. Chioibasu, D.; Achim, A.; Popescu, C.; Stan, G.E.; Pasuk, I.; Enculescu, M.; Iosub, S.; Duta, L.; Popescu, A. Prototype Orthopedic Bone Plates 3D Printed by Laser Melting Deposition. *Materials* **2019**, *12*, 906. [CrossRef] [PubMed]
35. Stan, G.E.; Marcov, D.A.; Popa, A.C.; Husanu, M.A. Polymer-like and diamond-like carbon coatings prepared by RF-PECVD for biomedical applications. *Dig. J. Nanomater. Biostruct.* **2010**, *5*, 705–718.
36. Owens, D.K.; Wendt, R.C. Estimation of the surface free energy of polymers. *J. Appl. Polym. Sci.* **1969**, *13*, 1741–1747. [CrossRef]
37. Popa, A.C.; Stan, G.E.; Husanu, M.A.; Pasuk, I.; Popescu, I.D.; Popescu, A.C.; Mihailescu, I.N. Multi-layer haemocompatible diamond-like carbon coatings obtained by combined radio frequency plasma enhanced chemical vapor deposition and magnetron sputtering. *J. Mater. Sci. Mater. Med.* **2013**, *24*, 2695–2707. [CrossRef] [PubMed]
38. Stan, G.E.; Morosanu, C.O.; Marcov, D.A.; Pasuk, I.; Miculescu, F.; Reumont, G. Effect of annealing upon the structure and adhesion properties of sputtered bio-glass/titanium coatings. *Appl. Surf. Sci.* **2009**, *255*, 9132–9138. [CrossRef]
39. Hench, L.L. Bioceramics: From Concept to Clinic. *J. Am. Ceram. Soc.* **1991**, *74*, 1487–1510. [CrossRef]
40. Pasteris, J.D.; Wopenka, B.; Freeman, J.J.; Rogers, K.; Valsami-Jones, E.; van der Houwen, J.A.M.; Silva, M.J. Lack of OH in nanocrystalline apatite as a function of degree of atomic order: Implications for bone and biomaterials. *Biomaterials* **2004**, *25*, 229–238. [CrossRef]
41. Cihlář, J.; Buchal, A.; Trunec, M. Kinetics of thermal decomposition of hydroxyapatite bioceramics. *J. Mater. Sci.* **1999**, *34*, 6121–6131. [CrossRef]
42. Komur, B.; Lohse, T.; Can, H.M.; Khalilova, G.; Geçimli, Z.N.; Aydoğdu, M.O.; Kalkandelen, C.; Stan, G.E.; Sahin, Y.M.; Sengil, A.Z.; et al. Fabrication of naturel pumice/hydroxyapatite composite for biomedical engineering. *Bio. Med. Eng. OnLine* **2016**, *15*, 81. [CrossRef]
43. Patterson, A.L. The Scherrer Formula for X-Ray Particle Size Determination. *Phys. Rev.* **1939**, *56*, 978. [CrossRef]
44. Markovic, M.; Fowler, B.O.; Tung, M.S. Preparation and Comprehensive Characterization of a Calcium Hydroxyapatite Reference Material. *J. Res. Natl. Inst. Stand. Technol.* **2004**, *109*, 553–568. [CrossRef] [PubMed]
45. Chirică, I.M.; Enciu, A.-M.; Tite, T.; Dudău, M.; Albulescu, L.; Iconaru, S.L.; Predoi, D.; Pasuk, I.; Enculescu, M.; Radu, C.; et al. The Physico-Chemical Properties and Exploratory Real-Time Cell Analysis of Hydroxyapatite Nanopowders Substituted with Ce, Mg, Sr, and Zn (0.5–5 at.%). *Materials* **2021**, *14*, 3808. [CrossRef] [PubMed]
46. Gadaleta, S.J.; Paschalis, E.P.; Betts, F.; Mendelsohn, R.; Boskey, A.L. Fourier transform infrared spectroscopy of the solution-mediated conversion of amorphous calcium phosphate to hydroxyapatite: New correlations between X-ray diffraction and infrared data. *Calcif. Tissue Int.* **1996**, *58*, 9–16. [CrossRef] [PubMed]
47. Querido, W.; Shanas, N.; Bookbinder, S.; Oliveira-Nunes, M.C.; Krynska, B.; Pleshko, N. Fourier transform infrared spectroscopy of developing bone mineral: From amorphous precursor to mature crystal. *Analyst* **2020**, *145*, 764–776. [CrossRef] [PubMed]
48. Sima, L.E.; Stan, G.E.; Morosanu, C.O.; Melinescu, A.; Ianculescu, A.; Melinte, R.; Neamtu, J.; Petrescu, S.M. Differentiation of mesenchymal stem cells onto highly adherent radio frequency-sputtered carbonated hydroxylapatite thin films. *J. Biomed. Mater. Res. A* **2010**, *95A*, 1203–1214. [CrossRef] [PubMed]
49. Visan, A.; Grossin, D.; Stefan, N.; Duta, L.; Miroiu, F.M.; Stan, G.E.; Sopronyi, M.; Luculescu, C.; Freche, M.; Marsan, O.; et al. Biomimetic nanocrystalline apatite coatings synthesized by matrix assisted pulsed laser evaporation for medical applications. *Mater. Sci. Eng. B* **2014**, *181*, 56–63. [CrossRef]
50. Spence, G.; Phillips, S.; Campion, C.; Brooks, R.; Rushton, N. Bone formation in a carbonate-substituted hydroxyapatite implant is inhibited by zoledronate. *J. Bone Joint Surg. Br.* **2008**, *90-B*, 1635–1640. [CrossRef]
51. Germaini, M.M.; Detsch, R.; Grünewald, A.; Magnaudeix, A.; Lalloue, F.; Boccaccini, A.R.; Champion, E. Osteoblast and osteoclast responses to A/B type carbonate-substituted hydroxyapatite ceramics for bone regeneration. *Biomed. Mater.* **2017**, *12*, 035008. [CrossRef]
52. Peng, G.W.; Chen, S.K.; Liu, H.S. Infrared Absorption Spectra and Their Correlation with the Ti-O Bond Length Variations for TiO_2 (Rutile), Na-Titanates, and Na-Titanosilicate (Natisite, $Na_2TiOSiO_4$). *App. Spectrosc.* **1995**, *49*, 1646–1651. [CrossRef]
53. Šupová, M. Substituted hydroxyapatites for biomedical applications: A review. *Ceram. Int.* **2015**, *41*, 9203–9231. [CrossRef]
54. Lescoute, E.; Hallo, L.; Hébert, D.; Chimier, B.; Etchessahar, B.; Tikhonchuk, V.T.; Chevalier, J.M.; Combis, P. Experimental observations and modeling of nanoparticle formation in laser-produced expanding plasma. *Phys. Plasmas* **2008**, *15*, 063507. [CrossRef]
55. Mihailescu, I.N.; Ristoscu, C.; Bigi, A.; Mayer, I. Advanced biomimetic implants based on nanostructured coatings synthesized by pulsed laser technologies. In *Laser-Surface Interactions for New Materials Production, Tailoring Structure and Properties*; Miotello, A., Ossi, M., Eds.; Springer: New York, NY, USA, 2010; pp. 235–268.
56. Hacking, S.A.; Boyraz, P.; Powers, B.M.; Sen-Gupta, E.; Kucharski, W.; Brown, C.A.; Cook, E.P. Surface roughness enhances the osseointegration of titanium headposts in non-human primates. *J. Neurosci. Methods* **2012**, *211*, 237–244. [CrossRef] [PubMed]
57. Dinda, G.P. Pulsed laser deposition of hydroxyapatite thin films on Ti–6Al–4V: Effect of heat treatment on structure and properties. *Acta Biomater.* **2009**, *5*, 1821–1830. [CrossRef] [PubMed]

58. Lu, Y.P.; Li, S.T.; Zhu, R.F.; Li, M.S.; Lei, T.Q. Formation of ultrafine particles in heat treated plasma-sprayed hydroxyapatite coatings. *Surf. Coating Technol.* **2003**, *165*, 65–70. [CrossRef]
59. Whitehouse, D.J. Surface Characterization and Roughness Measurement in Engineering. In *Photomechanics*; Rastogi, P.K., Ed.; Topics in Applied Physics; Springer: Berlin/Heidelberg, Germany, 2000; Volume 77. [CrossRef]
60. Kubiak, K.J.; Wilson, M.C.T.; Mathia, T.G.; Carval, P. Wettability versus roughness of engineering surfaces. *Wear* **2011**, *271*, 523–528. [CrossRef]
61. Duta, L.; Popescu, A.C.; Zgura, I.; Preda, N.; Mihailescu, I.N. Wettability of Nanostructured Surfaces. In *Wetting and Wettability*; Aliofkhazraei, M., Ed.; Intech: Vienna, Austria, 2015. [CrossRef]
62. US National Institute of Standards and Technology (NIST) e-Handbook of Statistical Methods. Available online: https://www.itl.nist.gov/div898/handbook (accessed on 2 November 2022).
63. Sedlacek, M.; Vilhena, L.M.S.; Podgornik, B.; Vizintin, J. Surface topography modelling for reduced friction. *J. Mech. Eng.* **2011**, *57*, 674–680. [CrossRef]
64. Stan, G.E.; Popescu, A.C.; Mihailescu, I.N.; Marcov, D.A.; Mustata, R.C.; Sima, L.E.; Petrescu, S.M.; Ianculescu, A.; Trusca, R.; Morosanu, C.O. On the bioactivity of adherent bioglass thin films synthesized by magnetron sputtering techniques. *Thin Solid Films* **2010**, *518*, 5955–5964. [CrossRef]
65. Gadelmawla, E.S.; Koura, M.M.; Maksoud, T.M.A.; Elewa, I.M.; Soliman, H.H. Roughness parameters. *J. Mater. Process Technol.* **2002**, *123*, 133–145. [CrossRef]
66. Liu, H.; Yazici, H.; Ergun, C.; Webster, T.J.; Bermek, H. An in vitro evaluation of the Ca/P ratio for the cytocompatibility of nano-to-micron particulate calcium phosphates for bone regeneration. *Acta Biomater.* **2008**, *4*, 1472–1479. [CrossRef]
67. León, B.; Jansen, J.A. *Thin Calcium Phosphate Coatings for Medical Implants*; Springer Science + Business Media: New York, NY, USA, 2009; pp. 1–328.
68. Surmenev, R.A.; Surmeneva, M.A.; Evdokimov, K.E.; Pichugin, V.F.; Peitsch, T.; Epple, M. The influence of the deposition parameters on the properties of an rf-magnetron-deposited nanostructured calcium phosphate coating and a possible growth mechanism. *Surf. Coat. Technol.* **2011**, *205*, 3600–3606. [CrossRef]
69. Sameer, R.P.; Zheng, C.; Wei, H.; Narendra, B.D. Wetting effects on in vitro bioactivity and in vitro biocompatibility of laser micro-textured Ca-P coating. *Biofabrication* **2010**, *2*, 025001. [CrossRef]
70. Wilson, C.J.; Clegg, R.E.; Leavesley, D.I.; Mark, A.; Pearcy, J. Mediation of biomaterial-cell interactions by adsorbed proteins: A review. *Tissue Eng.* **2005**, *11*, 1–18. [CrossRef] [PubMed]
71. Okabe, Y.; Kurihara, S.; Yajima, T.; Seki, Y.; Nakamura, I.; Takano, I. Formation of super-hydrophilic surface of hydroxyapatite by ion implantation and plasma treatment. *Surf. Coat. Technol.* **2005**, *196*, 303–306. [CrossRef]
72. Lotfi, M.; Nejib, M.; Naceur, M. Cell Adhesion to Biomaterials: Concept of Biocompatibility. In *Advances in Biomaterials Science and Biomedical Applications*; Pignatello, R., Ed.; Intech: Rijeka, Croatia, 2013; pp. 207–240. [CrossRef]
73. Gentleman, M.M.; Gentleman, E. The role of surface free energy in osteoblast-biomaterial interactions. *Int. Mater. Rev.* **2014**, *59*, 417–429. [CrossRef]
74. Pessková, V.; Kubies, D.; Hulejova, H.; Himmlova, L. The influence of implant surface properties on cell adhesion and proliferation. *J. Mater. Sci. Mater. Med.* **2007**, *18*, 465–473. [CrossRef]
75. Mitra, J.; Tripathi, G.; Sharma, A.; Basu, B. Scaffolds for bone tissue engineering: Role of surface patterning on osteoblast response. *RSC Adv.* **2013**, *3*, 11073–11094. [CrossRef]
76. *ISO 13779-2*; Implants for Surgery—Hydroxyapatite—Part 2: Coatings of Hydroxyapatite. ISO: Geneva, Switzerland, 2018.
77. Popa, A.C.; Stan, G.E.; Enculescu, M.; Tanase, C.; Tulyaganov, D.U.; Ferreira, J.M.F. Superior biofunctionality of dental implant fixtures uniformly coated with durable bioglass films by magnetron sputtering. *J. Mech. Behav. Biomed. Mater.* **2015**, *51*, 313–327. [CrossRef]
78. Stan, G.E.; Popa, A.C.; Galca, A.C.; Aldica, G.; Ferreira, J.M.F. Strong bonding between sputtered bioglass-ceramic films and Ti-substrate implants induced by atomic inter-diffusion post-deposition heat-treatments. *Appl. Surf. Sci.* **2013**, *280*, 530–538. [CrossRef]
79. Wang, Y.; Yang, X.; Gu, Z.; Qin, H.; Li, L.; Liu, J.; Yu, X. *In vitro* study on the degradation of lithium-doped hydroxyapatite for bone tissue engineering scaffold. *Mater. Sci. Eng. C Mater. Biol. Appl.* **2016**, *66*, 185–192. [CrossRef]

coatings

MDPI

Article

Bioactivity and Mechanical Properties of Hydroxyapatite on Ti6Al4V and Si(100) Surfaces by Pulsed Laser Deposition

Salizhan Kylychbekov [1], Yaran Allamyradov [1], Zikrulloh Khuzhakulov [1], Inomjon Majidov [1], Simran Banga [2,*], Justice ben Yosef [3], Liviu Duta [4,*] and Ali Oguz Er [1,*]

1 Department of Physics and Astronomy, Western Kentucky University, Bowling Green, KY 42101, USA; salizhan.kylychbekov837@topper.wku.edu (S.K.); yaran.allamyradov485@topper.wku.edu (Y.A.)
2 Department of Biology, Western Kentucky University, Bowling Green, KY 42101, USA
3 Department of Chemistry, Western Kentucky University, Bowling Green, KY 42101, USA
4 National Institute for Lasers, Plasma and Radiation Physics, 077125 Magurele, Romania
* Correspondence: simran.banga@wku.edu (S.B.); liviu.duta@inflpr.ro (L.D.); ali.er@wku.edu (A.O.E.)

Abstract: In this study, the effects of substrate temperature and ablation wavelength/mechanism on the structural, mechanical, and bioactivity properties of hydroxyapatite (HA) coatings were investigated. HA coatings were deposited on both Si(100) and Ti6Al4V surfaces. Substrate temperature varied from room temperature to 800 °C. Depositions were performed in Ar/H_2O and vacuum environments. X-ray diffraction, scanning electron microscopy, and atomic force microscopy techniques were used to analyze structural and morphological variations. The adherence of coatings to the substrates was assessed by the pull-out method. The obtained data indicated that with the temperature increase, the coatings steadily crystallized. However, temperatures above 700 °C adversely affected protein adsorption and adhesion properties. Similar trends were confirmed via pull-out testing, protein adsorption, and cell proliferation tests. The ablation mechanism was also proven to play an important role in the deposition process. Overall, this study provides further evidence that crystallinity is a vital factor in the functionality of the coatings and depends on the deposition conditions. However, all measurements directly indicated that beyond 700 °C, the morpho-structural, mechanical, and bioactivity properties degrade.

Keywords: hydroxyapatite coating; pulsed laser deposition; mechanical testing; bioactivity; protein adsorption; surface characterization

Citation: Kylychbekov, S.; Allamyradov, Y.; Khuzhakulov, Z.; Majidov, I.; Banga, S.; ben Yosef, J.; Duta, L.; Er, A.O. Bioactivity and Mechanical Properties of Hydroxyapatite on Ti6Al4V and Si(100) Surfaces by Pulsed Laser Deposition. *Coatings* **2023**, *13*, 1681. https://doi.org/10.3390/coatings13101681

Academic Editor: Michał Kulka

Received: 31 August 2023
Revised: 16 September 2023
Accepted: 18 September 2023
Published: 25 September 2023

1. Introduction

Biological responses to materials differ based on the differences in surface and interface properties. Conventional implants are commonly made from three families of metals: 316 L stainless steel, cobalt chromium, and titanium alloys. Although they have excellent tensile strength and stability over long periods, they are biologically inert [1]. Bones, in contrast, participate in vital biological processes in addition to structurally supporting the body. This drawback of metals results in complications with statistically significant effects [2–4]. To improve the biological performance of metal surfaces, multiple alternatives were proposed [5–12]. A good candidate for coatings should possess low degradation, high biocompatibility, antibacterial, and wear resistance properties [13,14]. Among them, coating metal surfaces with calcium phosphates (CaP), specifically with hydroxyapatite (HA, $Ca_{10}(PO_4)_6(OH)_2$), has been a main topic of interest for the last several decades for clinical applications such as prostheses, orthopedics, drug delivery, and dentistry [15–17]. The global implant market growth has been forecasted at a 6.8% compound annual growth rate (CAGR) for the period between 2022 and 2032 [18]. A similar trend pertains to the global HA market, with USD 2.3 billion in 2023 with a CAGR of 6.52% forecasted for 2023–2028. The main motivation for this industry is that coating the metal surface with CaP allows one to combine the metal's mechanical properties with the biocompatibility of CaP.

HA, being one of the main components of human bones, is biocompatible, bioactive, and thermodynamically stable in body fluid environments [19,20]. It also actively takes part in essential bone regeneration processes by osteogenic stimulation through the release of Ca^{2+}, PO^{4-}, and OH^- ions when in contact with blood plasma [21]. Coating the metal surface with HA offers numerous solutions to problems such as metal ion release from the metal surface, foreign body inflammation, and bone cell repair. Research progress has shown that crystalline HA is the most active constituent in human bones and is more stable in human body fluids that contain a variety of dissolving ions [15,21]. Among conventional coating methods, pulsed laser deposition (PLD) has been successful in producing crystalline HA coatings with applicable adhesion strength [22–24].

In addition, PLD can be used to synthesize various materials, including metals and semiconductors [25–30]. PLD is a thin film deposition technique where a high-power pulsed laser beam is focused inside a vacuum chamber to ablate the target of desired composition. Parallel to developments in lasers, PLD has become more popular in current research environments.

However, PLD is notorious for creating particulates during deposition. These droplets could later affect the photoemission performance of photocathodes. Particulate size ranges from the sub-micron level to several micrometers in size [31]. Particulate formation is affected by a number of parameters, such as the target, laser parameters, surface quality, and laser energy density. Multiple alternatives, such as using a shorter pulse width laser, or using bi-directional ablation were proposed to prevent particulate formation in PLD [11,32,33].

It has been reported that unmatched crystallinity and bonding strengths of HA could be obtained using the PLD technique [34]. Although DC Magnetron and RF sputtering have achieved relatable crystallinity, they showed weak adherence between the coating and metal surface [35]. Another widely suggested method is plasma spraying, but it uses high-temperature plasma leading to unwanted phases of CaP such as TTCP, α- and β-TCP, in addition to HA, which emerge at high temperatures [36–38]. In contrast, PLD has been used in the successful production of a variety of crystalline coatings with high adhesivity. Most of the studies reported post-deposition annealing as a necessary step to obtain crystalline coatings [39–43]. However, crystalline coatings at room temperature using a 532 nm Nd:YAG laser at high fluences were reported [44]. It was suggested that high fluences enabled more expelled elements and particulates and provided higher kinetic energy for the plume particles to find ideal nucleation sites. This was an unpredicted result compared to the general trend of using post-deposition annealing at high temperatures and providing enough H_2O vapor during the nucleation phase, as they are known to be critical for the successful production of HA crystals and maintenance of stoichiometry or Ca/P ratio [39–41]. Recently, the trend has shifted towards doped-HA coatings to merge additional functions through nanoparticles and elements such as Zn [45,46], Ag [47,48], F [49], and so forth [23,50,51].

One interesting aspect of PLD is the ablation mechanism. Historically, in the PLD of HA, two pulsed laser systems were used: Nd:YAG and excimer lasers. Nistor et al. [52] have reported coatings with 355 and 532 nm Nd:YAG lasers. Coating with a 532 nm laser produced pure HA phases, but 355 nm consisted of α-TCP and HA [52]. Although the observations are singular, they are relatively comparable. Recently, a Japanese group published several interesting reports on eliminating particulates using the so-called telescope model [53,54]. They reported that crystallinity was also significantly enhanced, as big particulates were unable to cross the obstacle, while atoms, electrons, and ions crossed and nucleated upon reaching the substrate. When considering the excimer system, ArF and KrF lasers were used. They were reported to induce columnar growth. In summary, while no single group studied all ablation systems at once, it is evident that the excimer lasers induce columnar growth while Nd:YAG lasers produce granular growth.

In the current work, we had two aims. Firstly, a complete analysis of substrate temperature effects was performed, as it is not fully understood yet. Although other groups have briefly touched on this subject, an extensive study with a full range is not yet available. Secondly, complete data on the unified set of Nd:YAG wavelengths: 1064, 532, and 355 nm in a single work was reported. This examination focused on three fronts, i.e., mechanical, morphological, and bioactivity assessments. Bioactivity by itself is a broad term, so the emphasis was placed on in vitro protein adsorption and cell proliferation.

2. Materials and Methods

2.1. Experimental Details

2.1.1. Substrate Preparation

Si(100) wafers and grade 5 Ti6Al4V plates were purchased from mtixtl.com and Titanium Industries Inc. Si(100) arrived with one side already polished, while Ti6Al4V were cut into 15 mm × 15 mm sheets and polished until a 0.05 μm mirror finish was obtained. First, sanding was performed using SiC pads sequentially with grit sizes of 180, 400, 600, and 1000. Then, 0.05 μm colloidal silica/alumina solution and colloidal silica suspension were used simultaneously for fine polishing. Finally, substrates were cleaned for 5 min each in both acetone and deionized water using an ultrasonic bath.

2.1.2. Coating Process

The PLD equipment consisted of the pulsed Nd:YAG nanosecond laser source (Continuum Surelite II), a stainless-steel ultrahigh vacuum chamber equipped with a resistive substrate heater, and a rotating target holder system. The laser had a fundamental wavelength of 1064 nm, a pulse width of 5 ns, a 10 Hz repetition rate, and a Gaussian pulse shape of 6 mm diameter when the beam was unfocused. Harmonic beams with 532 nm and 355 nm wavelengths were generated using KDP crystals. The fluence was adjusted at 7.4 J/cm^2. The substrate-to-target separation distance was set between 3 and 5 cm. The substrate temperature was varied from room temperature (RT) to 800 °C (i.e., RT, 300, 500, and 800 °C). Depositions were performed in vacuum at 10^{-6} Torr and 3.0×10^{-1} Torr water vapor, supplied to the vacuum chamber by bubbling the Ar gas through the water bath at RT. Commercially available HA (3D Biotek, LLC, Bridgewater, NJ, USA) dense discs, 9.5 mm in diameter, 1.6 mm thick, and 1.67 Ca/P ratio, were used as a target material. The targets were rotated at 8 rpm to ensure homogeneous ablation.

2.2. Surface Characterization: SEM, AFM, AmScope

The surface of the coatings was characterized by scanning electron microscopy (SEM), atomic force microscopy (AFM), and optical microscopy techniques. For SEM, a JEOL 6510LV 10–20 kV scanning electron microscope (JEOL Ltd., Tokyo, Japan) was used with energy-dispersive X-ray analysis. Thickness measurements were assessed by cracking the Si(100) sample and loading it vertically upward. The surface roughness measurements were performed with the Nanosurf FlexAFM atomic force microscope (Nanosurf AG, Liestal, Switzerland) operated in contact mode. Quantitative calculations were obtained using Gwyddion software (version 2.62). For bulk analysis, an AmScope optical microscope (United Scope LLC., Irvine, CA, USA) with the model number SM-1TSZZ-144S-10M was used.

2.3. Structural Analysis: XRD, FTIR

Phases of the coatings were identified with a 2nd generation Bruker X-ray Diffraction machine with Cu-Kα radiation source excited at 30 kV X-ray generation voltage and 1.54 Å wavelength. The bonding structure of the HA films was characterized by Fourier-transform infrared spectroscopy (FTIR Perkin Elmer Spectrum 100 (PerkinElmer Inc., Waltham, MA, USA)).

2.4. Mechanical Analysis and Adherence Strength: Pull-Out Tests

The pull-out bonding strength (tensile) at the film-substrate interface is considered of paramount importance in the process of high-quality implant fabrication [55] and the long-term functioning and stability of these medical devices [56] in situ. Thus, it was included as a quality factor in the International Standard 13779-2/2018 [57], which regulates the fabrication of load-bearing implant coatings.

The adherence of the synthesized HA coatings onto Ti substrates was estimated using the pull-out method. For these tests, a DFD Instruments® PATHandy MICRO AT101 (maximum pull force of 1 kN) adherence instrument, equipped with stainless-steel test elements (dollies) with diameters of 0.28 cm, was used. The dollies were glued to the coatings' surface with a special E1100S single-part epoxy glue.

First, the stub surface was polished, then ultrasonically cleaned in acetone and ethanol, and finally dried in nitrogen flow. After gluing, the HA coatings were dried in an oven (at 130 °C for 1 h). Prior to the investigations, three quality tests to control the bonding adhesive strength using uncoated Ti substrates were performed. The mean value of the adhesive strength, measured at the stainless-steel test element-Ti control interface, was 55 ± 4.3 MPa. Each dolly was pulled out vertically using a calibrated hydraulic pump. The extraction force was gradually increased until detachment. For each coating, the measurements were performed on quadruplicates and were in accordance with the ASTM D4541 [58] and ISO 4624 standards [59].

2.5. Biological Assessment

2.5.1. Cell Growth Experiment

HEK 293T cells were maintained in Dulbecco's modified eagle medium supplemented with 10% fetal bovine serum and antibiotics. For the assay, five different disc samples (i.e., control–uncoated and coated at RT, 300, 500, and 800 °C) were placed in 48-well plates in duplicate and seeded with 50,000 cells per well. The samples were incubated at 37 °C in 5% CO_2. At specific time points (i.e., 24, 72, and 120 h), the discs were washed gently with phosphate-buffered saline, and cells were collected to perform a cell count using a counting chamber. The cell counts were analyzed using ANOVA for statistical comparisons between different surfaces at each time point.

2.5.2. Protein Adsorption

Bovine Serum Albumin (BSA) is a relatively small protein with 66 kDa of mass and makes up approximately 60% of all proteins. Albumin is also the most common protein in human blood. Therefore, BSA was chosen for the protein adsorption study.

First, 400 μg/mL of BSA was prepared. HA coatings were submerged in ethanol and flamed for sterilization. Then, triplets from each sample were placed in 48-well plates, and 300 μL of BSA was added onto each sample. Every 24 h, 25 μL of BSA solution was taken from each sample and placed in a 96-well plate and examined under UV-Vis spectroscopy. For quantization, a PierceTM Protein Assay Kit was used. Optical Density was measured at 562 nm, and protein concentration in the liquid was calculated accordingly.

3. Results and Discussion

While multiple synthesis conditions affect the ablation process, one of the most significant conditions is the wavelength of the laser pulse. When the wavelength is shorter, the photon energies increase, allowing for bond breakage within the atomic lattice, thereby resulting in ionization [60]. This leads to the ejection of atoms and ions without resulting in adverse heating effects. Shorter wavelengths also result in shallower laser penetration depths and, therefore, lower the threshold fluences and ablation rates. In contrast, longer wavelengths often yield higher densities and larger sizes of particulates and/or droplets. Infrared or visible laser light usually processes materials by producing intense local heating that can melt or vaporize the material, which may cause serious damage to surrounding areas. This contributes to stronger compositional deviations in the film [61]. Therefore, UV irradiation at shorter wavelengths typically yields lower amounts of nano/microparticles as compared to irradiation at longer wavelengths.

During the ablation process, micro to nanostructures are frequently formed on the surface of the target material. These structures increase the surface area and, in turn, reduce the effective laser fluence. It has been reported that the wavelength of the laser pulse affects the size and frequency of the periodic structures that appear on the target surface [62]. Moreover, the wavelengths not only impact the absorption by the target material but also influence the absorption by the laser-generated plasma plume [61,63].

The useful range of laser wavelengths for ablation typically falls between 200 and 400 nm. Excimer lasers are often preferred for ablation due to their higher energies and relatively flat top energy profile. In contrast, Nd:YAG lasers demonstrate a Gaussian energy profile. Therefore, the variation in fluence inherent to the Gaussian profile must be considered when determining deposition parameters if using a Nd:YAG laser [64]. The ablation mechanism in PLD of HA with 532 nm using fast-gated ICCD cameras and spatio-temporal spectroscopic tools was studied. It was therefore reported that the plasma plume consisted of droplets, atoms, and ions-electrons [65].

While atoms and ions nucleate at the surface, droplets cause hardships in the seeding process. Therefore, minimizing amounts of droplets could enhance the crystallinity of the films [53,54].

Temperature is also known to affect many properties of the film during and after deposition. At higher temperatures, adatoms on the substrate's surface gain thermal energy, and surface diffusion is enhanced, thus promoting crystal growth. The transition from a textured microstructure to a densely packed crystal structure occurs over a range of temperatures.

To study the effect of wavelength and temperature, HA coatings were grown at three different wavelengths (i.e., 1064, 532, and 355 nm) and various temperatures (i.e., RT, 500, and 800 °C). The SEM images from Figure 1 show the clear effect of the temperature and wavelength.

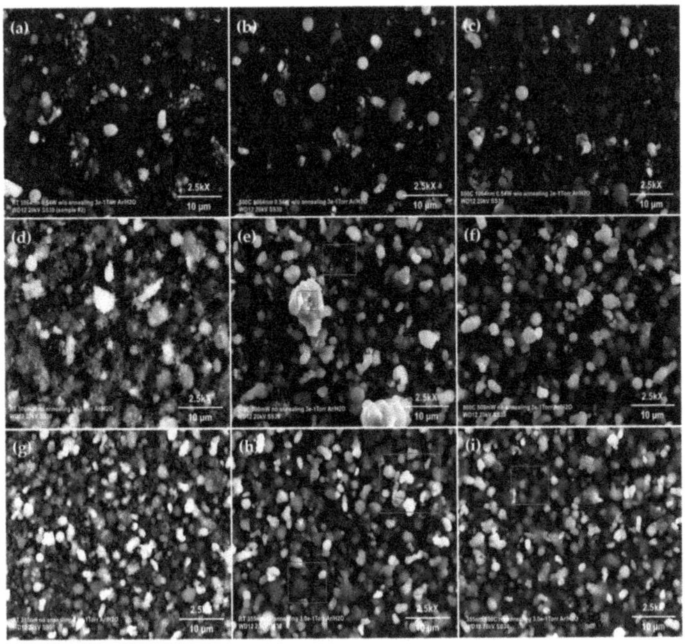

Figure 1. SEM images of hydroxyapatite coatings deposited at 1064 nm (**a–c**), 532 nm (**d–f**), and 355 nm (**g–i**) laser wavelengths at three different substrate temperatures: room temperature, 500 °C, and 800 °C. The red squares indicate the areas for which EDS analysis was performed.

As the wavelength increased from 355 to 1064 nm, the roughness of the coatings also increased. In addition, surface roughness initially increased when the temperature was raised from RT to 800 °C, but the magnitude of the increase between 500 °C and 800 °C is smaller as compared to the change between RT and 300 °C. The red squares on the SEM micrographs indicate the areas for which EDS analysis was performed. Based on the results, brighter particulates and the background had a similar Ca/P ratio.

Extensive EDS analysis was performed to understand the deposition stoichiometry, i.e., the Ca/P ratio. Results are provided in Supplementary Materials attached as Figures S4–S6. The first notable observation was that as the substrate temperature was raised, the Ca/P ratio increased in all wavelength regimes. A similar behavior was observed before by ArF excimer laser deposition, and their FTIR spectroscopic analysis revealed that PO^{4-} groups were gradually replaced with CO_3^{2-} groups as the temperature increased [66]. This result was also supported by our XPS measurements confirming the abundance of C1s excitations at higher temperatures.

Another observation point on the EDS analysis was about the effect of ablation wavelengths on the Ca/P ratio. 1064 nm and 532 nm produced HA coatings were had Ca/P ratios in the 1.6–1.7 range, while the 355 nm beam produced coatings with a Ca/P ratio higher than 1.7, for all substrate temperatures. Previously, Nistor et al. [52] reported that when identical deposition parameters were used but with two different wavelengths of 532 and 355nm, 355 nm one led to a smoother surface with non-HA phases [52]. Although it is not exactly known through which mechanism the stoichiometry is changing, it shows that one can get different stoichiometry and surface microstructures by varying the ablation wavelength.

Figure 2 shows the cross-sectional SEM image of HA coatings synthesized at 1064 nm. One can observe that an increase in the deposition temperature caused an increase in the surface roughness. With the temperature increase, the surface morphology became coarser due to faster surface-mass transport.

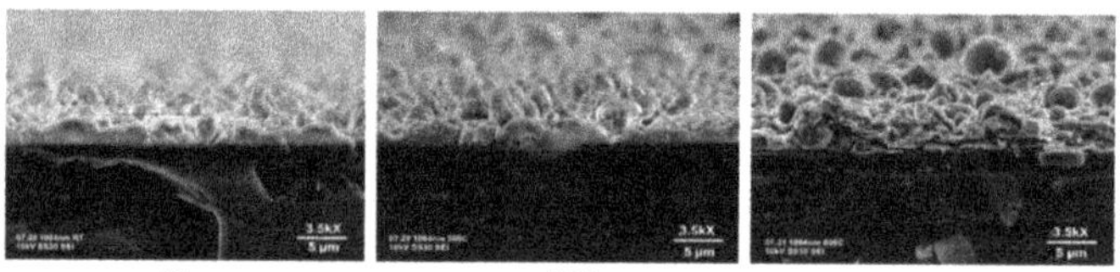

Figure 2. Cross-sectional view vs. substrate temperature in the case of hydroxyapatite coatings deposited at 1064 nm.

Figure 3 shows the cross-sectional view of SEM images of the deposited HA coatings at different wavelengths, along with the AFM line scan. From the cross-sectional view, one can observe that at 800 °C, the HA coating's microstructure is fragile. This effect had consequences that can be seen in protein adsorption, cell growth, and adherence strength tests. At this interface, the EDS results also suggested a lower Ca/P ratio, showing that the stoichiometry was not completely maintained, indicating some signs of degradation.

Another important aspect is the deposition rate. Among the three wavelengths, 532 nm gave the highest yield with 97.5 nm/min, then 355 nm with 38.5 nm/min, and 1064 nm had the lowest rate of 25.7 nm/min. It is understandable that the absorption wavelength of bulk hydroxyapatite is in the range of 300–500 nm; therefore, 532 nm had the better absorption, leading to a higher yield. However, when compared to Nd:YAG lasers, excimer lasers have relatively higher yields because of their repetition rate and flat-top pulse shape.

The AFM lines scan measurements indicated that the roughness values of the HA coatings increased with the used wavelength. When 355 nm was applied, RMS values of 0.65 ± 0.01 μm were inferred. RMS values increased to 1.01 ± 0.01 μm (for 532 nm) and 1.02 ± 0.01 μm (for 1064 nm). This result confirmed previous reports conducted at the same wavelength [44].

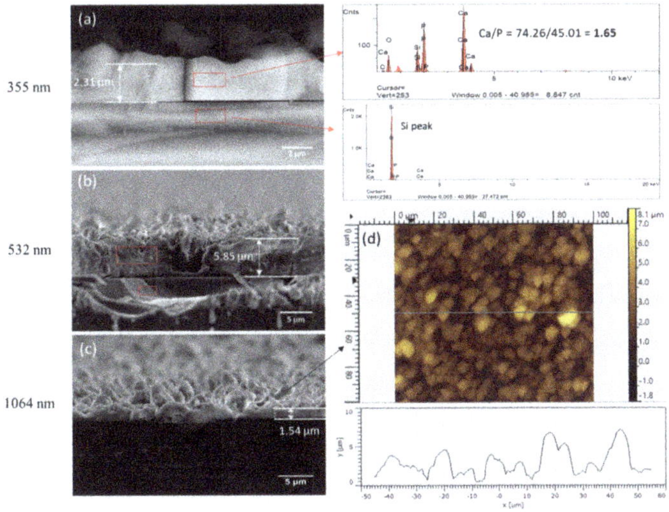

Figure 3. Cross-sectional SEM images of HA coatings deposited with (**a–c**) 355, 532, and 1064 nm laser beams. (**d**) AFM profile of (**c**).

Figure 4 shows the XRD peaks obtained at three different surface temperatures on Ti substrates (JCPDS 09-0432). At RT, two small peaks start to appear around θ = 32°. As the temperature increases, additional peaks appear, and the crystallinity of the sample increases with those additional peaks. At 800 °C, the XRD image shows extra peaks corresponding to tetracalcium phosphate (TTCP) in addition to the additional peaks of HA coatings. A small shift in the Ti substrate XRD peaks was observed. The apparent shift in the Ti substrate could be due to residual, thermal, and epitaxial strain. In addition, thermal strain may play a role due to variations in the cell parameters because of substrate and coating having different thermal expansion coefficients. In addition, some residual strain induced by the laser-ablated high-energy particle impingement may cause intrinsic residual strain, which may result in a slight change of XRD peaks of the substrate.

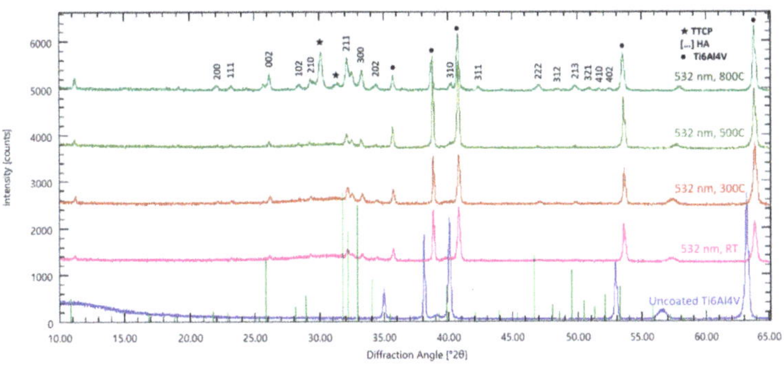

Figure 4. X-ray diffractograms of hydroxyapatite coatings synthesized onto Ti6Al4V substrates using a 532 nm beam. Referenced to JCPDS 09-0432.

The XRD images of HA coatings deposited onto Si(100) substrates are presented in Figure 5 (JCPDS 09-0432). The peak structure follows an analogous trend as in the case of HA coatings on Ti substrates. Similar to the case of HA coatings on Ti6Al4V, TTCP peaks emerged at 800 °C. It can, therefore, be concluded that temperature is a critical factor when synthesizing crystalline HA coatings on different types of substrates.

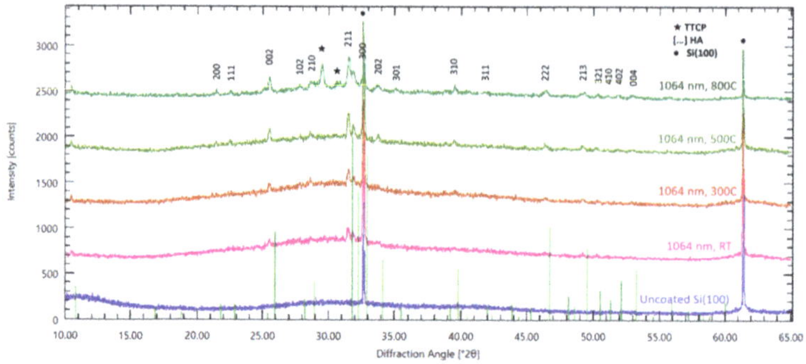

Figure 5. HA on Si(100) deposited with a 1064 nm beam. Referenced to JCPDS 09-0432.

Previously, it was demonstrated that the substrate temperature is a critical factor in determining the phase of the NbNx and Pb films. For a substrate temperature up to 450 °C, the film showed poor crystalline quality [17]. With temperature increase, the film became textured, and for a substrate temperature of 650–850 °C, a mix of cubic NbN and hexagonal phases was formed, and the surface roughness of the NbNx films increased as the temperature was raised from 450 °C to 850 °C. For Pb films, it was shown that when the substrate temperature was raised from RT to 300 °C, the film growth mechanism changed from Volmer-Weber to cluster formation mode. When the temperature was raised to values close to the melting temperature of Pb, the Ostwald ripening effect was observed [67]. In our case, the melting temperature of HA was far from the substrate temperature range; however, the results suggest that the variations in substrate temperature affected the nucleation-growth process, allowing smaller particles to conglomerate and possibly form larger crystals. The conglomeration process and growth of hexagonal-shaped crystals were observed at the powder-pellet level, further confirming the observations. As shown in Figures S2 and S3 in the Supplementary Materials, the crystal sizes steadily got bigger as the temperature was raised from 800 °C to 1200 °C, and crystallites melted at 1400 °C, at the temperature above the melting point of HA.

To estimate the change of crystallinity during various substrate temperatures, the crystallinity index (CI (%)) was calculated. The XRD deconvolution method was used for this purpose because it leads to more accurate values as compared to the easier and more popular Segal method [68]. The deconvolution of the X-ray diffraction patterns was performed with Origin Pro 2023b software, considering the Gaussian function as the shape of the resolved peaks. First, XRD patterns were baseline-corrected with user-defined points. Then, the CI (%) was calculated according to the following relation:

$$\mathrm{CI}(\%) = \frac{S_C}{S_t} \times 100\%$$

where S_c is the area of the crystalline domain, and S_t is the area under a total domain (crystalline and amorphous). The domain of the crystalline peak was identified according to FWHM.

The CI (%) values for the coatings deposited on Ti6Al4V were as follows: RT (16.6%), 300 °C (32.7%), 500 °C (51.1%), and 800 °C (81.5%). The CI (%) values for the coatings deposited on Si(100) were relatively lower: RT (6.2%), 300 °C (12.0%), 500 °C (32.1%), and 800 °C (61.4%). Corresponding calculations are provided in Figure S1 in the Supplementary Materials. In both cases, the crystalline-to-amorphous ratio steadily grew with respect to the substrate temperature.

It is important to note that the peaks of Si(100)-grown HA were shifted to lower angles, while the ones corresponding to Ti6Al4V-grown HA were shifted to higher angles. In principle, left-shift in XRD means lattice relaxation, while right-shift implies lattice strain. However, in the current situation, it is systematic, regardless of the applied temperatures.

Therefore, the origin of this effect is not evident. This imminent effect suggests an epitaxial-like behavior, but it does not follow the substrate crystal structure one-to-one. Murphy et al. [69] advanced an interesting observation regarding this issue. According to their results, once the initial layer of HA seeds, it grows identically regardless of the substrate. Therefore, the initial seeding process is determined by the substrate's thermodynamics, and the remaining growth process follows the primary nucleation regardless of the substrate structure.

The amorphous material lacks a long-range order, and XRD of amorphous materials has a broad hump, and sharp diffraction peaks are absent. Our results show that the HA coating is mostly crystalline, as confirmed by the presence of sharp diffraction peaks.

Coating Adherence

From a macroscopic point of view, all synthesized HA coatings were adherent to the Ti substrates, as demonstrated by preliminary adherence tests such as the finger test and application/removal of an adhesive tape (i.e., scotch test).

The measured values of adherence of the HA coatings synthesized by PLD at various substrate temperatures, i.e., RT–800 °C, are presented comparatively in Figure 6 and were calculated as mean ± standard deviation (SD). It should be emphasized that the events were considered only if the fracturing was of an adhesive type, i.e., when it occurred at the coatings-Ti substrate interface.

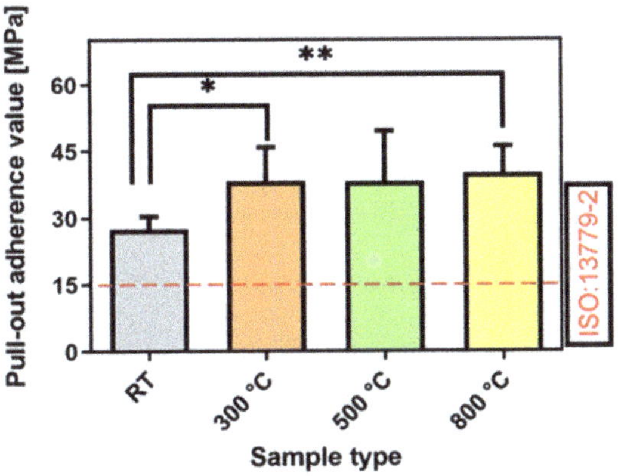

Figure 6. Mean bonding strength adherence values obtained in the case of hydroxyapatite coatings synthesized by pulsed laser deposition onto titanium substrates at RT, 300 °C, 500 °C, and 800 °C. Note: * $p < 0.05$, and ** $p < 0.01$, Student's *t*-test.

The pull-out test results (Figure 6) indicated dissimilar performances of the HA coatings fabricated at different substrate temperatures. HA coatings synthesized at RT were cleanly detached from the Ti substrate, yielding bonding strength values of ~27 MPa. The increase in the substrate deposition temperature seemed to result in an improvement in the films' adherence to the Ti substrate (Figure 6). Thus, in the case of HA coatings deposited at (300–800) °C, the pull-out detachment values were superior to the ones inferred for RT-synthesized coatings, being situated in a narrow range of 38–40 MPa.

The decreased value of the bonding strength inferred for HA coatings synthesized at RT could relate to a larger content of the amorphous counterpart, evident in this specific case using XRD. From the results presented in Figure 6, it can observed that the highest values of adherence corresponded to the HA coatings fabricated at 800 °C. This trend is similar to those previously reported in the literature for biological-origin HA coatings fabricated by PLD [22–24,70].

It is also important to note that only adherence values higher than the threshold imposed by ISO-137792 Part 2 Standard (i.e., >15 MPa) are considered acceptable for the coatings intended for applications with high biomechanical loads. Considering this criterion, the adherence values measured in the case of HA coatings (~three times higher than the threshold) should be emphasized.

Protein adsorption is a fundamental step in the interaction between cells and surface systems. When a biomaterial is exposed to biological fluids or tissues, proteins in the surrounding environment quickly adsorb onto its surface. The adsorbed proteins form a dynamic interface through noncovalent bonding that mediates subsequent cellular responses and influences the material's biocompatibility, which might be caused by the charged surface of the HA [71]. The adsorption of the protein to HA depends on the charged state of the surface. Given that we have a polycrystalline state, as is evident from the XRD image, the surface has multiple axis directions of domains. It is, therefore, expected that a positively charged surface (a-axis oriented domain) and a negatively charged surface (c-axis oriented domain) adsorb biological molecules such as carboxyl groups, phosphate groups, amino acids with charged side chains, and other charged particles.

We have measured the protein desorption of HA on Ti surface to evaluate its stability, as shown in Figure 7. Our results present that Ti with HA coatings shows better stability compared to a Ti-only surface. When the deposition temperature was situated between RT and 500 °C, no significant change in the protein desorption was observed. However, when the temperature was further increased to 800 °C, the protein desorption rate increased. It is normally expected that protein desorption should be the lowest at 800 °C due to its multiple peaks. However, it should be noted that the surface of the coating at 800 °C is quite coarse. In addition, surface roughness initially increases when the temperature is increased from RT to 800 °C, but the change in the increase between 500 °C and 800 °C is smaller compared to the change between RT and 300 °C.

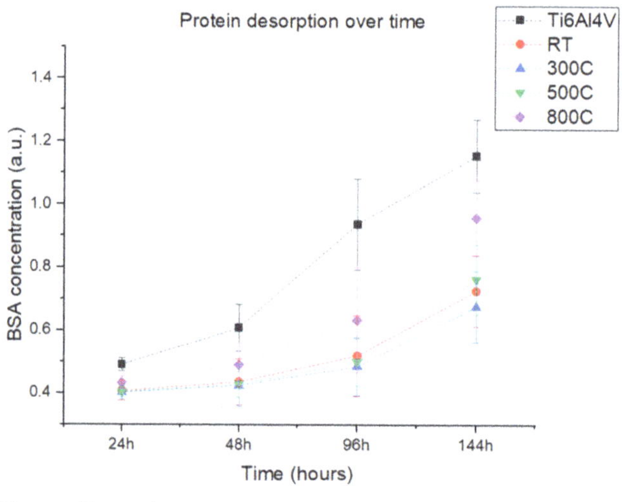

Figure 7. Protein desorption over time on the Ti6Al4V surface and hydroxyapatite coatings deposited at room temperature, 300 °C, 500 °C, and 800 °C.

Patterned surfaces have been shown to decrease the surface attachment of numerous microorganisms and to reduce the biofilm formation [11,72]. In the case of rough surfaces, the surface area of each island increases, which may lead to a decrease in the bonding site and the accessibility to the potential attachment sites, which leads to an energetically unstable surface to which biological molecules are unable to form a stable attachment.

As the wavelength increases from 355 to 1064 nm, the roughness of the film increases. Therefore, it is expected that there will be less protein absorption on the 1064 nm surface compared to the 355 nm surface. This is also in agreement with the observation of a higher number of peaks for the 532 nm samples, which have a higher protein absorption compared to 1064 nm.

It should be noted that there are many factors influencing the interaction between proteins and the HA surface, such as protein structures, morphology of HA surfaces, surroundings, and so on. Therefore, the interaction mechanism for various proteins might be different from each other.

A cell proliferation assay was performed to analyze the growth of mammalian cells on Ti6Al4V and HA-coated implant surfaces sintered at various temperatures. Our data shows that HA-coated surfaces support significantly more cells over a period of five days. Furthermore, an increase in temperature improves the biological activity of the HA-coated implants, as depicted in Figure 8.

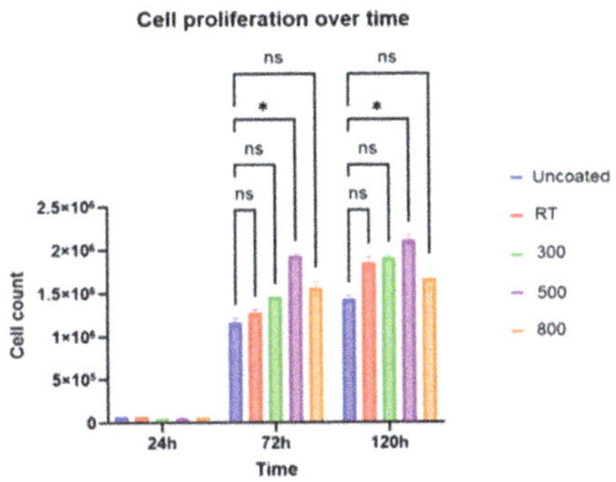

Figure 8. Proliferation of HEK293T cells on the Ti6Al4V substrate and hydroxyapatite coatings deposited at different temperatures with statistically significant (*) and no significance (ns).

One can observe from Figure 8 that the cell proliferation was significantly higher for all HA coatings and improved with the increase in the applied temperature, from RT to 500°C. As observed with protein absorption, higher temperature treatment performed at 800°C did not enhance the cell proliferation further, but it was significantly higher than the uncoated, bare Ti surfaces.

One should note that, for practical applications, the investigation of the long-term stability of HA coatings in model biofluids is of key importance and has to be considered in future studies.

4. Conclusions

HA coatings were grown onto silicon and titanium substrates using pulsed laser deposition (PLD) technique. The depositions were carried out at different surface temperatures and wavelengths. The effects of substrate temperature and ablation wavelength/mechanism on the structural, mechanical, and bioactivity properties of HA coatings were investigated using surface characterization tools, pull-out tests, and cell proliferation assays. Our results indicate that both surface temperature and wavelength are critical factors to obtain crystalline HA coatings. In addition, PLD was shown to be a feasible method to obtain HA coatings for the purpose of bioactivity measurements. The obtained results demonstrated that as the temperature increased, the coatings steadily crystallized. Further increase in the temperature above 700 °C adversely affected protein adsorption and mechanical properties. Similar trends were confirmed via cross-sectional SEM, pull-out testing, and cell proliferation tests performed on the HA coatings.

Supplementary Materials: The following supporting information can be downloaded at https://www.mdpi.com/article/10.3390/coatings13101681/s1, Figure S1. Representative calculations of crystallinity indices for (a,b) HA on Ti6Al4V at 800C and (c,d) HA on Si(100) at deposited RT. Crystallinity index: CI(%) = (S_c/S_t)*100%. Figure S2. SEM images of the HA pellets cold-pressed at 5000 psi, then annealed at different temperatures: (a) amorphous, (b) 800 °C, (c) 1000 °C, (d) 1200 °C, and (e) 1400 °C. Figure S3. XRD patterns of pellets shown in Figure S1. Figure S4. 1064 nm deposition, Ca/P ratio study through EDS analysis. Figure S5. 532 nm deposition, Ca/P ratio study through EDS analysis. Figure S6. 355 nm deposition, Ca/P ratio study through EDS analysis. Figure S7. XPS of HA coating on Ti6Al4V deposited at 500 °C with 532 nm beam.

Author Contributions: Conceptualization, A.O.E.; methodology, A.O.E.; formal analysis, S.K., Z.K., I.M., J.b.Y., S.B., Y.A., L.D. and A.O.E.; investigation, S.K., L.D. and A.O.E.; writing—original draft preparation, S.K. and A.O.E.; writing—review and editing, S.K., Z.K., I.M., J.b.Y., S.B., Y.A., L.D. and A.O.E.; supervision, A.O.E.; project administration, A.O.E. and L.D.; funding acquisition, A.O.E. and L.D. All authors have read and agreed to the published version of the manuscript.

Funding: This research was funded by KY IDeA Networks of Biomedical Research Excellence (ULRF12-1493C-01 via NIGMS grant# 8P20GM103436), KY NSF EPSCoR RA (3048115832-23-186) and RIE (3048115483-22-208), and WKU Research Foundation, Romanian Ministry of Research, Innovation, and Digitization under the Romanian National Core Program LAPLAS VII—contract no. 30N/2023.

Institutional Review Board Statement: Not applicable.

Informed Consent Statement: Not applicable.

Data Availability Statement: Not applicable.

Acknowledgments: We would like to thank John Andersland for SEM measurements.

Conflicts of Interest: The authors declare no conflict of interest.

References

1. Breme, H.; Biehl, V.; Reger, N.; Gawalt, E. Chapter 1a Metallic Biomaterials: Introduction. In *Handbook of Biomaterial Properties*; Springer: Berlin/Heidelberg, Germany, 2016; pp. 151–158.
2. Zhang, D.; Chen, Q.; Shi, C.; Chen, M.; Ma, K.; Wan, J.; Liu, R. Dealing with the Foreign-Body Response to Implanted Biomaterials: Strategies and Applications of New Materials. *Adv. Funct. Mater.* **2020**, *31*, 2007226. [CrossRef]
3. Espallargas, N.; Torres, C.; Muñoz, A. A metal ion release study of CoCrMo exposed to corrosion and tribocorrosion conditions in simulated body fluids. *Wear* **2015**, *332–333*, 669–678. [CrossRef]
4. Manivasagam, G.; Dhinasekaran, D.; Rajamanickam, A. Biomedical Implants: Corrosion and its Prevention—A Review. *Recent Patents Corros. Sci.* **2010**, *2*, 40–54. [CrossRef]
5. Wang, Q.; Wang, W.; Li, Y.; Li, W.; Tan, L.; Yang, K. Biofunctional magnesium coating of implant materials by physical vapour deposition. *Biomater. Transl.* **2021**, *2*, 248–256. [PubMed]
6. Harun, W.S.; Asri, R.I.; Sulong, A.B.; Ghani, S.A.; Ghazalli, Z. Hydroxyapatite-Based Coating on Biomedical Implant. *Hydroxyapatite Adv. Compos. Nanomater. Biomed. Appl. Its Technol. Facet.* **2018**, 6988. [CrossRef]
7. Asri, R.I.M.; Harun, W.S.W.; Samykano, M.; Lah, A.N.C.; Ghani, S.A.C.; Tarlochan, F.; Raza, M.R. Corrosion and surface modification on biocompatible metals: A review. *Mater. Sci. Eng. C* **2017**, *77*, 1261–1274. [CrossRef] [PubMed]
8. Li, Y.; Yang, W.; Li, X.; Zhang, X.; Wang, C.; Meng, X.; Pei, Y.; Fan, X.; Lan, P.; Wang, C.; et al. Improving Osteointegration and Osteogenesis of Three-Dimensional Porous Ti6Al4V Scaffolds by Polydopamine-Assisted Biomimetic Hydroxyapatite Coating. *ACS Appl. Mater. Interfaces* **2015**, *7*, 5715–5724. [CrossRef]
9. Li, Y.; Jahr, H.; Zhou, J.; Zadpoor, A.A. Additively manufactured biodegradable porous metals. *Acta Biomater.* **2020**, *115*, 29–50. [CrossRef]
10. Luo, Y.; Wang, J.; Ong, M.T.; Yung, P.S.; Wang, J.; Qin, L. Update on the research and development of magnesium-based biodegradable implants and their clinical translation in orthopaedics. *Biomater. Transl.* **2021**, *2*, 188–196.
11. Khuzhakulov, Z.; Kylychbekov, S.; Allamyradov, Y.; Majidov, I.; Ben Yosef, J.; Er, A.Y.; Kitchens, C.; Banga, S.; Badarudeen, S.; Er, A.O. Formation of picosecond laser-induced periodic surface structures on steel for knee arthroplasty prosthetics. *Front. Met. Alloy.* **2023**, *1*, 1090104. [CrossRef]
12. Papynov, E.K.; Shichalin, O.O.; Belov, A.A.; Buravlev, I.Y.; Mayorov, V.Y.; Fedorets, A.N.; Buravleva, A.A.; Lembikov, A.O.; Gritsuk, D.V.; Kapustina, O.V.; et al. $CaSiO_3$-HAp Metal-Reinforced Biocomposite Ceramics for Bone Tissue Engineering. *J. Funct. Biomater.* **2023**, *14*, 259. [CrossRef] [PubMed]
13. Dong, J.; Zhong, J.; Hou, R.; Hu, X.; Chen, Y.; Weng, H.; Zhang, Z.; Liu, B.; Yang, S.; Peng, Z. Polymer bilayer-Micro arc oxidation surface coating on pure magnesium for bone implantation. *J. Orthop. Transl.* **2023**, *40*, 27–36. [CrossRef] [PubMed]

14. Guo, Y.; Chen, C.; Zhang, S.; Ren, L.; Zhao, Y.; Guo, W. Mediation of mechanically adapted TiCu/TiCuN/CFR-PEEK implants in vascular regeneration to promote bone repair in vitro and in vivo. *J. Orthop. Transl.* **2022**, *33*, 107–119. [CrossRef] [PubMed]
15. Fiume, E.; Magnaterra, G.; Rahdar, A.; Verné, E.; Baino, F. Hydroxyapatite for Biomedical Applications: A Short Overview. *Ceramics* **2021**, *4*, 542–563. [CrossRef]
16. Pajor, K.; Pajchel, L.; Kolmas, J. Hydroxyapatite and Fluorapatite in Conservative Dentistry and Oral Implantology—A Review. *Materials* **2019**, *12*, 2683. [CrossRef] [PubMed]
17. Ielo, I.; Calabrese, G.; De Luca, G.; Conoci, S. Recent Advances in Hydroxyapatite-Based Biocomposites for Bone Tissue Regeneration in Orthopedics. *Int. J. Mol. Sci.* **2022**, *23*, 9721. [CrossRef]
18. Medical Implants Market Size to Surpass US$ 145.6 Bn by 2030. Available online: https://www.precedenceresearch.com/press-release/medical-implants-market (accessed on 28 August 2023).
19. Kim, H.; Camata, R.P.; Vohra, Y.K.; Lacefield, W.R. Control of phase composition in hydroxyapatite/tetracalcium phosphate biphasic thin coatings for biomedical applications. *J. Mater. Sci. Mater. Med.* **2005**, *16*, 961–966. [CrossRef]
20. Klein, C.P.; de Blieck-Hogemrst, J.; Wolket, J.; de Groot, K. Studies of the solubility of different calcium phosphate ceramic particles in vitro. *Biomaterials* **1990**, *11*, 509–512. [CrossRef]
21. Sobczak-Kupiec, A.; Drabczyk, A.; Florkiewicz, W.; Głąb, M.; Kudłacik-Kramarczyk, S.; Słota, D.; Tomala, A.; Tyliszczak, B. Review of the Applications of Biomedical Compositions Containing Hydroxyapatite and Collagen Modified by Bioactive Components. *Materials* **2021**, *14*, 2096. [CrossRef]
22. Duta, L.; Stan, G.E.; Popescu-Pelin, G.; Zgura, I.; Anastasescu, M.; Oktar, F.N. Influence of Post-Deposition Thermal Treatments on the Morpho-Structural, and Bonding Strength Characteristics of Lithium-Doped Biological-Derived Hydroxyapatite Coatings. *Coatings* **2022**, *12*, 1883. [CrossRef]
23. Mihailescu, N.; Stan, G.; Duta, L.; Chifiriuc, M.C.; Bleotu, C.; Sopronyi, M.; Luculescu, C.; Oktar, F.; Mihailescu, I. Structural, compositional, mechanical characterization and biological assessment of bovine-derived hydroxyapatite coatings reinforced with MgF 2 or MgO for implants functionalization. *Mater. Sci. Eng. C* **2016**, *59*, 863–874. [CrossRef] [PubMed]
24. Duta, L.; Oktar, F.; Stan, G.; Popescu-Pelin, G.; Serban, N.; Luculescu, C.; Mihailescu, I. Novel doped hydroxyapatite thin films obtained by pulsed laser deposition. *Appl. Surf. Sci.* **2013**, *265*, 41–49. [CrossRef]
25. Er, A.O.; Ren, W.; Elsayed-Ali, H.E. Low temperature epitaxial growth of Ge quantum dot on Si(100)-(2 $\times$ 1) by femtosecond laser excitation. *Appl. Phys. Lett.* **2011**, *98*, 013108. [CrossRef]
26. Farha, A.H.; Er, A.O.; Ufuktepe, Y.; Myneni, G.; Elsayed-Ali, H.E. Effects of substrate temperature on properties of NbNx films grown on Nb by pulsed laser deposition. *Appl. Surf. Sci.* **2011**, *258*, 1613–1618. [CrossRef]
27. Er, A.O.; Elsayed-Ali, H.E. Excitation-Induced Germanium Quantum Dot Formation on Si(100)−(2 $\times$ 1). *J. Appl. Phys.* **2010**, *108*, 034303–034310. [CrossRef]
28. Wang, X.; Jiang, J.; Tian, Y. A Review on Macroscopic and Microstructural Features of Metallic Coating Created by Pulsed Laser Material Deposition. *Micromachines* **2022**, *13*, 659. [CrossRef] [PubMed]
29. Er, A.O.; Elsayed-Ali, H.E. Femtosecond pulsed laser deposition of Ge quantum dot growth on Si (100)−(2 $\times$ 1). *Appl. Surf. Sci.* **2011**, *257*, 8078–8084. [CrossRef]
30. Er, A.O.; Elsayed-Ali, H.E. Electronically enhanced surface diffusion during Ge growth on Si(100). *J. Appl. Phys.* **2011**, *109*, 084320–084326.
31. Koren, G.; Gupta, A.; Baseman, R.J.; Lutwyche, M.I.; Laibowitz, R.B. Laser wavelength dependent properties of YBa2Cu3O7-d thin films deposited by laser ablation. *Appl. Phys. Lett.* **1989**, *55*, 2450–2452. [CrossRef]
32. György, E.; Mihailescu, I.N.; Kompitsasa, M.; Giannoudakosa, A. Particulates generation and solutions for their elimination in pulsed laser deposition. *J. Optoelectron. Adv. Mater.* **2004**, *6*, 39–46.
33. Mihailescu, I.N.; Teodorescu, V.S.; Gyorgy, E.; Luches, A.; Perrone, A.; Martino, M. About the nature of particulates covering the surface of thin films obtained by reactive pulsed laser deposition. *J. Phys. D Appl. Phys.* **1998**, *31*, 2236–2240. [CrossRef]
34. Hontsu, S. Preparation of Biomaterial Thin Films by Pulsed-Laser Deposition Technique. *Rev. Laser Eng.* **2000**, *28*, 407–412. [CrossRef]
35. Oladijo, S.; Akinlabi, E.; Mwema, F.; Stamboulis, A. An Overview of Sputtering Hydroxyapatite for Biomedical Application. In *IOP Conference Series: Materials Science and Engineering*; IOP Publishing: Bristol, UK, 2021.
36. Sun, L.; Berndt, C.C.; Gross, K.A.; Kucuk, A. Material fundamentals and clinical performance of plasma-sprayed hydroxyapatite coatings: A review. *J. Biomed. Mater. Res. Off. J. Soc. Biomater. Jpn. Soc. Biomater. Aust. Soc. Biomater. Korean Soc. Biomater.* **2001**, *58*, 570–592. [CrossRef] [PubMed]
37. Cheang, P.; Khor, K. Addressing processing problems associated with plasma spraying of hydroxyapatite coatings. *Biomaterials* **1996**, *17*, 537–544. [CrossRef] [PubMed]
38. Massaro, C.; Baker, M.A.; Cosentino, F.; Ramires, P.A.; Klose, S.; Milella, E. Surface and biological evaluation of hydroxyapatite-based coatings on titanium deposited by different techniques. *J. Biomed. Mater. Res. Off. J. Soc. Biomater. Jpn. Soc. Biomater. Aust. Soc. Biomater. Korean Soc. Biomater.* **2001**, *58*, 651–657. [CrossRef] [PubMed]
39. Cotell, C.M.; Chrisey, D.B.; Grabowski, K.S.; Sprague, J.A.; Gossett, C.R. Pulsed laser deposition of hydroxylapatite thin films on Ti-6Al-4V. *J. Appl. Biomater.* **1992**, *3*, 87–93. [CrossRef]
40. Baeri, P.; Torrisi, L.; Marino, N.; Foti, G. Ablation of hydroxyapatite by pulsed laser irradiation. *Appl. Surf. Sci.* **1992**, *54*, 210–214. [CrossRef]

41. Dinda, G.; Shin, J.; Mazumder, J. Pulsed laser deposition of hydroxyapatite thin films on Ti–6Al–4V: Effect of heat treatment on structure and properties. *Acta Biomater.* **2009**, *5*, 1821–1830. [CrossRef]
42. Blind, O.; Klein, L.H.; Dailey, B.; Jordan, L. Characterization of hydroxyapatite films obtained by pulsed-laser deposition on Ti and Ti-6AL-4v substrates. *Dent. Mater.* **2005**, *21*, 1017–1024. [CrossRef]
43. Arias, J.L.; Mayor, B.; Pou, J.; Rehman, I.; Knowles, J.; Best, S.; Bonfield, W.; García, F.; León, B.; Pérez-Amor, M. Effect of heat treatment on pulsed laser deposited amorphous calcium phosphate coatings. *J. Biomed. Mater. Res.* **1998**, *43*, 69–76.
44. Gomes, G.; Borghi, F.; Ospina, R.; López, E.; Borges, F.; Mello, A. Nd: YAG (532 nm) pulsed laser deposition produces crystalline hydroxyapatite thin coatings at room temperature. *Surf. Coat. Technol.* **2017**, *329*, 174–183. [CrossRef]
45. Yuan, B.; Chen, H.; Zhao, R.; Deng, X.; Chen, G.; Yang, X.; Xiao, Z.; Aurora, A.; Iulia, B.A.; Zhang, K.; et al. Construction of a magnesium hydroxide/graphene oxide/hydroxyapatite composite coating on Mg–Ca–Zn–Ag alloy to inhibit bacterial infection and promote bone regeneration. *Bioact. Mater.* **2022**, *18*, 354–367. [CrossRef] [PubMed]
46. Hidalgo-Robatto, B.; López-Álvarez, M.; Azevedo, A.; Dorado, J.; Serra, J.; Azevedo, N.; González, P. Pulsed laser deposition of copper and zinc doped hydroxyapatite coatings for biomedical applications. *Surf. Coat. Technol.* **2018**, *333*, 168–177. [CrossRef]
47. Ungureanu, E.; Dragomir, A.V.; Parau, A.C.; Mitran, V.; Cimpean, A.; Tarcolea, M.; Vranceanu, D.M.; Cotrut, C.M. In Vitro Evaluation of Ag- and Sr-Doped Hydroxyapatite Coatings for Medical Applications. *Materials* **2023**, *16*, 5428. [CrossRef] [PubMed]
48. Prodan, A.M.; Iconaru, S.L.; Predoi, M.V.; Predoi, D.; Motelica-Heino, M.; Turculet, C.S.; Beuran, M. Silver-Doped Hydroxyapatite Thin Layers Obtained by Sol-Gel Spin Coating Procedure. *Coatings* **2020**, *10*, 14. [CrossRef]
49. Cao, Z.; Li, L.; Yang, L.; Yao, L.; Wang, H.; Yu, X.; Shen, X.; Yao, L.; Wu, G. Osteoinduction Evaluation of Fluorinated Hydroxyapatite and Tantalum Composite Coatings on Magnesium Alloys. *Front. Chem.* **2021**, *9*, 727356. [CrossRef]
50. Zhang, L.; Liu, W.; Yue, C.; Zhang, T.; Li, P.; Xing, Z.; Chen, Y. A tough graphene nanosheet/hydroxyapatite composite with improved in vitro biocompatibility. *Carbon* **2013**, *61*, 105–115. [CrossRef]
51. Cotrut, C.M.; Ungureanu, E.; Ionescu, I.C.; Zamfir, R.I.; Kiss, A.E.; Parau, A.C.; Vladescu, A.; Vranceanu, D.M.; Saceleanu, A. Influence of Magnesium Content on the Physico-Chemical Properties of Hydroxyapatite Electrochemically Deposited on a Nanostructured Titanium Surface. *Coatings* **2022**, *12*, 1097. [CrossRef]
52. Nistor, L.; Ghica, C.; Teodorescu, V.; Nistor, S.; Dinescu, M.; Matei, D.; Frangis, N.; Vouroutzis, N.; Liutas, C. Deposition of hydroxyapatite thin films by Nd: YAG laser ablation: A microstructural study. *Mater. Res. Bull.* **2004**, *39*, 2089–2101. [CrossRef]
53. Yashiro, H.; Kakehata, M. A High-Quality Hydroxyapatite Layer Coated by Droplets Eliminated the Pulsed-Laser Deposition Scheme. In *Laser Applications in Microelectronic and Optoelectronic Manufacturing (LAMOM) XXVIII.*; SPIE: Bellingham, WA, USA, 2023.
54. Yashiro, H.; Kakehata, M. High crystalline hydroxyapatite coating by eclipse type pulsed-laser deposition for low annealing temperature. *Appl. Phys. Lett.* **2022**, *120*, 131602. [CrossRef]
55. Lin, K.C.; Chiu, Y.H.; Lin, J.H.; Pai, W.W. A quantitative analysis of the shape transition of Ge islands on Si(100) with NC-AFM. *Nanotechnology* **2005**, *16*, S63–S67. [CrossRef]
56. Popa, A.; Stan, G.; Enculescu, M.; Tanase, C.; Tulyaganov, D.; Ferreira, J. Superior biofunctionality of dental implant fixtures uniformly coated with durable bioglass films by magnetron sputtering. *J. Mech. Behav. Biomed. Mater.* **2015**, *51*, 313–327. [CrossRef] [PubMed]
57. *ISO 13779-2:2018*; Implants for surgery—Hydroxyapatite—Part 2: Thermally Sprayed Coatings of Hydroxyapatite. Available online: https://www.iso.org/standard/64617.html (accessed on 19 September 2023).
58. *ASTM D4541*; Standard Test Method for Pull-Off Strength of Coatings Using Portable Adhesion Testers. ASTM: West Conshohocken, PA, USA, 2017.
59. *ISO 4624*; Paints and varnishes—Pull-off test for adhesion. ISO: Geneva, Switzerland, 2016.
60. Bäuerle, D. *Laser Processing and Chemistry*; Springer Science & Business Media: Berlin/Heidelberg, Germany, 2013.
61. Chen, L.-C.; Chrisey, D.; Hubler, G. *Pulsed Laser Deposition of Thin Films*; Wiley: New York, NY, USA, 1994.
62. Kautek, W.; Roas, B.; Schultz, L. Formation of Y-Ba-Cu-oxide thin films by pulsed laser deposition: A comparative study in the UV, visible and IR range. *Thin Solid Film.* **1990**, *191*, 317–334. [CrossRef]
63. Liu, H.; Mao, X.; Yoo, J.; Russo, R. Early phase laser induced plasma diagnostics and mass removal during single-pulse laser ablation of silicon. *Spectrochim. Acta Part B At. Spectrosc.* **1999**, *54*, 1607–1624. [CrossRef]
64. Ojeda-G-P, A.; Döbeli, M.; Lippert, T. Influence of Plume Properties on Thin Film Composition in Pulsed Laser Deposition. *Adv. Mater. Interfaces* **2018**, *5*, 1701062. [CrossRef]
65. Agop, M.; Cimpoesu, N.; Gurlui, S.; Irimiciuc, S.A. Investigations of Transient Plasma Generated by Laser Ablation of Hydroxyapatite during the Pulsed Laser Deposition Process. *Symmetry* **2020**, *12*, 132. [CrossRef]
66. Mayor, B.; Arias, J.; Chiussi, S.; Garcia, F.; Pou, J.; Fong, B.L.; Pérez-Amor, M. Calcium phosphate coatings grown at different substrate temperatures by pulsed ArF-laser deposition. *Thin Solid Film.* **1998**, *317*, 363–366. [CrossRef]
67. Abdisatarov, B.; Ilhom, S.; Kholikov, K.; Loomis, D.; Dobrokhotov, V.; Khenner, M.; Er, A.O. Morphology and structure of Pb thin films grown on Si(111) by pulsed laser deposition. *Appl. Phys. A* **2020**, *126*, 1–7. [CrossRef]
68. Segal, L.; Creely, J.J.; Martin, A.E., Jr.; Conrad, C.M. An Empirical Method for Estimating the Degree of Crystallinity of Native Cellulose Using the X-Ray Diffractometer. *Text. Res. J.* **1959**, *29*, 786–794. [CrossRef]
69. Murphy, B.; Baez, J.; Morris, M.A. Characterizing Hydroxyapatite Deposited from Solution onto Novel Substrates in Terms of Growth Mechanism and Physical Chemical Properties. *Mater. Proc.* **2023**, *14*, 34.

70. Duta, L.; Mihailescu, N.; Popescu, A.C.; Luculescu, C.R.; Mihailescu, I.N.; Çetin, G.; Gunduz, O.; Oktar, F.N.; Popa, A.C.; Kuncser, A.; et al. Comparative physical, chemical and biological assessment of simple and titanium-doped ovine dentine-derived hydroxyapatite coatings fabricated by pulsed laser deposition. *Appl. Surf. Sci.* **2017**, *413*, 129–139. [CrossRef]
71. Dong, X.; Wang, Q.; Wu, T.; Pan, H. Understanding Adsorption-Desorption Dynamics of BMP-2 on Hydroxyapatite (001) Surface. *Biophys. J.* **2007**, *93*, 750–759. [CrossRef]
72. May, R.M.; Hoffman, M.G.; Sogo, M.J.; Parker, E.A.; O'toole, A.G.; Brennan, A.B.; Reddy, S.T. Micro-patterned surfaces reduce bacterial colonization and biofilm formation in vitro: Potential for enhancing endotracheal tube designs. *Clin. Transl. Med.* **2014**, *3*, 8. [CrossRef]

coatings

MDPI

Article

Lithium-Doped Biological-Derived Hydroxyapatite Coatings Sustain In Vitro Differentiation of Human Primary Mesenchymal Stem Cells to Osteoblasts

Paula E. Florian [1,*], Liviu Duta [2,*], Valentina Grumezescu [2], Gianina Popescu-Pelin [2], Andrei C. Popescu [3], Faik N. Oktar [4,5], Robert W. Evans [6] and Anca Roseanu Constantinescu [1]

1 Institute of Biochemistry, Romanian Academy, 060031 Bucharest, Romania; roseanua@gmail.com
2 National Institute for Lasers, Plasma and Radiation Physics, 077125 Magurele, Romania; valentina.grumezescu@inflpr.ro (V.G.); gianina.popescu@inflpr.ro (G.P.-P.)
3 Center for Advanced Laser Technologies (CETAL), National Institute for Lasers, Plasma and Radiation Physics, 077125 Magurele, Romania; andrei.popescu@inflpr.ro
4 Department of Bioengineering, Faculty of Engineering, Marmara University, Goztepe Campus, 34722 Istanbul, Turkey; foktar@marmara.edu.tr
5 Center of Nanotechnology & Biomaterials Application & Research, Marmara University, Goztepe Campus, 34722 Istanbul, Turkey
6 School of Engineering and Design, Brunel University, London UB8 3PH, UK; robertwevans49@gmail.com
* Correspondence: florian_paula@yahoo.com (P.E.F.); liviu.duta@inflpr.ro (L.D.); Tel.: +40-214574558 (L.D.)

Received: 27 October 2019; Accepted: 18 November 2019; Published: 21 November 2019

Abstract: This study is focused on the adhesion and differentiation of the human primary mesenchymal stem cells (hMSC) to osteoblasts lineage on biological-derived hydroxyapatite (BHA) and lithium-doped BHA (BHA:LiP) coatings synthesized by Pulsed Laser Deposition. An optimum adhesion of the cells on the surface of BHA:LiP coatings compared to control (uncoated Ti) was demonstrated using immunofluorescence labelling of actin and vinculin, two proteins involved in the initiation of the cell adhesion process. BHA:LiP coatings were also found to favor the differentiation of the hMSC towards an osteoblastic phenotype in the presence of osteoinductive medium, as revealed by the evaluation of osteoblast-specific markers, osteocalcin and alkaline phosphatase. Numerous nodules of mineralization secreted from osteoblast cells grown on the surface of BHA:LiP coatings and a 3D network-like organization of cells interconnected into the extracellular matrix were evidenced. These findings highlight the good biocompatibility of the BHA coatings and demonstrate that the use of lithium as a doping agent results in an enhanced osteointegration potential of the synthesized biomaterials, which might therefore represent viable candidates for future in vivo applications.

Keywords: biological-derived hydroxyapatite coatings; lithium doping; pulsed laser deposition; human mesenchymal stem cells; osteoblasts

1. Introduction

Surface properties of the biomaterials used for implants, such as chemistry and topography, are of crucial importance in establishing the tissue response to the success or failure of an implant [1–4]. Titanium (Ti) implants have been widely used in dental and orthopedic fields for bone replacement and tissue engineering due to their remarkable performances [5,6]. Current Ti surface functionalization focuses on simple and doped hydroxyapatite (HA) coatings as bioactive alternatives to autogenous bone for osseous remodeling processes [7–9].

Synthetic hydroxyapatite (HA) is a well-known biocompatible material used as an implantable ceramic for bone replacement due to its chemical-structural similarity with the inorganic part of human bone tissues [10]. It is stable in simulated biological fluids and exhibits an osteoconductive behavior [11,12]. Despite this capacity, HA ceramics possess weak mechanical properties [13]. To overpass this drawback, synthetic HA can be applied as a thin coating onto the surface of metallic implants, aiming to improve fixation at the living tissue—implant interface and to determine bone regeneration/formation [14].

It is worth noting that a great challenge in the biomedical field is the development of implants which could be easily integrated into the living body [15–17]. Biological-derived hydroxyapatite (BHA), which demonstrated exhibiting excellent biocompatibility, bioactivity, and osseoconduction characteristics [18], can serve as an alternative to conventional synthetic HA. Moreover, BHA materials doped with specific elements such as Ti, Li_2CO_3, Li_2O, Li_3PO_4, MgF_2, and/or MgO were reported to be a viable solution to improve the mechanical characteristics and to enhance the biological efficiency of an implant, due to their close resemblance with bone apatite [19]. One should stress upon the fact that the Earth's available mineral resources are threatened to become limited in the near future because of the rapid demographic increase and economic growth. Therefore, the access to low-cost sustainable natural resources (i.e., HA derived from animal bones, like in the current study) is critical, and consequently highly encouraged, for future economic growth and human welfare. The possibility to independently vary a large number of parameters (i.e., the laser fluence, pulse repetition rate, energy, target-to-substrate separation distance, substrate temperature, nature and pressure of the gas in the deposition chamber), on one hand, and the safety in functioning, fast processing, and low production costs, on the other hand, have advanced PLD, in the field of thin film growth, as a simple and versatile technique to produce layers with a high diversity of morphological and structural characteristics, which are superior to the ones pertaining to conventional deposition techniques [20–22]. For the difference of synthetic HA, natural apatites contain trace elements and new functional groups (i.e., HPO_4^{2-} and CO_3^{2-}) in their complex molecules. In this respect, the transfer of animal-origin HA by PLD is highly attractive due to the main advantage of this technique to grow stoichiometric films, with a controlled degree of crystallinity and thickness. One should note that these characteristics have a key influence over bio-resorption or dissolution, which are directly implicated in the process of films' osseointegration.

In our previous study [23], BHA:LiP coatings were shown both to exhibit a hydrophilic behavior as compared to bare Ti (control) and simple BHA coatings and to promote the human mesenchymal stem cells (hMSC) growth on the film surface, while maintaining an excellent adherence to the metallic substrate and evidencing a long-lasting anti-staphylococcal and -fungal biofilm activity [24]. The differentiation of hMSC to osteoblasts (OB) is a well-known model for bone regeneration used in the study of biocompatibility of materials proposed for implants [25,26].

Based on these observations, a leap forward in our research was performed by investigating the capacity of lithium-doped BHA (further denoted as BHA:LiP) coatings to promote the in vitro differentiation of primary hMSC from bone marrow aspirates to OB lineage. The expression of cell adhesion, cell morphology and OB specific markers were analyzed by fluorescence and Scanning Electron microscopy (SEM). Alkaline phosphatase (ALP) activity, an early key marker of osteogenesis, was quantitatively assessed to evaluate the capacity of BHA:LiP coatings to support the differentiation of hMSC to OB phenotype. To the best of our knowledge, this is the first evidence that synthesized BHA:LiP layers could offer proper conditions for hMSC adhesion and OB differentiation and sustain the achievement of an enhanced osseointegration potential for a new generation of metallic implants.

2. Materials and Methods

2.1. Powders Preparation

BHA powders were obtained from the cortical part of bovine femoral bones, following the protocol previously described in [20]. Briefly, the heads of femoral bones were cut off and the shafts processed. The bone marrows were then extracted and the soft tissue residues gently removed from the shafts. The femoral shafts were further cut into slices, cleaned and washed with distilled water and deproteinized for 14 days in an alkali media of 1% sodium hypochlorite. The resulting dry bone fragments were submitted to a calcination process (at 850 °C, 4 h in air), to remove any residual organic or biological hazardous components [27,28]. Before undergoing ball-milling to fine powders, the calcined bone specimens were crushed with a mortar and pestle. It should be emphasized that, for the fabrication of BHA powders, European Regulations [29] and ISO standard [30] were followed.

Batches of BHA powders were admixed with 1 wt.% of Li_3PO_4 (Sigma-Aldrich GmbH, St. Louis, MO, USA) and used for comparison.

2.2. *Target Preparation*

BHA and BHA:LiP powders were pressed at ~5 MPa in a 20 mm diameter mold. The resulting pellets were thermally treated in air, using an oven, for 4 h, at 700 °C. A heating rate of 20 °C/min and a cooling ramp of 5 °C/min were applied. Following this protocol, the fabrication of hard and compact targets, which are suited for PLD experiments, was achieved.

2.3. *PLD Experiment*

The PLD experiments were performed inside of a stainless-steel deposition chamber, in a pressure of 50 Pa water vapors. A KrF* excimer laser source, model COMPexPro 205 from Lambda Physics-Coherent, Göttingen, Germany (λ = 248 nm, $\tau_{FWHM} \leq 25$ ns), running at a repetition rate of 10 Hz, was used for the target ablation. The laser fluence was set at ~3.5 J/cm^2 (with a corresponding laser pulse energy of ~360 mJ). The laser beam was incident at 45° on the target surface. The ablated material was collected onto Ti substrates, with dimensions of (0.7×0.7) cm^2, which were placed parallel to the targets, at a separation distance of 5 cm. During the multi-pulse laser irradiation, the target was continuously rotated with 0.3 Hz and translated along two orthogonal axes to avoid piercing and to obtain unidirectional plasma. In addition, the substrates were heated and maintained at a constant temperature of 500 °C using a PID-EXCEL temperature controller (EXCEL Instruments, Mumbai, India), with a heating rate of 25 °C/min and a cooling ramp of 10 °C/min. Keeping the same parameters intact for the growth of one film, 15,000 consecutive laser pulses were applied.

Prior to introduction into the deposition chamber, the substrates were successively cleaned following a three-step protocol, implemented in our laboratory: acetone, ethylic alcohol, and deionized water, for 15 min each. Furthermore, to avoid any possible micro-contamination, the targets were submitted to a "cleaning" process with ~2000 laser pulses. To collect the flux of expulsed micro-impurities, a shutter was interposed between the target and the substrate.

To improve the crystallinity of the synthesized coatings, all samples were submitted to a 6 h post-deposition thermal treatment in water vapors enriched atmosphere, at a temperature of 600 °C.

2.4. *Cell Culture*

Primary hMSC were obtained from bone marrow aspirates and isolated by density gradient centrifugation, as previously described in Ref. [31]. Bone marrow was harvested from one healthy patient undergoing surgery for the orthopaedic implant procedure, with the approval of the Ethics Committee of the University of Medicine and Pharmacy of Craiova (Reference No. 68/11 July 2016). Human primary OB cells were obtained by differentiation of the hMSC.

The hMSC at low passage (2) were grown in complete Dulbecco's Modified Eagle Medium (DMEM) with low glucose (1 g/L), supplemented with 10% (*v*/*v*) fetal bovine serum (FBS) and 1% (*v*/*v*) streptomycin/penicillin (all from Gibco (Life Technologies, Paisley, UK)). For in vitro OB differentiation, the hMSC were grown in complete Minimum Essential Medium Alpha (α-MEM—Gibco, Thermo Fisher Scientific (Life Technologies, Paisley, UK)), supplemented with osteogenic factors: 82 µg/mL ascorbic acid (Sigma, Saint Louis, MO, USA), 100 nM dexamethasone (Sigma) and 10 mM β-glycerophosphate (Santa Cruz Biotechnology, (Dallas, Texas, USA)) [32]. The cells were maintained in culture at a density of 5×10^3 cells/cm^2 in 24-well plates, for 1, 3, and 21 days, respectively.

2.4.1. Cell Adhesion and Differentiation

Osteogenic Differentiation

The hMSC were cultivated in complete α-MEM or in complete α-MEM supplemented with osteogenic factors (osteoinduction medium, OIM), if differentiation to osteogenic phenotype was envisaged. Both hMSC and OB were cultivated in 24-well plates on the surface of Ti, BHA, and BHA:LiP coatings. All materials used for biocompatibility investigations were sterilized in a *Falcon 30 Autoclave* (LTE Scientific, (Greenfield, Oldham, UK)) using water vapors at 121 °C and 1 atm, for 30 min, before interaction with the cells. The cell adhesion and specific differentiation markers were assessed by immunofluorescence techniques.

The cells were cultivated for one and three days, then fixed with 4% paraformaldehyde (PFA) for 10 min and permeabilized with 0.2% Triton X-100 solution. Unspecific sites were blocked for 1 h with 0.5% bovine serum albumin solution. Cells were labeled with primary mouse anti-human vinculin antibodies (Sigma-Aldrich, Saint Louis, MO, USA) for 30 min before incubation with secondary antibodies goat anti-mouse conjugated with *Alexa Fluor 594* (Thermo Fisher Scientific, CA, USA). For actin filaments labeling, the samples were incubated for 30 min with phalloidin reagent conjugated with *Alexa Fluor 488* (Thermo Fisher Scientific). The nuclei were counterstained with Hoechst fluorescent dye (Life Technologies, Molecular Probes, Eugene, OR, USA). After the incubation period, the samples were mounted for microscopy investigation with *Medium Prolong Gold AntiFade* (Life Technologies). Samples were visualized using a *Zeiss AxioCam Erc5s* fluorescence microscope with *ApoTome 2* module and image acquisition was performed using *AxioCam MRm* camera (20× and 40× objectives). For image editing, the *Axio Vision Rel 4.8* software was utilized (Zeiss, Jena, Germany).

The differentiation capacity of the hMSC towards osteogenic phenotype was investigated using specific markers, such as osteocalcin and ALP.

ALP Activity Assay

The detection and quantification of ALP were performed using a colorimetric enzymatic assay (Quantitative Alkaline Phosphatase ES Characterization Kit, CHEMICON, Millipore (Darmstadt, Germany)). Briefly, the hMSC and OB were grown onto the surface of Ti, BHA, and BHA:LiP coatings for 14 days. The cell supernatant was then removed and the samples covered with cells were moved in a new 24-well plate for ALP activity measurements. After successive washing steps, using phosphate saline buffer, the cells were incubated with p-nitrophenyl phosphate substrate buffer to initiate the enzymatic hydrolysis. The absorbance level of yellow-colored by-product p-nitrophenol produced in the reaction was read at 405 nm in a spectrophotometer (Mithras, Berthold Technologies, Bad Wildbad, Germany) and was proportional with the amount of ALP present within the reaction. The ALP present in the sample was quantified using a standard curve of recombinant ALP present in the kit. All experiments were performed in triplicate.

Osteocalcin Detection

For osteocalcin protein detection in hMSC and OB grown on the surface of Ti, BHA and BHA:LiP coatings, a protocol for immunofluorescence—similar to the one for actin/vinculin detection described above—was applied. Briefly, the cells were cultivated for 1 and 21 days, then fixed and permeabilized before incubation with rabbit polyclonal primary anti-human osteocalcin antibodies (CHEMICON, Millipore). The detection was performed using a secondary antibody goat anti-rabbit conjugated with *Alexa Fluor 594* (Molecular Probes, Eugene, OR, USA).

2.4.2. Mineralization Assay

The extracellular matrix mineralization by OB was demonstrated by Alizarin Red staining [33,34]. Briefly, the cells were fixed with 4% PFA, and stained with a solution of 40 mM, pH 4.1 Alizarin Red S, for 30 min. Excess stain was removed with several distilled water washing steps. Finally, the cells were air-dried and the deposited Alizarin Red S dye was dissolved in a solution with 10% acetic acid and 20% methanol for 30 min. The absorbance of the resulting solution was recorded at a wavelength of 450 nm using a Berthold microplate reader (Bad Wildbad, Germany.), whilst the quantification was performed using a standard curve of the dye.

2.4.3. SEM

To investigate the discrete features of cellular morphology in the presence of biomaterials, the hMSC and OB grown onto the surfaces of Ti, BHA, and BHA:LiP coatings for 14 and 21 days were prepared for SEM imaging. Briefly, the cells were fixed with 2.5% glutaraldehyde solution and then dehydrated by sequential immersion in 70%, 90% and 100% ethanol. All specimens were washed with 50% and 75% hexamethyldisilazane (HMDS) solution prepared in ethanol and finally in 100% HMDS. The probes were subjected to HMDS evaporation in a Euroclone AURA 2000 M.A.C. (BioAir, Siziano, Pavia, Italy) fume hood. All samples were first air-dried and then metalized with a 100 Å thin silver layer, deposited using a BAL-TEC SCD 005 sputter coater (Schalksmühle, Germany). A *FEI Inspect S50* (Hillsboro, OR, USA) scanning electron microscope was used.

2.4.4. Statistical Analysis

All experiments were carried out in quadruplicates in order to achieve statistical significance. The statistical analysis was performed by using the unpaired Student's *t*-test and differences were considered significant for $p < 0.05$.

3. Results and Discussion

3.1. Cell Adhesion and Morphology

In vitro hMSC to OB phenotype differentiation is a long-lasting process (21 to 28 days), induced by specific osteogenic factors, which implies three cell stages: proliferation, differentiation, and maturation. Cell adhesion is an essential event that precedes the cell proliferation and differentiation. The process takes place before the formation of bone tissue and involves many proteins of the cytoskeleton that strongly interact with each other, such as integrins, actin filaments, and vinculin [35].

Cell adhesion is a parameter with a high specificity which characterizes relative sticking of the cell to a material or film. The evaluation of cell adhesion after longer incubation times provides relevant information regarding the behavior of cells on the surface of the biomaterials after implantation. In vivo, the biomaterial integration implies the formation of a stable interface between the extracellular matrix of the OB and the surface of the material. This excludes any intermediate fibrous tissue between the bone tissue and the surface of the material [36].

The hMSC and OB were grown onto the surface of Ti and BHA coatings for one and three days, and the samples were prepared for fluorescent microscopy imaging. The morphology and skeleton of the cells were investigated by labelling actin and vinculin, two proteins with important roles in the initiation of the cell adhesion process [37]. They are partially overlapping at the focal adhesion site, so the co-localization of the two proteins was investigated in all the tested samples.

After only one day in culture, in the presence of the OIM, the cells were already attached and presented an uniform pattern of spreading on the surface of both BHA and BHA:LiP coatings (Figure 1). After three days in culture, the cells revealed a pattern typical for actin (green) and vinculin (red) immunofluorescence labelling for all the tested samples. Thus, vinculin was easily spotted in focal contact sites, demonstrating the adhesion of the cells to the surface of the synthesized coatings, while the polymerized actin filaments network sustained the cells in this process. It is worth mentioning that the actin filaments form a dense cytoskeleton with a network-like appearance, indicating an optimum adhesion of the cells on the surface of the BHA and BHA:LiP coatings (Figure 1) Moreover, in the case of cells grown for three days onto the surface of the BHA:LiP coatings in the osteogenic medium, the actin filaments presented a circumferential pattern in the vicinity of the plasma membrane, while the vinculin was present at focal contacts. All these features suggest an optimum adhesion of the cells promoting a good differentiation towards osteogenic fate (Figure 1). Indeed, in our previous study, the viability assays revealed that the presence of lithium as a doping reagent promoted the in vitro growth of the hMSC cells on the coating surface [23]. Moreover, it was demonstrated that the synthesized lithium-doped coatings exhibited low cytotoxicity and an excellent biocompatibility on human osteosarcoma MG63, new initiated dermal fibroblasts, and immortalized keratinocytes HaCaT cells [24]. Additionally, a long-lasting anti-staphylococcal and -fungal biofilm activity of the lithium-doped coatings was proved on the tested species, which are representative for the etiology of medical devices biofilm-associated infections [24].

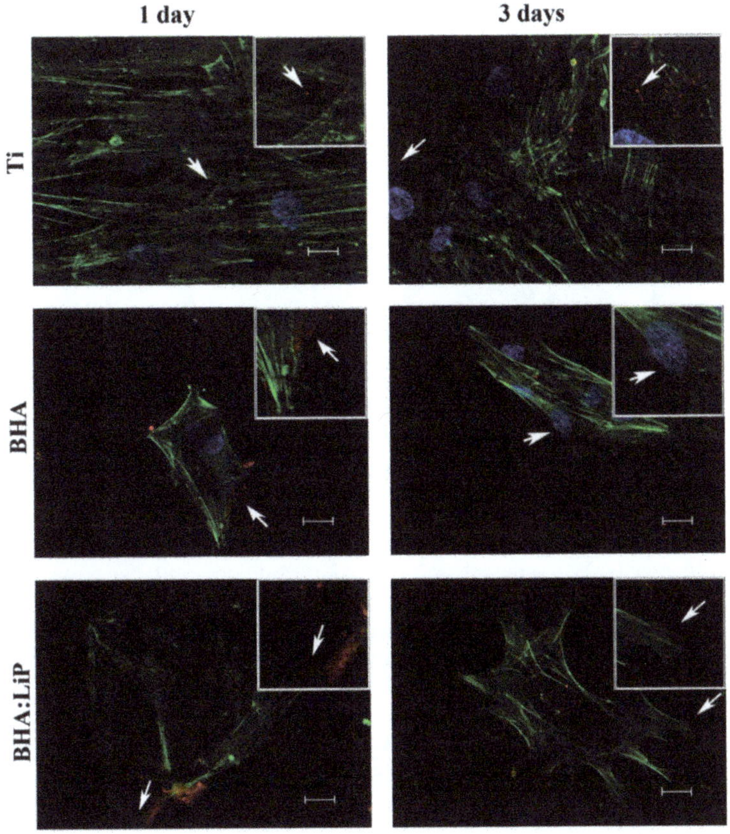

Figure 1. Cell morphology and cytoskeletal dynamic evaluation of human mesenchymal stem cells grown on the surface of Ti, biological-derived hydroxyapatite (BHA) and lithium-doped BHA (BHA:LiP) coatings after 1 and 3 days in culture. Green—actin; Red—vinculin; Blue—nucleus. Scale bar: 20 µm.

3.2. Analysis of ALP Activity

ALP is an enzyme present within all tissues of the body and is responsible for dephosphorization of a large number of molecules, such as deoxyribonucleic acid (DNA) (DNA), ribonucleic acid (RNA), and some proteins under alkaline levels of pH inside the cell [38]. ALP concentration is higher in some organs such as liver, kidney, bone, placenta, and embryo or in the case of specific disease states [39,40]. All pluripotent cells, including embryonic stem cells or carcinoma embryonic cells, express ALP protein. However, during the differentiation process towards the osteoblastic phenotype, ALP expression in bone progenitor cells is enhanced, thus constituting an important early marker of

OB differentiation [34,41,42]. To further investigate the characteristics of the differentiation process of primary hMSC to OB grown on the surface of BHA and BHA:LiP coatings, the enzymatic activity of ALP inside the cells cultivated for 14 days was measured (Figure 2).

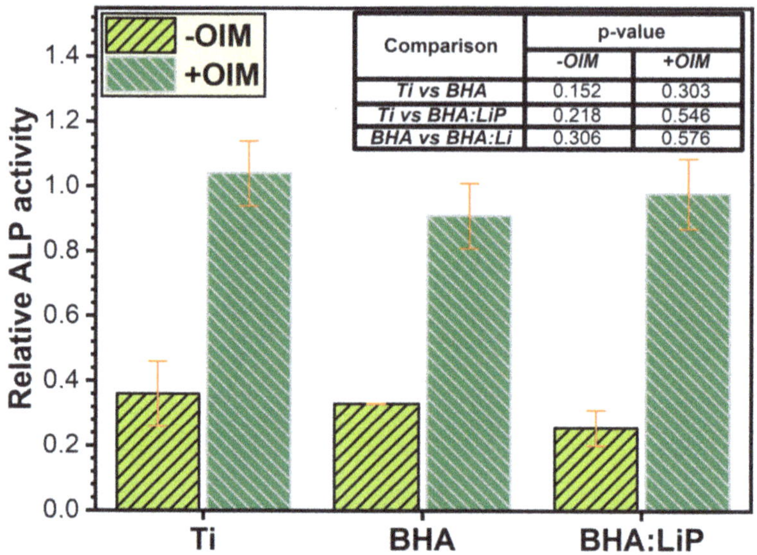

Figure 2. Alkaline phosphatase enzymatic activity detection in osteoblasts differentiated from primary human mesenchymal stem cells and grown on the surface of Ti, biological-derived hydroxyapatite (BHA) and lithium-doped BHA (BHA:LiP) coatings for 14 days, in the absence (−) or presence (+) of osteoinduction medium. Inset: table with the inferred *p*-values.

The in situ level of ALP, which is an early marker of osteogenic differentiation, in hMSC and differentiated OB after 14 days is depicted in Figure 2. The differences between the measured ALP levels were not found to be statistically significant for all the tested materials ($p > 0.05$—inset to Figure 2). However, the value registered for the enzymatic activity in cells grown on all the tested surfaces (Ti, BHA and BHA:LiP coatings) is higher in the presence of osteoinduction medium (+OIM) than in its absence (-OIM) (Figure 2). In the case of BHA:LiP surface, the obtained results could be explained by the characteristics of the coatings induced by the lithium doping agent (low roughness values and a hydrophilic behavior) [23,24]. One should emphasize that all these enhanced properties of the BHA:LiP coatings could indicate their potential in better supporting the bone tissue formation.

3.3. Expression of OB-Specific Differentiation Markers

The OB cells have the ability to synthesize the major components of their matrix, such as collagenous and non-collagenous proteins, offering support for bone mineral deposition. The correct assembly of bone extracellular matrix assures mechanical stress resistance of bone tissue and constitutes a good basis for the successful osteogenesis and osteoinduction of an implant [43,44].

The cells' capacity to differentiate towards osteogenic phenotype was investigated by fluorescence microscopy and enzymatic assay using specific markers, i.e., osteocalcin and ALP. The expression of the osteocalcin marker in extracellular matrix was investigated in the presence or absence of OIM, after 21 days of cell differentiation on the surface of Ti, BHA, and BHA:LiP coatings.

As shown in Figure 3, in all tested samples, a low level of osteocalcin marker expression detected in hMSC after one day in culture could be observed. On the contrary, after 21 days in culture, the expression of osteocalcin can be observed in hMSC and is characterized by the presence of discrete punctiform vesicles dispersed in an uniform manner throughout the cytoplasm. The pattern of dispersion of osteocalcin in the cell was observed for all tested materials and was previously reported by Sima et al. [32], in the case of OB cells cultivated on the surface of HA–sodium maleate-vinyl acetate copolymer coatings. These findings highlight a good biocompatibility of the BHA structures.

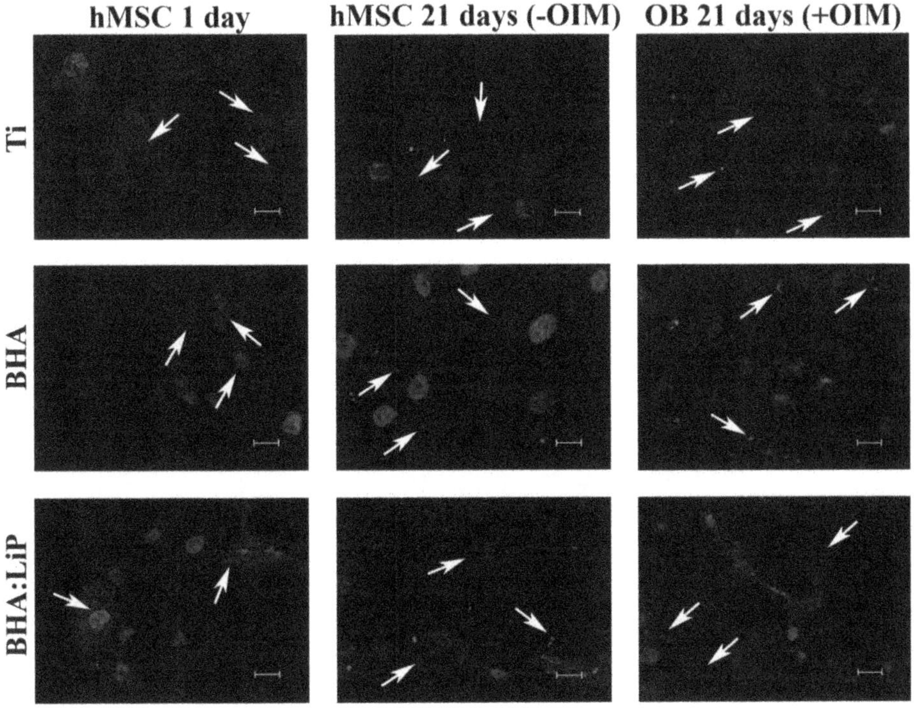

Figure 3. Investigation of osteocalcin expression in human mesenchymal stem cells and osteoblasts grown on the surface of Ti, biological-derived hydroxyapatite (BHA) and lithium-doped BHA (BHA:LiP) coatings after 1 and 21 days in culture. Red—osteocalcin; Blue—nucleus. Scale bar: 20 μm.

In the case of OB grown on the surface of BHA and BHA:LiP coatings, the presence of large granules was observed, in both the perinuclear region and near filamentous structures inside the cells. This pattern was more pronounced after 21 days of differentiation in the presence of OIM, suggesting that osteocalcin protein was engaged in the secretory pathway for the formation of extracellular matrix that surrounds the OB. Our results demonstrate that, in the presence of OIM, the cells grown on the surface of BHA and BHA:LiP coatings succeeded the differentiation towards osteoblastic phenotype.

3.4. Mineralization

As the cells differentiate from hMSC to OB, the expression of specific markers diminishes and the extracellular matrix mineralization process emerges. This specific morphological feature of calcium deposition and matrix mineralization characterizes the late stage of the osteogenesis process. Furthermore, the matrix mineralization was quantified by Alizarin Red dye detection (which specifically binds calcium ions) to evaluate the capacity of coatings to induce OB differentiation. The cells were grown for 28 days, in both the presence and absence of OIM.

As shown in Figure 4, the quantification of the bound dye revealed an enhanced secretion of calcium and thus a strong mineralization of the extracellular matrix by cells differentiated to OB and cultivated in OIM. One should note that the differences observed in the level of mineralization, expressed by the values of Alizarin Red dye, were found to be statistically significant for all three analyzed situations. In addition, the level of mineralization in the case of BHA:LiP coatings was significantly higher as compared to control Ti.

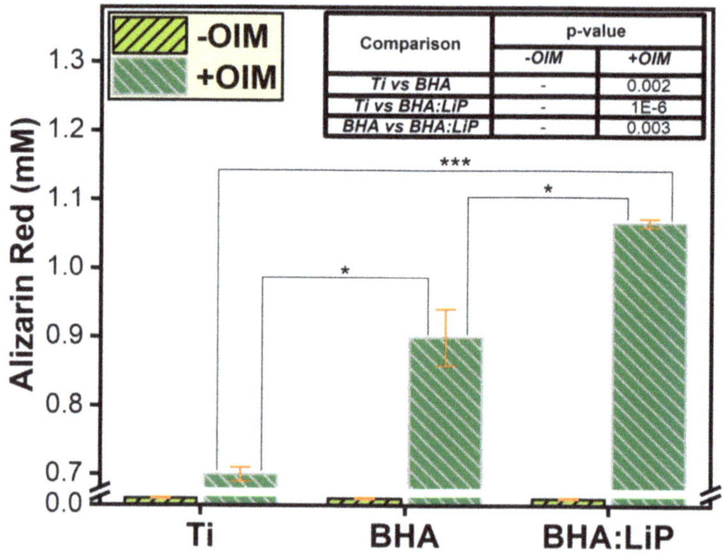

Figure 4. Extracellular matrix mineralization of osteoblasts differentiated from primary human mesenchymal stem cells grown on the surface of Ti, biological-derived hydroxyapatite (BHA), and lithium-doped BHA (BHA:LiP) coatings for 28 days, in the absence (−) or presence (+) of osteoinduction medium. Inset: table with the inferred p-values (* $p < 0.05$; *** $p < 0.0001$).

3.5. Structural Morphology of OB

The morphological investigation of attached cells offers important details on their interaction with the sample surfaces onto which they are grown. In this respect, SEM analysis was used to follow the dynamics of OB transformation from primary hMSC, over a period of 21 days. Moreover, the study of structural morphology details of OB helps build the picture of the progressive extracellular matrix mineralization and also elucidates the way the OB integrated in the mineralized matrix could maintain cellular connections with this natural biological support.

As shown in Figure 5, after 14 days in culture, the hMSC covered the surface of the materials, showing a good proliferation rate. Moreover, a typical fibroblast-like shape could be observed. In the case of cells grown in OIM, mineralization nodules were observed. However, a greater number of cells interconnected in the extracellular matrix was detected for BHA and BHA:LiP coatings. Visible differences could be observed between the morphology of cells grown on Ti and the ones cultivated on the surface of BHA and BHA:LiP coatings (Figure 5). Thus, in the latter case, the physical-chemical characteristics of the surface induced a proliferation in a tridimensional model.

The observed pattern was enhanced and could be easily observed after 21 days of differentiation on the surface of BHA and BHA:LiP coatings, in the presence of the OIM (Figure 6). The OB proliferated after three weeks in culture and secreted a mineralized extracellular matrix, the cells being inter-connected through dendritic processes and gap junctions, as previously described in literature [45–47]. This network of OB could constitute the premise for early differentiation of these cells grown onto the surfaces of BHA and BHA:LiP coatings in vitro and, consequently, for a better osteointegration in vivo [48,49]. This is in line with other studies showing that primary human osteoblastic cells were enabled to produce extracellular matrix and form an ex vivo 3D network of osteocytes, which are the terminally differentiated bone cells, via biomimetic assembly using microfluidic perfusion culture [48]. Chen et al. [49] reported on mineralized matrix morphology in 3D culture of osteoblasts and revealed by SEM a synergistic cross-talk between 1alpha, 25 $(OH)_2$ vitamin D3, and bone morphogenetic protein-2 that sustained improved osteogenesis and mineral deposition.

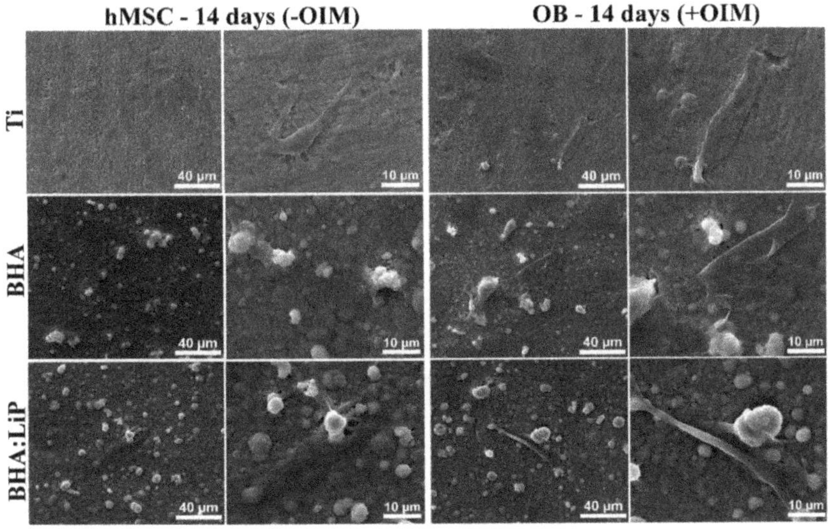

Figure 5. SEM micrographs of human mesenchymal stem cells differentiated to osteoblasts grown on the surface of Ti, biological-derived hydroxyapatite (BHA), and lithium-doped BHA (BHA:LiP) coatings for 14 days, in the absence or presence of osteogenic factors.

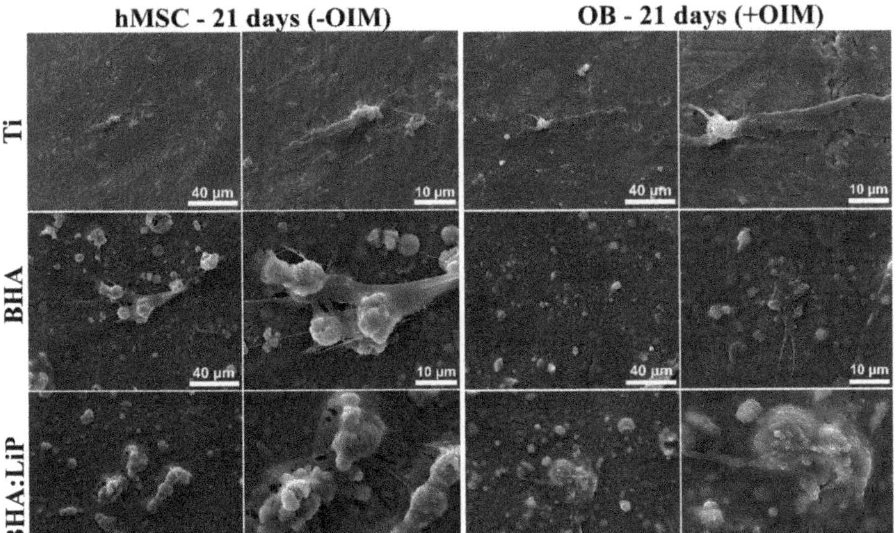

Figure 6. SEM micrographs of human mesenchymal stem cells differentiated to osteoblasts grown on the surface of Ti, biological-derived hydroxyapatite (BHA), and lithium-doped BHA (BHA:LiP) coatings for 21 days, in the absence or presence of osteogenic factors.

4. Conclusions

Biologically-derived hydroxyapatite (BHA) and lithium-doped BHA (BHA:LiP) coatings were synthesized by pulsed laser deposition onto medical-grade titanium (Ti) substrates. The morphological and cellular adhesion investigations performed in vitro showed that BHA:LiP coatings induced an enhanced adhesion of the primary human mesenchymal stem cells (hMSC) as compared to Ti or BHA coatings. The differentiation towards osteoblastic phenotype of hMSC grown on BHA and BHA:LiP coatings was demonstrated after cells cultured for a period of 21 days in osteogenic medium. In the case of cells grown on the surface of BHA:LiP coatings, large granules of osteocalcin protein were observed in the bone extracellular matrix, indicating a strong feature of the osteoblastic phenotype. In addition, the enzymatic activity of the alkaline phosphatase marker of osteogenic differentiation was increased in the case of cells grown on the surface of BHA:LiP coatings in the presence of osteogenic medium. In addition, the presence of numerous nodules of mineralization from osteoblast cells grown on the surface of BHA:LiP coatings, as well as a network-like organization of osteoblasts interconnected into the mineralized extracellular matrix, was evidenced. This network of osteoblasts could constitute the premise for early differentiation of cells grown on the surface of BHA:LiP coatings in vitro and, therefore, for an improved osteointegration in vivo.

Author Contributions: Conceptualization, P.E.F., L.D., and A.R.C.; Methodology, V.G.; Software, P.E.F, G.P.-P., and A.C.P.; Validation, P.E.F., L.D., F.N.O., R.W.E., and A.R.C.; Formal analysis, P.E.F., L.D., and A.R.C.; Investigation, G.P.-P. and P.E.F.; Resources, L.D., F.N.O., R.W.E., and A.R.C.; Data curation, P.E.F. and L.D.; Writing—original draft preparation, P.E.F., L.D., and A.R.C.; Writing—review and editing, P.E.F., L.D., V.G., G.P.-P., A.C.P., F.N.O., R.W.E., and A.R.C.; Visualization, P.E.F., L.D., V.G., G.P.P., A.C.P., R.W.E., and A.R.C.; Supervision, L.D.; Project administration, P.E.F., L.D., and A.R.C.; Funding acquisition, L.D. and A.R.C.

Funding: This work was supported by a grant from the Ministry of Research and Innovation, CNCS-UEFISCDI, No. PN-III-P1-1.1-PD-2016-1568 (PD 6/2018), within PNCDI III. L.D, A.P, V.G, and G.P-P acknowledge with thanks the partial support of the Romanian National Authority for Scientific Research and Innovation, CNCS-UEFISCDI, under Project No. PN-II-RU-TE-2014-1570 (TE 108/2015) and Core Programme–Contract 16N/2019. A.C.P. also acknowledges the support of CNCS-UEFISCDI in the framework of Contract PN-III-P1-1.1-TE-2016-2015 (TE136/2018). P.E.F. and A.R.C. acknowledge the support of the Institute of Biochemistry of the Romanian Academy under the Structural and Functional Proteomics Research Program.

Acknowledgments: All authors thank Livia Sima for useful discussions and for providing human mesenchymal stem cells for the experiments. All authors acknowledge George E. Stan for his useful comments and kind help.

Conflicts of Interest: The authors declare no conflict of interest.

References

1. Alvarez, K.; Nakajima, H. Metallic Scaffolds for Bone Regeneration. *Materials* **2009**, *2*, 790–832. [CrossRef]
2. Anselme, K. Biomaterials and interface with bone. *Osteoporos. Int.* **2011**, *22*, 2037–2042. [CrossRef] [PubMed]
3. Chen, S.; Guo, Y.; Liu, R.; Wu, S.; Fang, J.; Huang, B.; Li, Z.; Chen, Z.; Chen, Z. Tuning surface properties of bone biomaterials to manipulate osteoblastic cell adhesion and the signaling pathways for the enhancement of early osseointegration. *Colloids Surf. B Biointerfaces* **2018**, *164*, 58–69. [CrossRef]
4. Dohan Ehrenfest, D.M.; Coelho, P.G.; Kang, B.S.; Sul, Y.T.; Albrektsson, T. Classification of osseointegrated implant surfaces: Materials, chemistry and topography. *Trends Biotechnol.* **2010**, *28*, 198–206. [CrossRef] [PubMed]
5. Gaviria, L.; Salcido, J.P.; Guda, T.; Ong, J.L. Current trends in dental implants. *J. Korean Assoc. Oral Maxillofac. Surg.* **2014**, *40*, 50–60. [CrossRef]
6. Shah, F.A. Fluoride-containing bioactive glasses: Glass design, structure, bioactivity, cellular interactions, and recent developments. *Mater. Sci. Eng. C* **2016**, *58*, 1279–1289. [CrossRef]
7. Bouler, J.; Pilet, P.; Gauthier, O.; Verron, E. Biphasic calcium phosphate ceramics for bone reconstruction: A review of biological response. *Acta Biomater.* **2017**, *53*, 1–12. [CrossRef]
8. Narayanan, R.; Seshadri, S.K.; Kwon, T.Y.; Kim, K.H. Calcium phosphate-based coatings on titanium and its alloys. *J. Biomed. Mater. Res. Part B Appl. Biomater.* **2008**, *85*, 279–299. [CrossRef]
9. Yang, Y.; Dennison, D.; Ong, J.L. Protein adsorption and osteoblast precursor cell attachment to hydroxyapatite of different crystallinities. *Int. J. Oral Maxillofac. Implant* **2005**, *20*, 187–192.

10. Graziani, G.; Boi, M.; Bianchi, M. A Review on ionic substitutions in hydroxyapatite thin films: towards complete biomimetism. *Coatings* **2018**, *8*, 269. [CrossRef]
11. Vlădescu, A.; Pârâu, A.; Pană, I.; Cotruț, C.M.; Constantin, L.R.; Braic, V.; Vrânceanu, D.M. In vitro activity assays of sputtered HAp coatings with SiC addition in various simulated biological fluids. *Coatings* **2019**, *9*, 389. [CrossRef]
12. Jonauske, V.; Stanionyte, S.; Chen, S.-W.; Zarkov, A.; Juskenas, R.; Selskis, A.; Matijosius, T.; Yang, T.C.K.; Ishikawa, K.; Ramanauskas, R.; et al. Characterization of sol-gel derived calcium hydroxyapatite coatings fabricated on patterned rough stainless steel surface. *Coatings* **2019**, *9*, 334. [CrossRef]
13. León, B.; Jansen, J.A. *Thin Calcium Phosphate Coatings for Medical Implants*; Springer: New York, NY, USA, 2009.
14. Batebi, K.; Khazaei, B.A.; Afshar, A. Characterization of sol-gel derived silver/fluor-hydroxyapatite composite coatings on titanium substrate. *Surf. Coat. Technol.* **2018**, *352*, 522–528. [CrossRef]
15. Roseti, L.; Parisi, V.; Petretta, M.; Cavallo, C.; Desando, G.; Bartolotti, I.; Grigolo, B. Scaffolds for Bone Tissue Engineering: State of the art and new perspectives. *Mater. Sci. Eng. C Mater. Biol. Appl.* **2017**, *78*, 1246–1262. [CrossRef] [PubMed]
16. Mariani, E.; Lisignoli, G.; Borzì, R.M.; Pulsatelli, L. Biomaterials: Foreign Bodies or Tuners for the Immune Response? *Int. J. Mol. Sci.* **2019**, *20*, 636. [CrossRef] [PubMed]
17. Chen, Z.; Klein, T.; Murray, R.Z.; Crawford, R.; Chang, J.; Wu, C.; Xiao, Y. Osteoimmunomodulation for the development of advanced bone biomaterials. *Mater. Today* **2016**, *19*, 304–321. [CrossRef]
18. Tite, T.; Popa, A.C.; Balescu, L.M.; Bogdan, I.M.; Pasuk, I.; Ferreira, J.M.F.; Stan, G.E. Cationic Substitutions in Hydroxyapatite: Current Status of the Derived Biofunctional Effects and Their In Vitro Interrogation Methods. *Materials* **2018**, *11*, 2081. [CrossRef]
19. Duta, L.; Popescu, A.C. Current Status on Pulsed Laser Deposition of Coatings from Animal-Origin Calcium Phosphate Sources. *Coatings* **2019**, *9*, 335. [CrossRef]
20. Chrisey, D.B.; Hubler, G.K. *Pulsed Laser Deposition of Thin Films*, 1st ed.; John Wiley & Sons: Hoboken, NJ, USA, 1994; pp. 1–649.
21. Eason, R. *Pulsed Laser Deposition of Thin Films-Applications-Led Growth of Functional Materials*; Wiley-Interscience: Hoboken, NJ, USA, 2006; pp. 1–682.
22. Christen, H.M.; Eres, G. Recent advances in pulsed-laser deposition of complex oxides. *J. Phys. Condens. Matter* **2008**, *20*, 264005. [CrossRef]
23. Popescu, A.C.; Florian, P.E.; Stan, G.E.; Popescu-Pelin, G.; Zgura, I.; Enculescu, M.; Oktar, F.N.; Trusca, R.; Sima, L.E.; Roseanu, A.; et al. Physical-chemical characterization and biological assessment of simple and lithium-doped biological-derived hydroxyapatite thin films for a new generation of metallic implants. *Appl. Surf. Sci.* **2018**, *439*, 724–735. [CrossRef]
24. Duta, L.; Chifiriuc, M.C.; Popescu-Pelin, G.; Bleotu, C.; (Pircalabioru) Gradisteanu, G.; Anastasescu, M.; Achim, A.; Popescu, A. Pulsed Laser Deposited Biocompatible Lithium-Doped Hydroxyapatite Coatings with Antimicrobial Activity. *Coatings* **2019**, *9*, 54. [CrossRef]
25. Vats, A.; Bielby, R.C.; Tolley, N.S.; Nerem, R.; Polak, J.M. Stem cells. *Lancet* **2005**, *366*, 592–602. [CrossRef]
26. Vohra, S.; Hennessy, K.M.; Sawyer, A.A.; Zhuo, Y.; Bellis, S.L. Comparison of mesenchymal stem cell and osteosarcoma cell adhesion to hydroxyapatite. *J. Mater. Sci. Mater. Med.* **2008**, *19*, 3567–3574. [CrossRef] [PubMed]
27. Duta, L.; Mihailescu, N.; Popescu, A.C.; Luculescu, C.R.; Mihailescu, I.N.; Cetin, G.; Gunduz, O.; Oktar, F.N.; Popa, A.C.; Kuncser, A.; et al. Comparative physical, chemical and biological assessment of simple and titanium-doped ovine dentine-derived hydroxyapatite coatings fabricated by pulsed laser deposition. *Appl. Surf. Sci.* **2017**, *413*, 129–139. [CrossRef]
28. Duta, L.; Oktar, F.N.; Stan, G.E.; Popescu-Pelin, G.; Serban, N.; Luculescu, C.; Mihailescu, I.N. Novel doped hydroxyapatite thin films obtained by pulsed laser deposition. *Appl. Surf. Sci.* **2013**, *265*, 41–49. [CrossRef]
29. Eisenhart, S. EU Regulation 722. In *New EU Animal Tissue Regulations in Effect for Some Medical Devices*; Emergo: Hong Kong, China, 2013; Available online: https://www.emergobyul.com/blog/2013/09/new-eu-animaltissue-regulations-effect-some-medical-devices (accessed on 29 August 2019).
30. ISO 22442-1. *Medical Devices Utilizing Animal Tissues and Their Derivatives—Part 1: Application of Risk Management*; International Organization for Standardization: Berlin, Germany, 2015.
31. Negroiu, G.; Piticescu, R.M.; Chitanu, G.C.; Mihailescu, I.N.; Zdrentu, L.; Miroiu, M. Biocompatibility evaluation of a novel hydroxyapatite-polymer coating for medical implants (in vitro tests). *J. Mater. Sci.* **2008**, *19*, 1537–1544. [CrossRef]

32. Sima, L.E.; Filimon, A.; Piticescu, R.M.; Chitanu, G.C.; Suflet, D.M.; Miroiu, M.; Socol, G.; Mihailescu, I.N.; Neamtu, J.; Negroiu, G. Specific biofunctional performances of the hydroxyapatite–sodium maleate copolymer hybrid coating nanostructures evaluated by in vitro studies. *J. Mater. Sci. Mater. Med.* **2009**, *20*, 2305–2316. [CrossRef]
33. Gregory, C.A.; Gunn, W.G.; Peister, A.; Prockop, D.J. An Alizarin red-based assay of mineralization by adherent cells in culture: Comparison with cetylpyridinium chloride extraction. *Anal. Biochem.* **2004**, *329*, 77–84. [CrossRef]
34. Sima, L.E.; Stan, G.E.; Morosanu, C.O.; Melinescu, A.; Ianculescu, A.; Melinte, R.; Neamtu, J.; Petrescu, S.M. Differentiation of mesenchymal stem cells onto highly adherent radio frequency-sputtered carbonated hydroxylapatite thin films. *J. Biomed. Mater. Res. A* **2010**, *95*, 1203–1214. [CrossRef]
35. Brakebusch, C.; Fassler, R. The integrin-actin connection, an eternal love affair. *EMBO J.* **2003**, *22*, 2324–2333. [CrossRef]
36. Anselme, K. Osteoblast adhesion on biomaterials. *Biomaterials* **2000**, *21*, 667–681. [CrossRef]
37. Okumura, A.; Goto, M.; Goto, T.; Yoshinari, M.; Masuko, S.; Katsuki, T.; Tanaka, T. Substrate affects the initial attachment and subsequent behavior of human osteoblastic cells (Saos-2). *Biomaterials* **2001**, *22*, 2263–2271. [CrossRef]
38. Hoylaerts, M.F.; Manes, T.; Millan, J.L. Mammalian alkaline phosphatases are allosteric enzymes. *J. Biol. Chem.* **1997**, *272*, 22781–22787. [CrossRef] [PubMed]
39. Sharma, U.; Pal, D.; Prasad, R. Alkaline phosphatase: An overview. *Indian J. Clin. Biochem.* **2014**, *29*, 269–278. [CrossRef]
40. Weiss, M.J.; Henthorn, P.S.; Lafferty, M.A.; Slaughter, C.; Raducha, M.; Harris, H. Isolation and characterization of a cDNA encoding a human liver/bone/kidney-type alkaline phosphatase. *Proc. Natl. Acad. Sci. USA* **1986**, *83*, 7182–7186. [CrossRef]
41. Gronthos, S.; Zannettino, A.C.; Graves, S.E.; Ohta, S.; Hay, S.J.; Simmons, P.J. Differential cell surface expression of the STRO-1 and alkaline phosphatase antigens on discrete developmental stages in primary cultures of human bone cells. *J. Bone Miner. Res.* **1999**, *14*, 47–56. [CrossRef]
42. Dos Santos, E.A.; Farina, M.; Soares, G.A.; Anselme, K. Chemical and topographical influence of hydroxyapatite and beta-tricalcium phosphate surfaces on human osteoblastic cell behavior. *J. Biomed. Mater. Res. A* **2009**, *89*, 510–520. [CrossRef]
43. Hamilton, D.W.; Chehroudi, B.; Brunette, D.M. Comparative response of epithelial cells and osteoblasts to microfabricated tapered pit topographies in vitro and in vivo. *Biomaterials* **2007**, *28*, 2281–2293. [CrossRef]
44. Kunzler, T.P.; Drobek, T.; Schuler, M.; Spencer, N.D. Systematic study of osteoblast and fibroblast response to roughness by means of surface-morphology gradients. *Biomaterials* **2007**, *28*, 2175–2182. [CrossRef]
45. Van de Peppel, J.; van Leeuwen, J.P.T.M. Vitamin D and gene networks in human osteoblasts. *Front. Physiol.* **2014**, *5*, 137. [CrossRef]
46. Dallas, S.L.; Prideaux, M.; Bonewald, L.F. The osteocyte: An endocrine cell... and more. *Endocr. Rev.* **2013**, *34*, 658–690. [CrossRef] [PubMed]
47. Popescu, A.C.; Sima, F.; Duta, L.; Popescu, C.; Mihailescu, I.N.; Capitanu, D.; Mustata, R.; Sima, L.E.; Petrescu, S.M.; Janackovic, D. Biocompatible and bioactive nanostructured glass coatings synthesized by pulsed laser deposition: In vitro biological tests. *Appl. Surf. Sci.* **2009**, *255*, 5486–5490. [CrossRef]
48. Sun, Q.; Choudhary, S.; Mannion, C.; Kissin, Y.; Zilberberg, J.; Lee, W.Y. Ex vivo construction of human primary 3D-networked osteocytes. *Bone* **2017**, *105*, 245–252. [CrossRef] [PubMed]
49. Chen, J.; Dosier, C.R.; Park, J.H.; De, S.; Guldberg, R.E.; Boyan, B.D.; Schwartz, Z. Mineralization of three-dimensional osteoblast cultures is enhanced by the interaction of 1α,25-dihydroxyvitamin D3 and BMP2 via two specific vitamin D receptors. *J. Tissue Eng. Regen. Med.* **2016**, *10*, 40–51. [CrossRef]

coatings

MDPI

Article

Magnesium Doped Hydroxyapatite-Based Coatings Obtained by Pulsed Galvanostatic Electrochemical Deposition with Adjustable Electrochemical Behavior

Diana Maria Vranceanu [1], Ionut Cornel Ionescu [1,2,*], Elena Ungureanu [1], Mihai Ovidiu Cojocaru [1], Alina Vladescu [3,4,*] and Cosmin Mihai Cotrut [1]

1 Faculty of Materials Science and Engineering, University Politehnica of Bucharest, RO060042 Bucharest, Romania; diana.vranceanu@upb.ro (D.M.V.); ungureanu.elena14@yahoo.com (E.U.); cojocarumihaiovidiu@yahoo.co.uk (M.O.C.); cosmin.cotrut@upb.ro (C.M.C.)
2 Faculty of Dental Medicine, University of Medicine and Pharmacy "Carol Davila", RO020021 Bucharest, Romania
3 Department for Advanced Surface Processing and Analysis by Vacuum Technologies, National Institute of Research and Development for Optoelectronics-INOE 2000, RO77125 Magurele, Romania
4 National Research Tomsk Polytechnic University, 634050 Tomsk, Russia
* Correspondence: ionescu.ionut.cornel@gmail.com (I.C.I.); alinava@inoe.ro (A.V.); Tel.: +4-021-457-57-59 (A.V.)

Received: 25 June 2020; Accepted: 22 July 2020; Published: 24 July 2020

Abstract: The aim of this study was to adapt the electrochemical behavior in synthetic body fluid (SBF) of hydroxyapatite-based coatings obtained by pulsed galvanostatic electrochemical deposition through addition of Mg in different concentrations. The coatings were obtained by electrochemical deposition in a typical three electrodes electrochemical cell in galvanic pulsed mode. The electrolyte was obtained by subsequently dissolving $Ca(NO_3)_2{\cdot}4H_2O$, $NH_4H_2PO_4$, and $Mg(NO_3)_2{\cdot}6H_2O$ in ultra-pure water and the pH value was set to 5. The morphology consists of elongated and thin ribbon-like crystals for hydroxyapatite (HAp), which after the addition of Mg became a little wider. The elemental and phase composition evidenced that HAp was successfully doped with Mg through pulsed galvanostatic electrochemical deposition. The characteristics and properties of hydroxyapatite obtained electrochemically can be controlled by adding Mg in different concentrations, thus being able to obtain materials with different properties and characteristics. In addition, the addition of Mg can lead to the control of hydroxyapatite bioactive ceramics in terms of dissolution rate.

Keywords: coatings; electrochemical deposition; doped hydroxyapatite; magnesium; electrochemical behavior

1. Introduction

Intense and regular use of medical devices has resulted in hostile factors, including wearing out, micro-motions, and increase in corrosion. Thus, the challenges in the medical field are to design implantable materials with tunable properties in terms of bioactivity, biodegradation, and bactericidal effect, but the difficulties are redoubtable since these surfaces' characteristics tend to deteriorate in time.

Recently, the coatings field has attracted much interest and the ideal implant coating will likely be multifunctional, combining different technologies to promote a rapid osseointegration while inhibiting adverse tissue reactions, foreign body response or other events [1–5]. In addition to optimization of mechanical and chemical compatibilities, another important aspect for implant materials is the establishment of a suitable surface state in terms of a tailored surface topography and chemistry that stimulates bone cell growth and activates an optimum bone osseointegration [6,7]. To achieve a successful fixation, it is important to create a stable implant-bone interface [1].

To overcome these problems, the main inorganic component in natural bone-hydroxyapatite (HAp, $Ca_{10}(PO_4)_6(OH)_2$)-is widely used as a bioactive coating material [8–10]. Moreover, it is possible to tailor the properties of HAp-based materials by addition of a wide variety of substitutions and ion doping found in natural bones (e.g., Mg, Zn, Sr) into its crystal structure [11–20] in search for the promotion of bone remodeling, antibacterial activity, and enhanced bio-integration. It is reasonable to deduce that by incorporation of different ions into HAp coating simultaneously could be beneficial for the improvement of both bioactivity and long-term stability such as biodegradation [21] and electrochemical behavior of the coating in physiological media [22,23].

Hydroxyapatite is a virtually ubiquitous bio-ceramic and is extensively used as a bioactive coating on Ti-based implants due to its chemical and structural similarity to hard tissues, and consequently it can be bound directly to bone. In the HAp structure, the PO_4^{3-} anions can be replaced by carbonate groups while the calcium ions by Mg^{2+}, Na^+, K^+, etc., and trace elements like Pb^{2+}, Cu^{2+}, and Fe^{2+} [24].

Magnesium (Mg) is also essential for metabolism, being the fourth most abundant element in the human body. Especially in the medical field, Mg has gained more and more attention from the scientific community, where it is employed either as a main or alloying element or as a substitution/doping element [25–28]. By doping HAp with Mg, the mineral metabolism can be influenced, modifying the dissolution rate of crystals and the biodegradation of the corresponding materials [29]. Landi et al. [30] showed in an in vivo study that Mg-HAp enhanced osseoconductivity and resorption, compared to commercial HAp. The substitution of Ca^{2+} by Mg^{2+} in HAp is limited due to the large size difference between Mg^{2+} and Ca^{2+} (~0.28 Å difference in radius), which leads to strong distortions of the HAp lattice, reducing its crystallinity. These changes have a direct impact on the properties of Mg-HAp, compared to their non-substituted analogues [31], by increasing their solubility. Zhao et al. reported that Mg/HAp-coated surface promotes osteogenic differentiation of pre-osteoblasts and osseointegration during the early stage of bone healing [32].

Since 1990, the interest in electrochemical deposition (ED) has evolved, getting ascendant attention as a superior method for the preparation of different protective coatings [33,34]. ED is ideal for the deposition of doped HAp coatings [13,18,35–37] due to: (i) the low temperatures involved, which enable formation of highly crystalline deposits with low solubility in body fluids and low residual stresses; (ii) the ability to coat porous and geometrically complex materials; (iii) the possible improvement of the substrate/coating bond strength; and (iv) the ability to control the thickness, composition, and microstructure of the deposit.

The aim of this study is to adapt the electrochemical behavior in synthetic body fluid (SBF) of hydroxyapatite-based coatings through addition of Mg in different concentrations obtained by pulsed galvanostatic electrochemical deposition. Magnesium was selected because it increases the solubility of HAp and thus, properties, such as the electrochemical behavior can be controlled and/or adjusted with respect to the intended medical applications. Even though indirectly, the electrochemical tests carried out in this study in synthetic body fluid (SBF) at human body temperature can be considered as a prediction tool which can indicate the potential biodegradable behavior of the proposed coatings.

2. Materials and Methods

2.1. Preparation of the Coatings

Titanium alloy (Ti6Al4V, purchased from Bibus Metals AG, Essen, Germany) discs (20 mm × 2 mm) were used as substrates. One side of each disc was metallographically prepared on silicon carbide paper of different grit (320 ÷ 800 grit). The prepared discs were cleaned with ultra-pure water, ASTM I, and iso-propylic alcohol. Before deposition, the Ti6Al4V discs were sonicated in acetone for 20 min to remove all residues and grease and cleaned with ultra-pure water.

The electrolyte used for the coating preparation is presented in Table 1 and was obtained by subsequently dissolving the chemical reagents (Table 2) in ultra-pure water. The pH was adjusted to 5 by dropwise addition of 1 M NaOH. The electrolyte was de-aerated with N_2 for 30 min to minimize the formation of $CaCO_3$ deposits. The temperature was maintained at 75 °C using a heating plate (KA RCT Basic Safety Control Hotplate/Stirrer and ETS-D6 Temp, IKA, Staufen, Germany) and a magnetic stirrer was used to keep the concentration of the electrolyte uniform during deposition.

Table 1. Chemical composition of the used electrolytes and sample codification.

Coating Type	Sample Codification	Chemical Composition (mM)			Molar Ratio (Ca + Mg)/P
		$Ca(NO_3)_2{\cdot}4H_2O$	$NH_4H_2PO_4$	$Mg(NO_3)_2{\cdot}6H_2O$	
HAp	H	10.500	6.300	0	1.67
HAp with Mg	H-Mg1	9.450		1.050	
	H-Mg1.5	8.925		1.575	
	H-Mg2	8.400		2.100	

Table 2. Grade and purity of the chemical reagents employed in electrolytes preparation.

Chemical Formula	Reagent Grade	Purity	Manufacturer
$Ca(NO_3)_2{\cdot}4H_2O$	ACS reagent	≥99.00%	Sigma Aldrich, Munich, Germany
NH_4H_2PO	Trace metal basis	99.99%	
$Mg(NO_3)_2{\cdot}6H_2O$	ACS reagent	99.00%	
NaOH	ACS reagent	≥98.00%	

Pulsed galvanostatic deposition was carried out in a conventional three electrode electrochemical cell using a multichannel Potentiostat/Galvanostat (Parstat MC, Princeton Applied Research–Ametek, Oak Ridge, TN, USA). The deposition was performed for 40 min on Ti6Al4V discs in pulsed galvanostatic mode with a constant current density of −0.85 mA/cm^2 [12,19,38]. The activation and relaxation stages were set as follows: $i_{ON} = -0.85$ mA/cm^2 for $t_{ON} = 1$ s, followed by $i_{OFF} = 0$ mA/cm^2 for $t_{OFF} =$ 1 s. After the electrochemical deposition, the samples were removed from the electrolyte and gently rinsed with ultra-pure water and stored in a desiccator. Figure 1 shows a schematic illustration of the electrochemical deposition process and the sample codification.

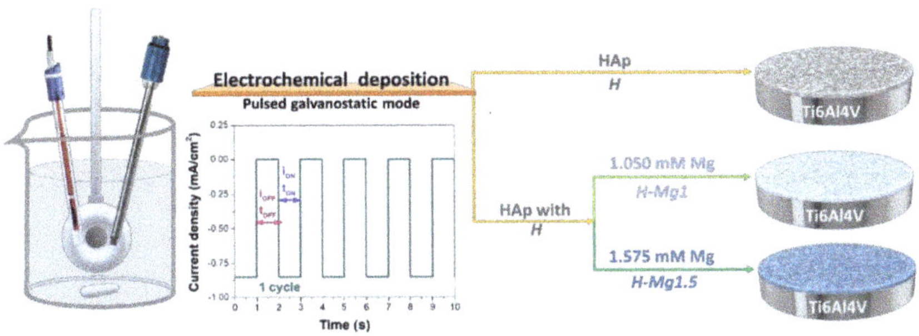

Figure 1. Schematic illustration of the electrochemical deposition process and samples codification.

As a function of Mg concentration, it was observed that up to 1.5 mM $Mg(NO_3)_2{\cdot}6H_2O$ the coatings cover the entire surface exposed to the electrochemical deposition process with a uniform and compact layer which does not present defects. By increasing the Mg concentration to 2 mM, the uniformity and compactness of the coating begins to alter indicating that an increased quantity of Mg (higher than 1.5 mM) can lead to a porous, irregular, and heterogenous coating which does not meet the minimal criteria selection (for electrochemical investigation) and thus, this coating was not further investigated (H-Mg2).

2.2. Characterization of Coatings

The coatings' morphology and elemental composition were investigated by scanning electron microscopy equipped with an X-ray energy dispersive spectrometer (SEM-EDS, Phenom ProX, Phenom World, Eindhoven, Netherlands).

The phase composition of the investigated samples was examined using a SmartLab X-ray diffractometer (Rigaku, city, Japan) with a CuKα radiation (λ = 1.5405 Å). Grazing incidence measurements were performed in the θ/2θ range of 20–60° with a 0.02° step, at an incidence angle of 3°. For this purpose, a medium resolution parallel beam optical system was used.

Surface roughness was measured over a length of 4000 μm using DEKTAK 150 (Veeco Instruments, Plainview, NY, USA) stylus profilometer. The thickness of the coatings was determined by Dektak 150 stylus profilometer (stylus radius of 12.5 μm), measuring the forming ledge between the deposited film and the uncoated substrate.

The adhesion between the Ti6Al4V substrate and the undoped and doped HAp coatings was measured by performing a "tape test" according to ASTM D3359-17 standard [39] with an Elcometer 107 Cross Hatch Adhesion Tester kit (Ulmer, Aalen, Germany). The test consists of 6 × 6 mm^2 parallel cuts with gaps of 1 mm between them made with a cutting edge that will generate a lattice pattern. It is important to apply sufficient pressure to ensure that the cutting edge will go through the coating to the substrate's surface. The adhesive tape (ISO/ASTM tape) was placed over the lattice pattern indentation, and after 90 s the tape was removed by pulling in a single smooth action at angle of 180° to the coating surface. After the adhesion tests, the coatings were examined by scanning electron microscopy (SEM Phenom Pro X, Phenom World, Eindhoven, Netherlands). The coating adhesion was assessed in terms of area removed by viewing the lattice pattern indentation and was classified in terms of percentages (from highest to weakest): 5B: 0%; 4B: ≤5% (not large than 5%); 3B: 5% ÷ 15%; 2B: 15% ÷ 35%; 1B: 35% ÷ 65%; 0B: ≥65% (larger than 65%).

Electrochemical behavior was investigated at 37 ± 0.5 °C in synthetic body fluid with a pH of 7.4 using a PARSTAT 4000 Potentiostat/Galvanostat (Princeton Applied Research, Oak Ridge, TN, USA). The chemical composition of the testing media (SBF) is presented in Table 3 and was made in accordance with the receipt proposed by Kokubo [40]. For electrochemical behavior, an area of 1 cm^2 was exposed to the electrolyte (SBF). A typical three-electrode cell with the following set-up: sample as working electrode (WE), a platinum electrode was used as a counter electrode (CE), and saturated calomel (SCE) as reference electrode (RE), was used for all electrochemical measurements. The open circuit potential (OCP) was monitored for 1 h, starting right after the sample's immersion in the electrolyte and the Tafel plots were recorded from −150 (vs. OCP) to +150 mV (vs. SCE) at a scanning rate of 1 mV/s. All measurements were achieved according to the ASTM G5-94 standard (reapproved 2011) [41]. Experiments were performed in triplicate and the obtained data are presented as mean ± SD.

Table 3. Chemical composition and reagents grade and purity used for preparation of synthetic body fluid (SBF) media (1000 mL, pH = 7.4).

No.	Chemical Formula	Amount	Reagent Grade	Purity	Manufacturer
#1	NaCl	8.035 g/L	ACS reagent	>99.99%	Sigma Aldrich, Munich, Germany
#2	$NaHCO_3$	0.355 g/L	Bio Reagent	≥99.50%	
#3	KCl	0.225 g/L	Purris p.a.	≥99.50%	
#4	$K_2HPO_4{\cdot}3H_2O$	0.231 g/L	Reagent Plus	≥99.00%	
#5	$MgCl_2{\cdot}6H_2O$	0.311 g/L	Purum p.a.	≥98.00%	
#6	HCl	39 mL	ACS reagent	≥37.00%	
#7	$CaCl_2$	0.292 g/L	ACS reagent	≥99.00%	
#8	Na_2SO_4	0.072 g/L	ACS reagent	≥99.00%	
#9	$(CH_2OH)_3CNH_2$	6.118 g/L	Standard & Buffer	≥99.90%	

3. Results and Discussion

3.1. Morphology

The surface morphology and crystallinity of Hap-based coatings are in strong correlation with the current density and the electrolyte ion concentration [13,42]. Nevertheless, the cell performance and biological activity are dependent on the morphology and crystal orientation [13,43,44].

Figure 2 presents the SEM images revealing the morphology of the obtained coatings. As it can be seen, the entire investigated surface has been fully covered regardless of the electrolyte composition with a uniform layer, which also due to its morphology has a high surface to volume ratio. Particularly, for undoped HAp coatings, the morphology consists of elongated and thin ribbon-like crystals which was also reported in other studies [11,45,46].

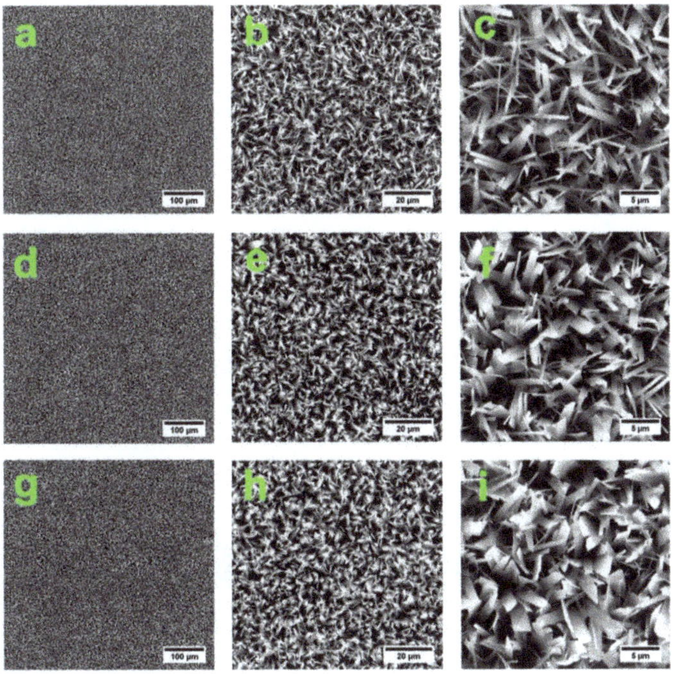

Figure 2. Scanning electron microscopy (SEM) images at different magnifications evidencing the morphology of the H (**a–c**), H-Mg1 (**d–f**), and H-Mg1.5 (**g–i**) based coatings.

At small magnifications it was observed that, regardless of its concentration, the addition of Mg did not lead to substantial alteration of the ribbon-like morphology, while at higher magnification some dimensional variations were noted. Thus, it can be said that, compared to undoped HAp, Mg addition made the ribbons a little wider. Similar results were also found by Bakin et al. [47].

3.2. Elemental Composition

The EDS coatings analysis confirmed the presence of Mg in the H-Mg1 and H-Mg1.5 coatings and the presence of Ca and P in all coatings. The EDS mapping analysis was used to investigate elemental distribution (Ca, P, and Mg) and the results are presented in Figure 3.

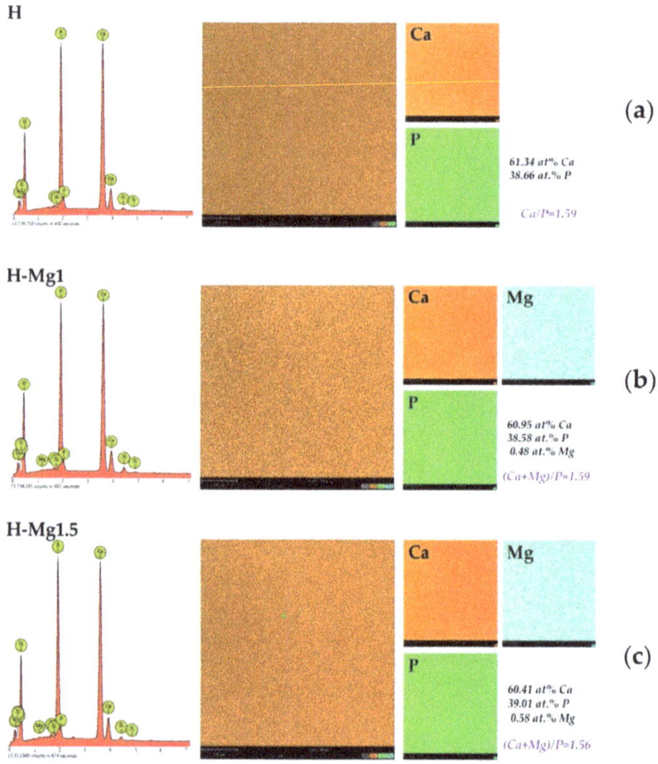

Figure 3. X-ray energy dispersive spectroscopy (EDS), elemental distribution, and chemical composition of the undoped (**a**) and Mg-doped hydroxyapatite (Hap) (**b**,**c**) based coatings.

According to this investigation, all elements were uniformly distributed in all three types of coatings. The (Ca + Mg)/P ratio is lower than the stoichiometric one (Figure 4), suggesting that the doping process causes Ca-deficient apatite, probably due to the generation of crystal defects from the ion substitution [16]. A ratio of 1.59 was noted for H and H-Mg1 coatings, while for H-Mg1.5 the ratio was 1.56. According to some studies [48,49], Ca-deficient apatite is beneficial for bone formation, but further investigations are required.

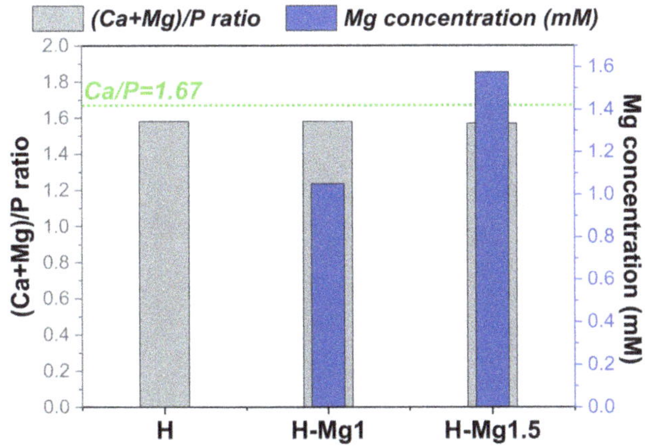

Figure 4. Dependency between (Ca+Mg)/P ratio and the Mg concentration.

3.3. *Phase Composition*

The X-ray diffractograms of undoped and doped HAp coatings are presented in Figure 5.

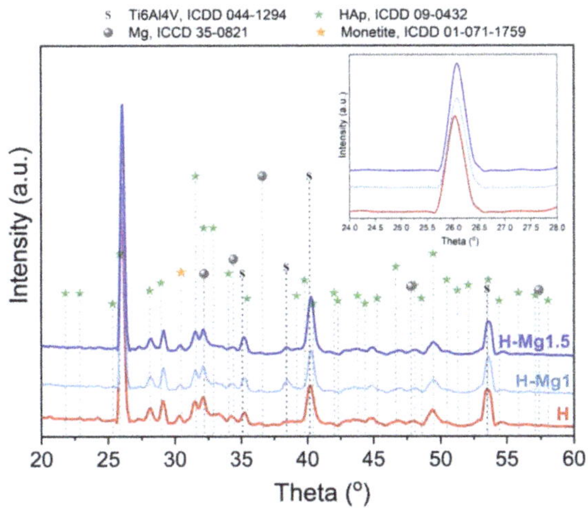

Figure 5. The X-ray diffractograms for undoped and Mg-doped HAp.

All of the coatings showed the characteristic diffraction peaks of HAp (ICDD-The International Centre for Diffraction Data, #09-0432). Nevertheless, the strongest relative intensity was registered for the plan (002) at 25.95° which is typical for electrochemical deposition [36,50,51], suggesting that the coatings are grown perpendicular on the Ti6Al4V substrate. According to the literature, oriented HAp may enable control over cells' behavior [52], and in particular the (002) preferred orientation promotes cell adhesion [53,54]. All of the coatings present a small peak which was assigned to monetite (ICDD, #01-071-1759), a precursor of hydroxyapatite. Monetite phase can be converted into hydroxyapatite by alkaline treatment [55,56]. According to the literature [57], besides its ability to regenerate bone, monetite resorbs in vivo faster than other calcium phosphates enabling the implant replacement by a newly formed tissue.

In the detailed diffractogram from Figure 5, it can be observed that after addition of Mg into HAp, the peak associated with (002) has slightly shifted towards higher angles from 25.95° registered for H samples to 26.06° and 26.10° for H-Mg1 and H-Mg1.5, respectively. These findings are in good correlation with other papers [19,58,59] and indicate that Mg has replaced Ca in the HAp lattice. This result is consistent with the smaller ionic radius of Mg^{2+} compared with Ca^{2+} (0.69 and 0.99 Å, respectively). No other impurities were detected.

The average crystal size was calculated using the Debye–Scherrer approximation considering the (002) peak. In the literature, it is suggested that this peak should be selected for quantitative estimations because apart from others it does not show interferences [60]. According to our results, the H coatings has a crystal size of 8.4 ± 0.1 nm, while the addition of Mg led to a slight increment of the crystal size reaching a value of 8.8 ± 0.1 and 10.2 ± 0.1 nm for H-Mg1 and H-Mg1.5, respectively; the addition of a small amount of Mg did not significantly affect the crystal size.

In the literature, there are few modalities for quantifying the degree of crystallinity. In this paper, the degree of crystallinity was calculated based on the intensity of (002) reflection peak using the equation reported in previous studies [60–62]. It was found that the crystallinity increases with respect to the Mg quantity. Thus, compared to H coatings, which presented a crystallinity of 17.8%, by increasing the Mg amount the crystallinity reaches a maximum of 29.6% for H-Mg1.5 coating, closely followed by H-Mg1 coating which had a value of 27.1%.

Based on the XRD results, it can be concluded that the hydroxyapatite-based coating become more crystalline as a function of Mg concentration. Usually, both crystallinity degree and crystal size are related with the impurity amount and the stoichiometry (Ca/P ratio). Note that for the H and H-Mg1 coatings, the Ca/P is similar, while the both H-Mg1 and H-Mg1.5 exhibited closer Mg content and probably this is the reason that the crystal size or crystallinity degree has closer values.

3.4. Roughness

The coating roughness was carried out by stylus profilometry, and the main results are presented in Table 4. As it can be seen, the average roughness and root mean square roughness parameters have increased after Mg addition, being also dependent on the quantity which has been added into the electrolyte. Figure 6 presents the representative profiles for each coating. By increasing the Mg quantity from 1 to 1.5 mM, the average roughness has increased ~8 times, passing from nanometric scale in the case of H-Mg1 to micrometric scale for the H-Mg1.5 coatings. According to the literature, the average roughness values of the commercial coated dental implants is found to be between 0.1 and 10 μm [20]. Thus, it can be noted that the developed coatings are found to be in this limitation.

The skewness parameter offers information regarding the height distribution (skew <0 for a surface with deeper valleys, skew >0 for a surface with higher peaks) [63]. Thus, it can be noted that all coatings presented positive skewness suggesting that the surface is predominantly with peaks that protrude above the mean line. By comparing the values among them, it is visible that compared to the undoped HAp, the lowest skew was registered for H-Mg1 coatings, suggesting that in this concentration, Mg reduces the peaks height and the surface tends towards a symmetrical topography, whereas increasing the Mg content leads to a very high skewness value which suggests that the surface presents very high peaks.

Table 4. Roughness parameters (Ra-average roughness, Rq-root mean square, Skew-skewness) and thickness of the investigated coatings.

Coating	Ra (nm)	Rq (nm)	Skew	Thickness (μm)
H	439.7 (±33.9)	587.9 (±66.7)	0.84 (±0.1)	11.1 (±0.9)
H-Mg1	558.2 (±37.9)	710.9 (±47.0)	0.38 (±0.1)	12.2 (±0.7)
H-Mg1.5	4396.9 (±895.4)	6609.1 (±1071.4)	2.44 (±0.25)	10.5 (±0.7)

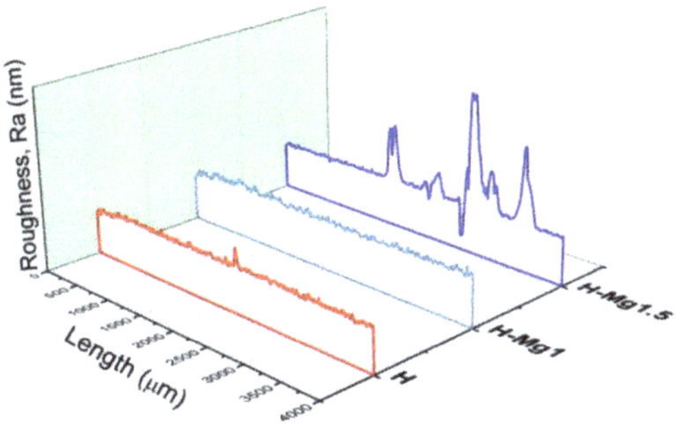

Figure 6. Representative profiles of the obtained coatings.

By measuring the forming ledge between the deposited film and the uncoated substrate, the coating thickness was obtained (Figure 7). Compared to the undoped HAp which has registered a thickness of approximately 11 µm, the H-Mg1 coatings registered an increment, until a value of ~12 µm, while for the H-Mg1.5, the thickness decreased, reaching the lowest thickness of 10.5 µm. The thickness differences can be associated with the deposition rate which is influenced by the Mg quantity added in the electrolyte. Considering that all the parameters involved in the electrochemical deposition were maintained constant, it can be assumed that the quantity of Mg affects the deposition rate/HAp nucleation and growth. It can be seen that for a small amount of magnesium added (1 mM), the thickness of the HAp layer increases, while after this concentration the thickness decreases. Nevertheless, Mg^{2+} ions inhibit HAp crystallization, avoiding the formation of large crystals and promoting the formation of more apatite nuclei. Thus, the presence of strong stresses inside the formed layer could be the reason why the HAp coating doped with 2.1 Mm of Mg is porous, irregular, and heterogenous. This finding can also be correlated with the fact that magnesium can inhibit apatite crystallization [64].

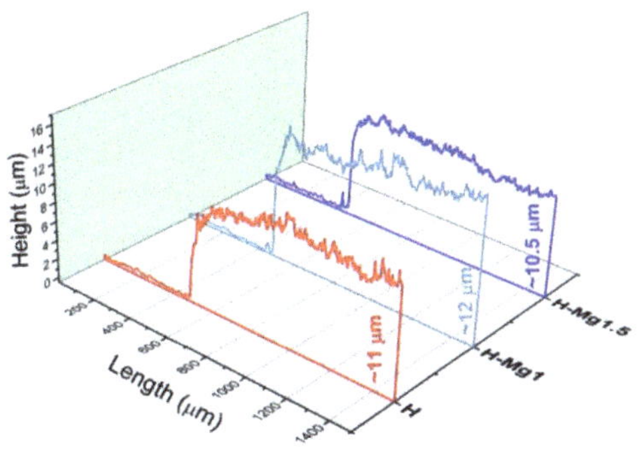

Figure 7. Coating thickness.

3.5. Adhesion

Figure 8 depicts the coatings surfaces after adhesion assays by "tape test" according to ASTM procedure. It is worth mentioning that within the adhesive tape removal, from all investigated coatings, a very thin, superficial layer was removed, without affecting the coating structural integrity, leaving behind a very consistent layer of HAp. Similar observations were also made by Parcharoen et al. [65], who investigated the adhesion between HAp-based coatings electrochemically deposited on anodized titania nanotubes of different lengths.

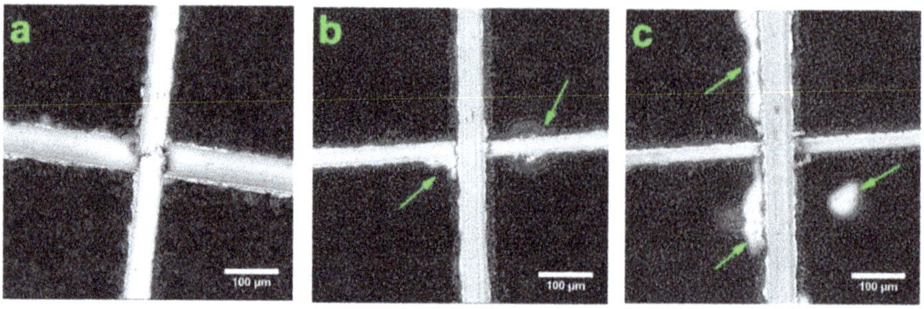

Figure 8. Representative SEM images of H (**a**), H-Mg1 (**b**), and H-Mg1.5 (**c**) coatings on Ti6Al4V substrate after the adhesion assay by the "tape test" (green arrow indicating the delaminated/affected areas).

Based on the obtained results after the adhesion assay, it was noted that the coatings codified with H had the best adhesion to the substrate followed closely by the coatings with 1 mM of Mg (H-Mg1). These coatings were classified according to ASTM D3359-17 [39] in the following categories: 5B for H coatings because the edges of the cuts were completely smooth with none of the lattice squares detached and 4B for H-Mg1, where some detachments/flakes, of areas not greater than 5% at the intersection of the cuts were observed. By increasing the Mg amount, it was noted that some delamination/detachment were found to be not just at the intersection of the cuts, but also along the cutting line and inside the squares of the lattice, though the affected area is not greater than 35%. Thus, the H-Mg1.5 coatings were classified into a 3B category.

Therefore, it can be concluded that even though after the adhesion assay most of the coating remained on the substrate on all of investigated samples, by increasing the Mg amount, the adhesion begins to weaken. Accordingly, the best results were obtained for H and H-Mg1 coatings, indicating that for a higher concentration of Mg (1.5 mM) the substrate may require some surface modification treatment before deposition to enhance the bonding strength at the coating–substrate interface.

3.6. Electrochemical Behaviour

Coating synthesis on titanium-based substrate enhances the biomineralization ability and acts as protection barrier against ions (especially Al and V) that can be released from the substrate improving the overall in vivo behavior and features [66]. The electrochemical techniques are a very important and useful tool in predicting a material's behavior in vivo by analyzing the reactions at the coating–substrate interface. Figure 9 presents the open circuit potential curves and the Tafel plots of the undoped and Mg-doped coatings deposited on Ti6Al4V substrate for only one set of samples.

The polarization resistance (Rp) was calculated by using the Stern-Geary equation:

$$R_p = \frac{1}{2.3 \cdot i_{corr}} \cdot \frac{\beta_a \cdot |\beta_c|}{\beta_a + |\beta_c|} \quad (1)$$

where Rp is the polarization resistance, β_a and β_c are the anodic and cathodic Tafel slops respectively, and i_{corr} is the corrosion current density.

Based on Tafel extrapolation the main electrochemical parameters were extracted and presented in Table 5.

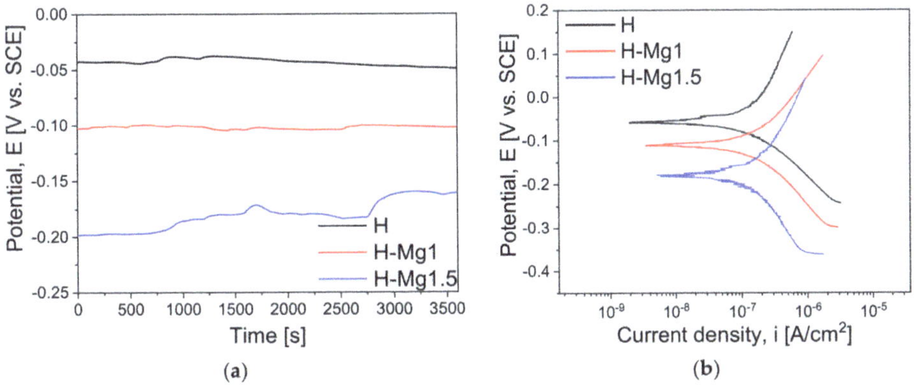

Figure 9. Open circuit potential curves (**a**) and Tafel plots (**b**) of the investigated coatings.

Table 5. Electrochemical parameters of the investigated samples (E_{oc}-open circuit potential; E_{corr}-corrosion potential; i_{corr}-corrosion current density; R_p-polarization resistance).

Sample	E_{oc} (mV)	E_{corr} (mV)	i_{corr} (nA/cm^2)	Bc (mV)	βa (mV)	R_p (kΩ·cm^2)
H	−48.9 (±5.2)	−56.9 (±6.3)	22.45 (±3.82)	34.23 (±5.51)	35.98 (±6.15)	339.70 (±10.42)
H-Mg1	−102.2 (±7.8)	−110.3 (±9.1)	53.70 (±4.56)	63.22 (±9.47)	60.67 (±8.77)	250.67 (±11.13)
H-Mg1.5	−160.7 (±9.4)	−180.5 (±10.7)	90.43 (±6.75)	100.33 (±12.34)	105.39 (±11.92)	247.11 (±12.58)

It is known that a material is resistant to the corrosive attack if it has an electropositive E_{corr}, low i_{corr}, and high R_p [67]. If we take into consideration the corrosion potential (E_{corr}), it was noted that the most electropositive value was registered for H coatings, followed by H-Mg1 and H-Mg1.5, suggesting that by increasing the Mg content the coatings' resistance to the corrosive attack of the media is poorer. Considering the corrosion current density (i_{corr}), the smallest value was registered also for H coatings, followed by H-Mg1 and H-Mg1.5 and indirectly suggesting that by adding Mg, the solubility of HAp can increase. Based on the values obtained for anodic (β_a) and cathodic (β_c) Tafel slops of each investigated coating it can be said that the kinetics related to the electron transfer and the number and nature of electrons involved in the two reactions are also similar, indicating a mixed controlled. If we consider the obtained polarization resistance values, it was observed that the highest R_p was obtained for H followed by H-Mg1 and H-Mg1.5 coatings, suggesting the order of dissolution/biodegradability as well.

The present study highlights that the addition of Mg tailors the electrochemical behavior and so, the solubility/biodegradability of HAp. Overall, the findings are in good correlation with a study performed by Cai et al. [68], in which they found that the by increasing the Mg content a higher dissolution rate can be obtained due to the incorporation of Mg into HAp.

4. Conclusions

In this study, Mg in different concentrations was incorporated in HAp structure by electrochemical deposition method using galvanic pulses technique on Ti6Al4V substrate. All coatings presented a morphology consisting of ribbon-like crystals.

Compared with undoped HAp coatings, the ribbon-like crystals appear to increase their dimension as a function of Mg concentration added in the electrolyte. The coatings were preferentially oriented in the (002) direction which is typical for electrochemical deposition and after addition of Mg some small shift towards higher angles were noted indicating that the substitution occurred in the HAp lattice. Furthermore, Mg addition led to a higher degree of crystallinity and crystal size, compared to undoped

HAp. The (Ca + Mg)/P ratio was 1.59 for H and H-Mg1 and 1.56 for H-Mg1.5, suggesting that by increasing the Mg amount, the ratios tend towards lower values. Adhesion assays evidenced that a concentration of 1.5 mM of Mg in the electrolyte leads to a weaker adhesion, while using just 1 mM the adhesion is closer to the undoped HAp-based coatings. The electrochemical behavior highlighted that undoped HAp presented the most electropositive corrosion potential and the highest polarization resistance. By comparing the coatings with Mg addition, it was observed that the H-Mg1.5 coatings presented the highest corrosion current density and the lowest polarization resistance suggesting the weakest electrochemical behavior in SBF media. Thus, it can be observed that by doping with Mg the electrochemically deposited HAp, the electrochemical behavior can be tuned and so, the dissolution rate as well.

In conclusion, the properties and characteristics of HAp obtained by electrochemical deposition can be controlled through addition of Mg in different concentrations, thus obtaining biomaterials with tunable properties suitable for medical applications.

Author Contributions: Conceptualization and methodology, C.M.C. and D.M.V.; investigation, E.U., I.C.I., and A.V.; data analysis and interpretation, D.M.V., A.V., and C.M.C.; writing—original draft preparation, I.C.I., C.M.C., and D.M.V.; writing—review and editing, D.M.V., M.O.C., A.V., and C.M.C. All authors have read and agreed to the published version of the manuscript.

Funding: This research has been funded by the Operational Program Human Capital of the Ministry of European Funds through the Financial Agreement 51668/09.07.2019, SMIS code 124705; by grants of the Romanian Ministry of Research and Innovation, CCCDI-UEFISCDI, projects number COFUND-ERANET-RUS-PLUS-CoatDegraBac no.68/2018 as well the Core Program of 2019 (no. 18N/2019). The XRD investigations were performed using the systems modernized by the PROINSTITUTIO Project-contract No. 19PFE/17.10.2018.

Conflicts of Interest: The authors declare no conflicts of interest.

References

1. Goodman, S.B.; Yao, Z.; Keeney, M.; Yang, F. The future of biologic coatings for orthopaedic implants. *Biomaterials* **2013**, *34*, 3174–3183. [CrossRef]
2. Surmenev, R.A.; Surmeneva, M.A.; Ivanova, A.A. Significance of calcium phosphate coatings for the enhancement of new bone osteogenesis—A review. *Acta Biomater.* **2014**, *10*, 557–579. [CrossRef]
3. Raphel, J.; Holodniy, M.; Goodman, S.B.; Heilshorn, S.C. Multifunctional coatings to simultaneously promote osseointegration and prevent infection of orthopaedic implants. *Biomaterials* **2016**, *84*, 301–314. [CrossRef] [PubMed]
4. Canillas, M.; Pena, P.; de Aza, A.H.; Rodríguez, M.A. Calcium phosphates for biomedical applications. *Boletín Soc. Española Cerámica Vidr.* **2017**, *56*, 91–112. [CrossRef]
5. Furko, M.; Havasi, V.; Kónya, Z.; Grünewald, A.; Detsch, R.; Boccaccini, A.R.; Balázsi, C. Development and characterization of multi-element doped hydroxyapatite bioceramic coatings on metallic implants for orthopedic applications. *Bol. La Soc. Esp. Ceram. Y Vidr.* **2018**, *57*, 55–65. [CrossRef]
6. Damiati, L.; Eales, M.G.; Nobbs, A.H.; Su, B.; Tsimbouri, P.M.; Salmeron-Sanchez, M.; Dalby, M.J. Impact of surface topography and coating on osteogenesis and bacterial attachment on titanium implants. *J. Tissue Eng.* **2018**, *9*, 1–19. [CrossRef]
7. Souza, J.C.M.; Sordi, M.B.; Kanazawa, M.; Ravindran, S.; Henriques, B.; Silva, F.S.; Aparicio, C.; Cooper, L.F. Nano-scale modification of titanium implant surfaces to enhance osseointegration. *Acta Biomater.* **2019**, *94*, 112–131. [CrossRef] [PubMed]
8. Mitran, V.; Ion, R.; Miculescu, F.; Necula, M.; Mocanu, A.-C.; Stan, G.; Antoniac, I.; Cimpean, A. Osteoblast Cell Response to Naturally Derived Calcium Phosphate-Based Materials. *Materials* **2018**, *11*, 1097. [CrossRef]
9. Dascalu, C.-A.; Miculescu, F.; Mocanu, A.-C.; Constantinescu, A.E.; Butte, T.M.; Pandele, A.M.; Ciocoiu, R.-C.; Voicu, S.I.; Ciocan, L.T. Novel Synthesis of Core-Shell Biomaterials from Polymeric Filaments with a Bioceramic Coating for Biomedical Applications. *Coatings* **2020**, *10*, 283. [CrossRef]
10. Miculescu, F.; Mocanu, A.C.; Stan, G.E.; Miculescu, M.; Maidaniuc, A.; Cîmpean, A.; Mitran, V.; Voicu, S.I.; Machedon-Pisu, T.; Ciocan, L.T. Influence of the modulated two-step synthesis of biogenic hydroxyapatite on biomimetic products' surface. *Appl. Surf. Sci.* **2018**, *438*, 147–157. [CrossRef]

11. Bhattacharjee, A.; Gupta, A.; Verma, M.; Anand, M.P.; Sengupta, P.; Saravanan, M.; Manna, I.; Balani, K. Antibacterial and magnetic response of site-specific cobalt incorporated hydroxyapatite. *Ceram. Int.* **2020**, *46*, 513–522. [CrossRef]
12. Thanh, T.T.; Cotrut, C.M.; Vranceanu, M.D.; Ungureanu, E.; Tarcolea, M. Studies of microstructure and composition of modified hydroxyapatite coatings via SEM investigations. *U.P.B. Sci. Bull. Ser. B* **2020**, *82*, 145–154.
13. Huang, Y.; Hao, M.; Nian, X.; Qiao, H.; Zhang, X.; Zhang, X.; Song, G.; Guo, J.; Pang, X.; Zhang, H. Strontium and copper co-substituted hydroxyapatite-based coatings with improved antibacterial activity and cytocompatibility fabricated by electrodeposition. *Ceram. Int.* **2016**, *42*, 11876–11888. [CrossRef]
14. Li, T.; Ling, L.; Lin, M.; Peng, H.; Ren, H.; Lou, C.-W.; Lin, J.-H. Recent advances in multifunctional hydroxyapatite coating by electrochemical deposition. *J. Mater. Sci.* **2020**, *55*, 6352–6374. [CrossRef]
15. Sharma, M.; Nagar, R.; Meena, V.K.; Singh, S. Electro-deposition of bactericidal and corrosion-resistant hydroxyapatite nanoslabs. *RSC Adv.* **2019**, *9*, 11170–11178. [CrossRef]
16. Munirathinam, B.; Jaladurgam, N.R.; Magesh, J.; Narayanan, R.; Mol, J.M.C.; Neelakantan, L. Improved corrosion protection of titanium implant material by crystallographic texturing of Sr doped calcium phosphate electrodeposits. *Thin Solid Films* **2019**, *675*, 115–121. [CrossRef]
17. Morejón-Alonso, L.; Mochales, C.; Nascimento, L.; Müller, W. Electrochemical deposition of Sr and Sr/Mg-co-substituted hydroxyapatite on Ti-40Nb alloy. *Mater. Lett.* **2019**, *248*, 65–68. [CrossRef]
18. Huang, Y.; Song, G.; Chang, X.; Wang, Z.; Zhang, X.; Han, S.; Su, Z.; Yang, H.; Yang, D.; Zhang, X. Nanostructured Ag+-substituted fluorhydroxyapatite-TiO_2 coatings for enhanced bactericidal effects and osteoinductivity of Ti for biomedical applications. *Int. J. Nanomed.* **2018**, *13*, 2665–2684. [CrossRef]
19. Vranceanu, D.M.; Tran, T.; Ungureanu, E.; Negoiescu, V.; Tarcolea, M.; Dinu, M.; Vladescu, A.; Zamfir, R.; Timotin, A.C.; Cotrut, C.M. Pulsed electrochemical deposition of Ag doped hydroxyapatite bioactive coatings on Ti6Al4V for medical purposes. *UPB Sci. Bull. Ser. B Chem. Mater. Sci.* **2018**, *80*, 173–184.
20. Yajing, Y.; Qiongqiong, D.; Yong, H.; Han, S.; Pang, X. Magnesium substituted hydroxyapatite coating on titanium with nanotublar TiO_2 intermediate layer via electrochemical deposition. *Appl. Surf. Sci.* **2014**, *305*, 77–85. [CrossRef]
21. Lin, W.-C.; Chuang, C.-C.; Wang, P.-T.; Tang, C.-M. A Comparative Study on the Direct and Pulsed Current Electrodeposition of Cobalt-Substituted Hydroxyapatite for Magnetic Resonance Imaging Application. *Materials* **2018**, *12*, 116. [CrossRef] [PubMed]
22. Vranceanu, D.M.; Parau, A.C.; Cotrut, C.M.; Kiss, A.E.; Constantin, L.R.; Braic, V.; Vladescu, A. In vitro evaluation of Ag doped hydroxyapatite coatings in acellular media. *Ceram. Int.* **2019**, *45*, 11050–11061. [CrossRef]
23. Vlădescu, A.; Pârâu, A.; Pană, I.; Cotruț, C.M.; Constantin, L.R.; Braic, V.; Vrânceanu, D.M. In Vitro Activity Assays of Sputtered HAp Coatings with SiC Addition in Various Simulated Biological Fluids. *Coatings* **2019**, *9*, 389. [CrossRef]
24. Ergun, C.; Webster, T.J.; Bizios, R.; Doremus, R.H. Hydroxylapatite with substituted magnesium, zinc, cadmium, and yttrium. I. Structure and microstructure. *J. Biomed. Mater. Res.* **2002**, *59*, 305–311. [CrossRef]
25. Antoniac, I.; Miculescu, F.; Cotrut, C.; Ficai, A.; Rau, J.V.; Grosu, E.; Antoniac, A.; Tecu, C.; Cristescu, I. Controlling the Degradation Rate of Biodegradable Mg-Zn-Mn Alloys for Orthopedic Applications by Electrophoretic Deposition of Hydroxyapatite Coating. *Materials* **2020**, *13*, 263. [CrossRef]
26. Antoniac, I.; Popescu, D.; Zapciu, A.; Antoniac, A.; Miculescu, F.; Moldovan, H. Magnesium Filled Polylactic Acid (PLA) Material for Filament Based 3D Printing. *Materials* **2019**, *12*, 719. [CrossRef]
27. Vladescu, A.; Mihai Cotrut, C.; Ak Azem, F.; Bramowicz, M.; Pana, I.; Braic, V.; Birlik, I.; Kiss, A.; Braic, M.; Abdulgader, R.; et al. Sputtered Si and Mg doped hydroxyapatite for biomedical applications. *Biomed. Mater.* **2018**, *13*, 025011. [CrossRef]
28. Rau, J.V.; Antoniac, I.; Filipescu, M.; Cotrut, C.; Fosca, M.; Nistor, L.C.; Birjega, R.; Dinescu, M. Hydroxyapatite coatings on Mg-Ca alloy prepared by Pulsed Laser Deposition: Properties and corrosion resistance in Simulated Body Fluid. *Ceram. Int.* **2018**, *44*, 16678–16687. [CrossRef]
29. Bertinetti, L.; Drouet, C.; Combes, C.; Rey, C.; Tampieri, A.; Coluccia, S.; Martra, G. Surface characteristics of nanocrystalline apatites: Effect of Mg surface enrichment on morphology, surface hydration species, and cationic environments. *Langmuir* **2009**, *25*, 5647–5654. [CrossRef]

30. Landi, E.; Logroscino, G.; Proietti, L.; Tampieri, A.; Sandri, M.; Sprio, S. Biomimetic Mg-substituted hydroxyapatite: From synthesis to in vivo behaviour. *J. Mater. Sci. Mater. Med.* **2008**, *19*, 239–247. [CrossRef]
31. Chen, W.; Liu, Y.; Courtney, H.S.; Bettenga, M.; Agrawal, C.M.; Bumgardner, J.D.; Ong, J.L. In vitro anti-bacterial and biological properties of magnetron co-sputtered silver-containing hydroxyapatite coating. *Biomaterials* **2006**, *27*, 5512–5517. [CrossRef]
32. Zhao, S.; Jiang, Q.; Peel, S.; Wang, X.; He, F. Effects of magnesium-substituted nanohydroxyapatite coating on implant osseointegration. *Clin. Oral Implant. Res.* **2013**, *24*, 34–41. [CrossRef] [PubMed]
33. Bucur, A.I.; Linul, E.; Taranu, B.-O. Hydroxyapatite coatings on Ti substrates by simultaneous precipitation and electrodeposition. *Appl. Surf. Sci.* **2020**, *527*, 146820. [CrossRef]
34. Dorozhkin, S.V. Calcium orthophosphate deposits: Preparation, properties and biomedical applications. *Mater. Sci. Eng. C* **2015**, *55*, 272–326. [CrossRef]
35. Eliaz, N.; Sridh, T.M. Electrocrystallization of hydroxyapatite and its dependence on solution conditions. *Cryst. Growth Des.* **2008**, *8*, 3965–3977. [CrossRef]
36. Cotrut, C.M.; Vladescu, A.; Dinu, M.; Vranceanu, D.M. Influence of deposition temperature on the properties of hydroxyapatite obtained by electrochemical assisted deposition. *Ceram. Int.* **2017**, *44*, 669–677. [CrossRef]
37. Vladescu, A.; Vranceanu, D.M.; Kulesza, S.; Ivanov, A.N.; Bramowicz, M.; Fedonnikov, A.S.; Braic, M.; Norkin, I.A.; Koptyug, A.; Kurtukova, M.O.; et al. Influence of the electrolyte's pH on the properties of electrochemically deposited hydroxyapatite coating on additively manufactured Ti64 alloy. *Sci. Rep.* **2017**, *7*. [CrossRef]
38. Oriň, R.; Oriň, A.; Kupková, M.; Hrubovčáková, M.; Škantárová, L.; Arlinghaus, H.F. Study of Electrochemical Deposition and Degradation of Hydroxyapatite Coated Iron Biomaterials. *Int. J. Electrochem. Sci.* **2015**, *10*, 659–670.
39. ASTM International. *Standard Test Methods for Rating Adhesion by Tape Test*; ASTM D3359-17; ASTM International: West Conshohocken, PA, USA, 2017; pp. 1–9.
40. Kokubo, T.; Takadama, H. How useful is SBF in predicting in vivo bone bioactivity? *Biomaterials* **2006**, *27*, 2907–2915. [CrossRef]
41. ASTM International. *Standard Reference Test Method for Making Potentiostatic and Potentiodynamic Anodic Polarization Measurements*; ASTM G5-94(2011)e1; ASTM International: West Conshohocken, PA, USA, 2011.
42. Li, T.-T.; Ling, L.; Lin, M.-C.; Jiang, Q.; Lin, Q.; Lou, C.-W.; Lin, J.-H. Effects of ultrasonic treatment and current density on the properties of hydroxyapatite coating via electrodeposition and its in vitro biomineralization behavior. *Mater. Sci. Eng. C* **2019**, *105*, 110062. [CrossRef]
43. Chakraborty, R.; Seesala, V.S.; Sengupta, S.; Dhara, S.; Saha, P.; Das, K.; Das, S. Comparison of Osteoconduction, cytocompatibility and corrosion protection performance of hydroxyapatite-calcium hydrogen phosphate composite coating synthesized in-situ through pulsed electro-deposition with varying amount of phase and crystallinity. *Surf. Interfaces* **2018**, *10*, 1–10. [CrossRef]
44. Huang, Y.; Zhou, G.; Zheng, L.; Liu, H.; Niu, X.; Fan, Y. Micro-/Nano- sized hydroxyapatite directs differentiation of rat bone marrow derived mesenchymal stem cells towards an osteoblast lineage. *Nanoscale* **2012**, *4*, 2484. [CrossRef] [PubMed]
45. Wang, H.; Eliaz, N.; Xiang, Z.; Hsu, H.P.; Spector, M.; Hobbs, L.W. Early bone apposition in vivo on plasma-sprayed and electrochemically deposited hydroxyapatite coatings on titanium alloy. *Biomaterials* **2006**, *27*, 4192–4203. [CrossRef]
46. Kanamoto, K.; Imamura, K.; Kataoka, N.; Oshitani, J.; Imanaka, H.; Nakanishi, K. Formation characteristics of calcium phosphate deposits on a metal surface by H2O2-electrolysis reaction under various conditions. *Colloids Surf. A Physicochem. Eng. Asp.* **2009**, *350*, 79–86. [CrossRef]
47. Bakin, B.; Koc Delice, T.; Tiric, U.; Birlik, I.; Ak Azem, F. Bioactivity and corrosion properties of magnesium-substituted CaP coatings produced via electrochemical deposition. *Surf. Coatings Technol.* **2016**, *301*, 29–35. [CrossRef]
48. Biemond, J.E.; Hannink, G.; Jurrius, A.M.G.; Verdonschot, N.; Buma, P. In Vivo Assessment of Bone Ingrowth Potential of Three-Dimensional E-Beam Produced Implant Surfaces and the Effect of Additional Treatment by Acid Etching and Hydroxyapatite Coating. *J. Biomater. Appl.* **2012**, *26*, 861–875. [CrossRef]
49. Šponer, P.; Strnadová, M.; Urban, K. In vivo behaviour of low-temperature calcium-deficient hydroxyapatite: Comparison with deproteinised bovine bone. *Int. Orthop.* **2011**, *35*, 1553–1560. [CrossRef]

50. Thanh, D.T.M.; Nam, P.T.; Phuong, N.T.; Que, L.X.; Van Anh, N.; Hoang, T.; Lam, T.D. Controlling the electrodeposition, morphology and structure of hydroxyapatite coating on 316L stainless steel. *Mater. Sci. Eng. C* **2013**, *33*, 2037–2045. [CrossRef]
51. Li, T.-T.; Ling, L.; Lin, M.-C.; Jiang, Q.; Lin, Q.; Lin, J.-H.; Lou, C.-W. Properties and Mechanism of Hydroxyapatite Coating Prepared by Electrodeposition on a Braid for Biodegradable Bone Scaffolds. *Nanomaterials* **2019**, *9*, 679. [CrossRef]
52. Dong, X.-L.; Zhou, H.-L.; Wu, T.; Wang, Q. Behavior Regulation of Adsorbed Proteins via Hydroxyapatite Surface Texture Control. *J. Phys. Chem. B* **2008**, *112*, 4751–4759. [CrossRef]
53. Liu, X.; He, D.; Zhou, Z.; Wang, G.; Wang, Z.; Wu, X.; Tan, Z. Characteristics of (002) Oriented Hydroxyapatite Coatings Deposited by Atmospheric Plasma Spraying. *Coatings* **2018**, *8*, 258. [CrossRef]
54. Kim, H.; Camata, R.P.; Lee, S.; Rohrer, G.S.; Rollett, A.D.; Hennessy, K.M.; Bellis, S.L.; Vohra, Y.K. Calcium Phosphate Bioceramics with Tailored Crystallographic Texture for Controlling Cell Adhesion. *MRS Proc.* **2006**, *925*. [CrossRef]
55. Zou, Z.; Liu, X.; Chen, L.; Lin, K.; Chang, J. Dental enamel-like hydroxyapatite transformed directly from monetite. *J. Mater. Chem.* **2012**, *22*, 22637. [CrossRef]
56. Mišković-Stanković, V.B. Electrophoretic Deposition of Ceramic Coatings on Metal Surfaces. In *Electrodeposition and Surface Finishing: Fundamentals and Applications*; Stojan, S.D., Ed.; Springer: New York, NY, USA, 2014; pp. 133–216. ISBN 978-1-4939-4830-7.
57. Tamimi, F.; Le Nihouannen, D.; Eimar, H.; Sheikh, Z.; Komarova, S.; Barralet, J. The effect of autoclaving on the physical and biological properties of dicalcium phosphate dihydrate bioceramics: Brushite vs. monetite. *Acta Biomater.* **2012**, *8*, 3161–3169. [CrossRef] [PubMed]
58. Batra, U.; Kapoor, S.; Sharma, S. Influence of Magnesium Ion Substitution on Structural and Thermal Behavior of Nanodimensional Hydroxyapatite. *J. Mater. Eng. Perform.* **2013**, *22*, 1798–1806. [CrossRef]
59. Gopi, D.; Karthika, A.; Nithiya, S.; Kavitha, L. In vitro biological performance of minerals substituted hydroxyapatite coating by pulsed electrodeposition method. *Mater. Chem. Phys.* **2014**, *144*, 75–85. [CrossRef]
60. Rusu, V.M.; Ng, C.H.; Wilke, M.; Tiersch, B.; Fratzl, P.; Peter, M.G. Size-controlled hydroxyapatite nanoparticles as self-organized organic-inorganic composite materials. *Biomaterials* **2005**, *26*, 5414–5426. [CrossRef]
61. Landi, E.; Tampieri, A.; Celotti, G.; Sprio, S. Densification behaviour and mechanisms of synthetic hydroxyapatites. *J. Eur. Ceram. Soc.* **2000**, *20*, 2377–2387. [CrossRef]
62. Poralan, G.M.; Gambe, J.E.; Alcantara, E.M.; Vequizo, R.M. X-ray diffraction and infrared spectroscopy analyses on the crystallinity of engineered biological hydroxyapatite for medical application. *IOP Conf. Ser. Mater. Sci. Eng.* **2015**, *79*, 012028. [CrossRef]
63. Leach, R. Surface Topography Characterisation. In *Fundamental Principles of Engineering Nanometrology*; Elsevier: Amsterdam, The Netherlands, 2014; pp. 241–294.
64. Mayer, I.; Schlam, R.; Featherstone, J.D. Magnesium-containing carbonate apatites. *J. Inorg. Biochem.* **1997**, *66*, 1–6. [CrossRef]
65. Parcharoen, Y.; Kajitvichyanukul, P.; Sirivisoot, S.; Termsuksawad, P. Hydroxyapatite electrodeposition on anodized titanium nanotubes for orthopedic applications. *Appl. Surf. Sci.* **2014**, *311*, 54–61. [CrossRef]
66. Zhang, Z.; Dunn, M.F.; Xiao, T.D.; Tomsia, A.P.; Saiz, E. Nanostructured Hydroxyapatite Coatings for Improved Adhesion and Corrosion Resistance for Medical Implants. *MRS Proc.* **2001**, *703*. [CrossRef]
67. Baboian, R.; Scully, J.R.; Dean, S.W.J. *Corrosion Tests and Standards*; Fontana, M.G., Staehle, R.W., Eds.; Springer US Plenum Press: New York, NY, USA, 2005; Volume 6, ISBN 978-1-4684-8988-0.
68. Cai, Y.; Zhang, S.; Zeng, X.; Wang, Y.; Qian, M.; Weng, W. Improvement of bioactivity with magnesium and fluorine ions incorporated hydroxyapatite coatings via sol–gel deposition on Ti6Al4V alloys. *Thin Solid Films* **2009**, *517*, 5347–5351. [CrossRef]

coatings

Article

Effect of Deposition Temperature on the Structure, Mechanical, Electrochemical Evaluation, Degradation Rate and Peptides Adhesion of Mg and Si-Doped Hydroxyapatite Deposited on AZ31B Alloy

Anca Constantina Parau [1], Mihaela Dinu [1], Cosmin Mihai Cotrut [2], Iulian Pana [1], Diana Maria Vranceanu [2], Lidia Ruxandra Constantin [1], Giuseppe Serratore [3], Ioana Maria Marinescu [1], Catalin Vitelaru [1], Giuseppina Ambrogio [3], Dennis Alexander Böhner [4], Annette G. Beck-Sickinger [4] and Alina Vladescu (Dragomir) [1,5,*]

[1] Department for Advanced Surface Processing and Analysis by Vacuum Technologies, National Institute of Research and Development for Optoelectronics—INOE 2000, 77125 Magurele, Romania; lidia.constantin@inoe.ro (L.R.C.)
[2] Department of Metallic Materials Science, Physical Metallurgy, Faculty of Materials Science and Engineering, University Politehnica of Bucharest, 60042 Bucharest, Romania
[3] Department of Mechanical Energy and Management Engineering, University of Calabria, 87036 Rende, Italy
[4] Institute of Biochemistry, Faculty of Life Sciences, Leipzig University, D04316 Leipzig, Germany; dennis.boehner@uni-leipzig.de (D.A.B.)
[5] Physical Materials Science and Composite Materials Centre, Research School of Chemistry & Applied Biomedical Sciences, National Research Tomsk Polytechnic University, 634050 Tomsk, Russia
* Correspondence: alinava@inoe.ro; Tel.: +40-21-457-57-59

Citation: Parau, A.C.; Dinu, M.; Cotrut, C.M.; Pana, I.; Vranceanu, D.M.; Constantin, L.R.; Serratore, G.; Marinescu, I.M.; Vitelaru, C.; Ambrogio, G.; et al. Effect of Deposition Temperature on the Structure, Mechanical, Electrochemical Evaluation, Degradation Rate and Peptides Adhesion of Mg and Si-Doped Hydroxyapatite Deposited on AZ31B Alloy. *Coatings* **2023**, *13*, 591. https://doi.org/10.3390/coatings13030591

Academic Editor: Devis Bellucci

Received: 8 February 2023
Revised: 28 February 2023
Accepted: 3 March 2023
Published: 9 March 2023

Abstract: Degradable and non-degradable biomaterials are two categories that can be used to classify the existing biomaterials, being a solution for eliminating a second surgical intervention of the implant when the tissue has properly recovered. In the present paper, the effect of deposition temperature on the structure, morphology, hardness, electrochemical evaluation, degradation properties and functional peptides adhesion of Mg and Si-doped hydroxyapatite was investigated. The coatings were obtained by RF magnetron sputtering technique at room temperature (RT) and 200 °C on AZ31B alloy substrate. Results showed that an increase in deposition temperature led to an improvement in hardness and reduced modulus of about 47%. From an electrochemical point of view, a comparative assessment of corrosion resistance was made as a function of the immersion medium used, highlighting the superior behaviour revealed by the coating deposited at elevated temperature when immersed in DMEM medium (i_{corr}~12 μA/cm^2, R_{coat} = 705 Ω cm^2, R_{ct} = 7624 Ω cm^2). By increasing the deposition temperature up to 200 °C, the degradation rate of the coatings was slowed, more visible in the case of DMEM, which had a less aggressive effect after 14 days of immersion. Both deposition temperatures are equally suitable for further bio-inspired coating with a mussel-derived peptide, to facilitate biointegration.

Keywords: magnetron sputtering; hydroxyapatite; hardness; corrosion; degradation; peptides adhesion

1. Introduction

Degradable and non-degradable biomaterials are two categories that can be used to classify existing biomaterials. Researchers have shown a greater interest in biodegradable materials in recent years compared to more typical biologically inert metal materials, such as stainless steel, titanium alloy, cobalt-based alloy, and so on [1–6]. The necessity for a second surgical intervention needed to remove the implant when the tissue has properly recovered is eliminated when using biodegradable implants, which is one of the most significant advantages of these types of devices [7,8].

In the topic of magnesium and magnesium alloys, more than 4000 papers were published in 2021 alone, illustrating that this is a hot spot in the materials science and engineering field [9]. Also, there has been an increasing amount of focus in terms of developing magnesium alloys as a viable treatment solution for orthopaedic injuries and cardiovascular diseases. Nevertheless, the quick rate of magnesium breakdown continues to be a significant barrier to the widespread utilisation of magnesium. Therefore, the development of biodegradable magnesium alloys with tuneable degradation rates is crucial [4,10–12].

As a potential material for use in biomedicine, magnesium alloy possesses several benefits. It is one of the primary metal elements found in the human body, and its levels are second only to those of calcium, sodium, and potassium, being a crucial component for bone development, favouring cell proliferation [3,10,13]. Also, it is involved in 300 enzymatic reactions [5]. Alloys made of magnesium have mechanical characteristics that are comparable to those of human bones [4,10,14]. In the long-axis direction, the tensile strength, young's modulus, and density of human cortical bone are as follows: 120–150 MPa, 20 GPa, and 1.8 g/cm^3, respectively. In comparison, the mechanical properties of magnesium alloy are as follows: 200–300 MPa, 40–45 GPa, and 1.74 g/cm^3, respectively. In addition, the elastic modulus is somewhere in the range of 41–45 GPa, which is very close to that of the human bone [7,10,15]. Also, the low density of the engineering metals of magnesium is approximately 65% as dense as aluminium alloys, 38% as dense as titanium, and 25% as dense as steel [16,17]. As the structural metal with the lowest density, magnesium is widely regarded as the ideal candidate for use in modern alloys [18].

The fundamental principles of degradation mechanism and corrosion products have been largely acknowledged during the many years of study conducted worldwide on magnesium alloys. When it comes to magnesium alloys, the primary areas of focus are still the control of the degradation rate and the behaviour of ion release [12,18,19].

In 1878, vascular ligation with magnesium alloys was performed for the first time, and since then, the benefit of using biodegradable materials in medicine was continuously demonstrated [7,10,15,18,20]. Recent research has resulted in the development of new magnesium zinc composites with further characteristics since zinc is another vitally important trace element for the human body. It is second only to iron in terms of its content in the human body, which comes in at roughly 2 g. According to the studies carried out so far, the presence of zinc within magnesium alloys has a significant impact on their resistance to corrosion [3,8,10].

The most difficult aspect of this situation is figuring out how to customise the degrading process in a manner that is appropriate for a biological setting. Surface treatment is one of the primary tactics that has been extensively researched as a way of modifying the mechanical characteristics of magnesium and its alloys in order to slow down the degradation rate [5,19].

If researchers and engineers could forecast the rates of corrosion, they would be better able to develop materials with appropriate corrosion rates without significantly compromising the material's mechanical qualities [21]. Despite the continuous efforts undertaken to make the metallic materials more resistant to corrosion, which is an electrochemical process that includes both reduction and oxidation reactions, the underlying problem persists [19]. One suitable possibility to control the corrosion process and to prevent the body fluids from directly interacting with magnesium-based alloys is the usage of coatings, which act as protective layers at the body-magnesium implant interface [10,20].

Surface coatings can either drastically cut down the rate of localised degradation or, at the very least, postpone the rate at which magnesium-based materials are attacked locally. There are a few different coating processes available for magnesium and its alloys, through which ceramic and/or polymeric-based films can be obtained [6,12,22–26].

Among the material of choice as a coating for magnesium alloy substrate, one can find hydroxyapatite (HAp, $Ca_{10}(PO_4)_6(OH)_2$, which is a ceramic biomaterial, also allowing the addition of different doping elements to further favour the osteointegration and control the alloys corrosion rate [27,28]. In addition to this, it is known that calcium is one of the most crucial nutrients, involved in several processes and mechanisms of the human body [13]. Because HAp possesses strong bioactivity and osteoconductivity, it can rapidly integrate with the bones and stimulate the formation of new hard tissue. This ability is critical for bone regeneration, as was reported in Refs. [22,29].

Coatings made of calcium phosphate-based materials, such as HAp, are non-toxic, osteoconductive, and have high biocompatibility [30]. As a result, a significant number of studies have concentrated on calcium phosphate coatings for use in biomedical applications such as bone substitutes and orthopaedics. A layer of hydroxyapatite, which is the primary component of natural bone, can occur through the presence of calcium and phosphorus elements [2,15,20]. There are a few review articles that discuss several coating solutions, including HAp coatings, that can be used on magnesium-based materials [16,22,30–32].

Surface modification of biomaterials can be accomplished by several different procedures, one of which is called magnetron sputtering. Using this method, hydroxyapatite coating properties like topography, Ca/P ratio, density, thickness, etc., can be modified by altering the sputtering parameters such as air pressure, applied voltage, substrate-to-target distance, and deposition time [1,14,16,33]. Magnetron sputtering results in coatings that are dense, have a strong adherence to metallic surfaces and have an elemental composition that can be controlled and tuned [34,35].

The interest in the addition of biocompatible elements has increased in medical applications, especially in the coatings field. For this study, silicon (Si) was used as a doping element in order to increase HAp's corrosion resistance since it acts as a barrier between the substrate and the body fluids. Silicon is a biocompatible element that sustains the function of osteogenic cells [14,22] and is essential for the formation and growth of bone, teeth, and other skeletal elements. Studies suggest that the addition of Si to HAp facilitates the precipitation of an apatite layer on the materials' surface in an artificial physiological solution [36,37]. Used as a doping element, magnesium helps the osteointegration process [37]. According to the findings of in-vitro degradation investigations, the presence of the coating has significantly improved the corrosion resistance of magnesium alloy and has increased the bioactivity [38].

The objective of this study was to investigate the effect of deposition temperature (room temperature vs. 200 °C) on the structure, morphology, hardness, electrochemical evaluation, degradation properties and peptides adhesion of Mg and Si-doped hydroxyapatite coatings obtained by RF magnetron sputtering to improve AZ31B alloy properties. The deposition temperature was selected based on the previous results published in Ref. [34], where it was reported that the best deposition temperature was 200 °C. According to the previously published results [34], the deposition temperature has a significant effect on the crystallinity, mechanical and degradation rate of HAp coatings. Based on this paper, it was found that the coatings deposited at RT have an amorphous structure, while those deposited at 200 °C were crystalline with good corrosion resistance in simulated body fluid (SBF). Both coatings proved to have proper mechanical properties compared to those deposited at 100, 300 or 400 °C. Thus, based on the reported results, the present research is focused on the coatings prepared at RT and 200 °C.

2. Experimental Details

2.1. Substrate

AZ31B was used as substrates for the present paper, being supplied in flat sheets with a nominal thickness of 1 mm made by cold rolling and final annealing treatment. The chemical composition is reported in Table 1.

Table 1. AZ31B chemical composition (wt.%) of the as-received alloy.

Al	Zn	Mn	Fe	Cu	Si	Ni	Ca	Mg
2.98	1.03	0.34	0.0022	0.0067	0.0089	0.00047	-	Rest

Prior to coating deposition, the specimens were processed via Single Point Incremental forming (SPIF) [39] to obtain samples with mechanical and superficial conditions more equivalent to those that could be obtainable in a real prosthetic usecase. A standard geometry, i.e., a pyramid with a 154 square base characterised by an inclined sidewall of 40°, a major base equal to 149 mm 155 and a final depth of 30 mm, was manufactured, and a set of small samples (10 × 10 mm^2) were extracted by each side of the pyramid for the subsequent coating. The SPIF process was achieved by a hemispherical head punch tool with a diameter of 10 mm and a step depth of 0.1 mm on a Mazak Nexus 410 milling machine equipped with a heating chamber able to keep constant the sheet temperature up to about 250 °C during the forming process [40]. A feed rate of 0.5 m/min and a spindle speed of 4000 rpm were used. A D321R Molykote spray was employed as a lubricant. The substrates were obtained by the side faces of the above-described geometry.

2.2. Coatings

RF magnetron sputtering deposition unit was used for preparing the coatings (AJA International, Scituate, MA, USA). The system was equipped with three cathodes (Φ 20.4 mm) made of HAp (99.99% purity), MgO (99.99% purity) and SiC (99.99% purity), positioned in a confocal geometry. To achieve a similar deposition rate as of the HAp, the addition of Mg and Si into HAp-based coatings was carried out with cathodes made of oxides. In the present study, the coatings were prepared on two types of substrates: silicon wafers with <111> orientation and AZ31B alloy, depending on the type of investigation technique. To ensure the uniformity of the coatings, the samples were positioned on a rotating holder (15 rpm) at a distance of 120 mm from the targets. Prior to each deposition run, the substrates were ultrasonically cleaned with isopropyl alcohol for 15 min in an ultrasonic bath, and consequently, they were sputter cleaned in Ar^+ plasma for 15 min (f = 13.56 MHz, p = 50 W, U_{bias} = −310 V, p = 0.67 Pa, without any intentional heating on substrates holder) and the vacuum chamber was evacuated down to 1.4 × 10^{-4} Pa. The deposition conditions are presented in Table 2.

Table 2. The deposition conditions: p—pressure, V_{RF}—RF bias on substrates, P—power applied on the cathode, T—deposition temperature.

Coatings	p (Pa)	V_{RF} (V)	P_{MgO} (W)	P_{SiC} (W)	T (°C)
HAp + Mg + Si_RT	0.67	−60	25	15	RT
HAp + Mg + Si_200	0.67	−60	25	15	200

2.3. Investigation Techniques

An energy-dispersive X-ray spectrometer (EDS, Bruker, Billerica, MA, USA) was used to determine the elemental composition of coatings.

The coatings topography was evaluated by atomic force microscopy (AFM, Veeco-Bruker, Billerica, MA, USA), using tapping mode 1 × 1 μm^2 with a scan speed of 0.3 Hz.

Grazing Incidence X-ray diffraction (GIXRD) was carried out to evaluate the phase composition of the coating using a SmartLab diffractometer (Rigaku, Tokyo, Japan) at an incident angle of the X-ray beam to 3°.

The nanoindentation test was used for determining the hardness and elastic modulus of the coatings using a Hysitron Premier TI nanomechanical characterisation system (Bruker, Billerica, MA, USA). The nanoindenter was equipped with a 142.3° blunt Berkovich tip with a 100 nm curvature radius. The surface of the sample was also scanned with the same indenter tip before testing to ensure a smooth surface for nanoindentation. Before any indentation testing, potential sources of uncertainty and errors such as thermal drift, initial penetration depth, and machine compliance were considered. A standard fused quartz sample (H = 9.25 GPa ± 10%, E = 69.6 GPa ± 10%) was used to calibrate the load force. The contact depth of every single indentation point was larger than 40 nm to overcome the limitation given by the geometrical characteristics of the used Berkovich tip. A minimum physical distance of 5 μm was established between at least 25 indentation points, while the time intervals for load, hold and unload were 7 s, 2 s and 7 s, respectively.

In vitro electrochemical tests in SBF (simulated body fluid) and DMEM (Dulbecco's Modified Eagle Medium) solutions were used for evaluating the corrosion resistance of the coatings using a PARSTAT 4000 Potentiostat/Galvanostat (Princeton Applied Research—Ametek, Oak Ridge, TN, USA). A conventional three-electrode electrochemical cell was employed: working electrode (WE, coated samples), calomel reference electrode (SCE) and Pt foil as a counter electrode (CE). The tests were performed in SBF (pH = 7.4, [41]) at 37 ± 0.5 °C, being held constant with a heated circulating bath (Jeio Tech, CW-05G, Yuseong-gu, Daejeon, Republic of Korea). The electrochemical tests were performed in good agreement with a protocol described in ASTM standard G5-94 (2014) as follows. The open circuit potential (E_{OC}) was monitored for 1 h, and the Tafel plots were carried out between ±0.200 mV vs. E_{OC} at a scanning rate of 1 mV/s. The main electrochemical parameters—corrosion potential (E_{corr}) and corrosion current density (i_{corr}), were estimated by Tafel plots extrapolation, according to ASTM G59-97 standard (reapproved 2020) [42].

In vitro, electrochemical impedance spectroscopy (EIS, Princeton Applied Research—Ametek, Oak Ridge, TN, USA) measurements were performed over the frequency range of $0.1 \div 10^5$ Hz by applying a sinusoidal signal with an amplitude of 10 mV RMS vs. E_{OC}. The data were recorded by VersaStudio software (version 2.62.2, Princeton Applied Research, Oak Ridge, TN, USA), and the EIS fitting procedure was performed using ZView software (version 12136-4, Scribner Associates Inc., Southern Pines, NC, USA).

In vitro degradation tests were performed in SBF and DMEM for 1, 3, 7, and 14 days at 37 ± 0.5 °C, using an incubator (Memmert IF 55, Memmert GmbH, Büchenbach, Germany). The testing solutions were changed everytwo days to ensure the necessary ionic content and to prevent the growth/formation of bacteria or other microorganisms. After each period of immersion, each sample was washed with distilled water and then dried for 24 h in a desiccator to remove any remaining water from its surface. Then, the samples were weighed with an analytical balance to evaluate the sample's mass evolution [41,43]. Five samples of each coating were used for immersion assays, and the mass measurements were conducted five times for each sample. The mass evolution was calculated based on the formula: $\Delta m = mf - mi$ (mg), where: Δm represents the mass variation; mf is the sample's mass after immersion; mi is the sample's initial mass.

A scanning electron microscope (SEM, TableTop 3030PLUS, Tokyo, Japan) was used to evaluate the coatings aspect after the electrochemical and biodegradation tests.

Decorating biomaterial surfaces with functional peptides is a promising approach to improve implant-host interactions. Immobilisation of the cell-adhesive motifs c [RGDfK] or the heparin-binding sequence FHRRIKA was successfully achieved by short bio-inspired peptides derived from mussel foot proteins [44,45]. The effect of the HAp-based coating on the surface binding ability of a similar mussel-derived peptide (MP) was investigated. To protect the uncoated backside from solvent exposure, samples were glued into a 12-well plate using silicone. For surface coating, samples were incubated with 1 μM peptide solution in TBS buffer (25 mM Tris, 137 mM NaCl, 2.7 mM KCl, pH 7.6) for 238 16 h shaking at room temperature. For SEM imaging, samples were incubated with unmodified MP, whereas a biotinylated peptide was used for the investigation of the surface binding affinity. The next day, the coating solution was aspired, and samples were washed twice with TBS-T buffer (25 mM Tris, 137 mM NaCl, 2.7 mM KCl, 0.1% Tween20, pH 7.6) as well as TBS buffer. After transferring and glueing samples into new wells, surfaces were first blocked with 10% BSA in TBS buffer for 20 min and then incubated with horseradish peroxidase-conjugated streptavidin (1:2000 in TBS containing 1% BSA) for 1 h. Subsequently, samples were washed four times with TBS-T buffer and detection of bound HRP-streptavidin conjugate was carried out using 3,3′,5,5′-tetramethylbenzidine (TMB). After 3 min the reaction was stopped with 1 M HCl and the absorption at 450 nm of the solution was measured (Infinite M200, Tecan). The result is shown as mean + standard error of the mean and represents two independent experiments performed in duplicates. Data were analysed using GraphPad

Prism 5.03 (GraphPad Software). Statistical significances were determined by one-way ANOVA following Dunnett's multiple comparison test.

3. Results

3.1. Elemental and Phase Compositions and Morphology

The EDS investigations were performed on the coatings deposited on the Si wafers substrate for determining the Mg content and on the Mg alloy substrate for determining the Si content. This research design was made in order to eliminate the influence of the Mg or Si element from each substrate. The chemical composition is presented in Figure 1a. The EDS spectrum of each coating is presented in Figure 1c. One may observe that Si and Mg content is up to ~7 at.%, indicating the formation of HAp-doped coatings. The Ca/P ratio was calculated at about 1.53 for the coating deposited at RT and 1.68 for those prepared at 200 °C. This finding showed that an increment of the deposition temperature assured the formation of a stoichiometric HAp structure. Usually, for the doped HAp coatings, the (Ca + Si + Mg)/P ratio is calculated, but in the case of the present paper, it was impossible to be evaluated because the substrates were Si or Mg alloy, having at least one element found in the coating. However, based on the EDS spectrum (Figure 1b), there are seen a small amount of some elements of bare Mg alloy, such as Zn. Other elements, such as Mn, Fe, Cu, and Ni, are too small to be detected by the EDS system.

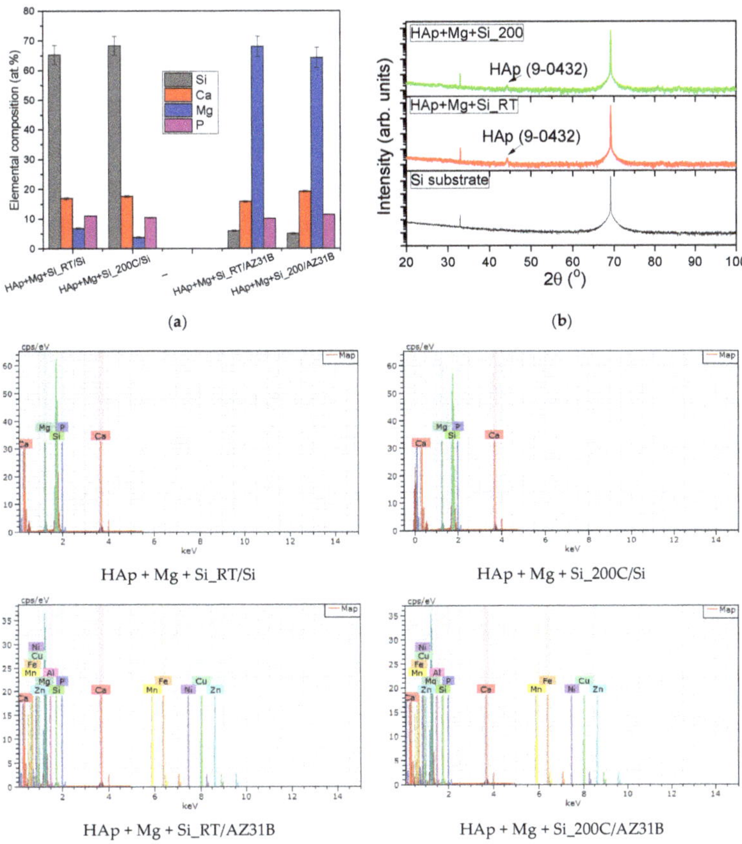

Figure 1. (**a**) Elemental composition; (**b**) GIXRD diffraction patterns of the coatings; and (**c**) EDS spectrum of each coating.

The GIXRD diffraction patterns of the coatings deposited on Si wafers are presented in Figure 1b. In both coatings, a small peak located at 44.32° can be found, being attributed to the HAp phase (JCPDS card no. 09-0432). This result indicates that the Si addition does not affect the formation of the HAp phase.

The morphology was evaluated based on AFM images (Figure 2). One can note that the morphology is quite similar for both coatings. Those prepared at 200 °C exhibited a denser structure with fine grains. It should be mentioned that the AFM images were recorded on coatings deposited on Si wafers to avoid disturbances from the roughness of the Mg alloy.

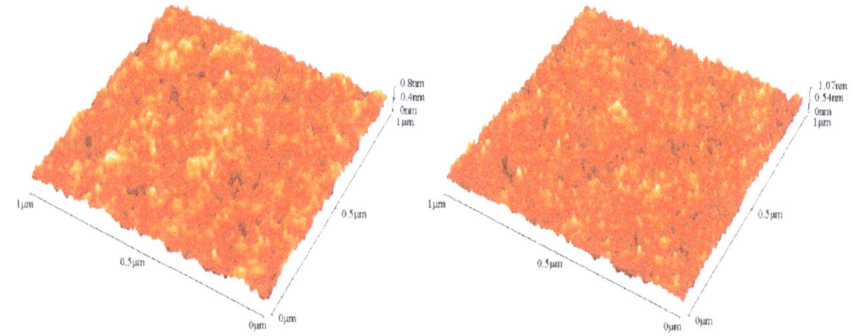

(**a**) (**b**)

Figure 2. AFM images: (**a**) HAp + Mg + Si_RT and (**b**) HAp + Mg + Si_200 coatings deposited on Si waffers.

The 2D SEM images acquired for the coatings deposited directly on AZ31B alloys are presented in Figure 3. In both cases, the grooves' presence at 100× magnification is a common feature expected after the polishing process of AZ31B alloys before coating.

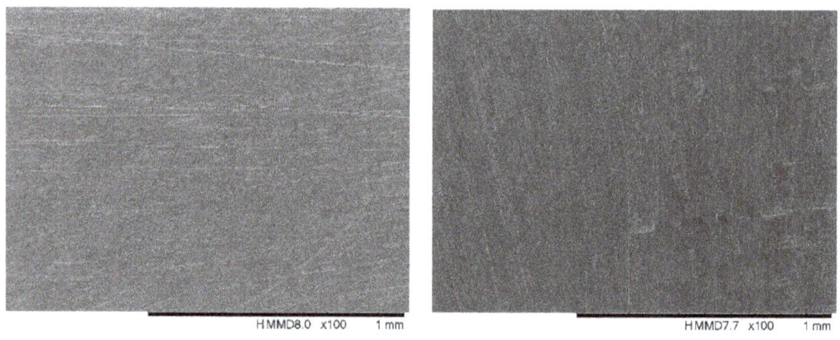

(**a**) (**b**)

Figure 3. 2D SEM images: (**a**) HAp + Mg + Si_RT and (**b**) HAp + Mg + Si_200 coatings deposited on AZ31B substrates.

3.2. Nanoindentation

The mechanical properties of the coatings can significantly differ from those of bulk materials due to the confinement of the material at the nanometric scale. The unique properties of these thin layers are highly dependent on several factors such as the deposition method, thickness, structure, substrate type, or interface between substrate and layer. Therefore, it is important to adapt the deposition methods to obtain coatings that meet the specific technological requirements for smart functional and multifunctional biomaterials.

The hardness and reduced elastic modulus of the coatings were evaluated at penetration depths comparable to the coating's thicknesses. The results are presented in Figure 4. The substrate effect on the mechanical properties of the investigated layers was considered by employing a model that estimates the dependencies of hardness and reduced modulus as a function of substrate characteristics, maximum indentation depth and thickness [46]. The used model was only valid between the minimum 40 nm contact depth limit of the used nanomechanical system and a maximum penetration depth of ~90% of the film thickness. The maximum applied force used for nanoindentations was selected based on the coating's thicknesses and, in this case, had not exceeded a value of 2.5 mN.

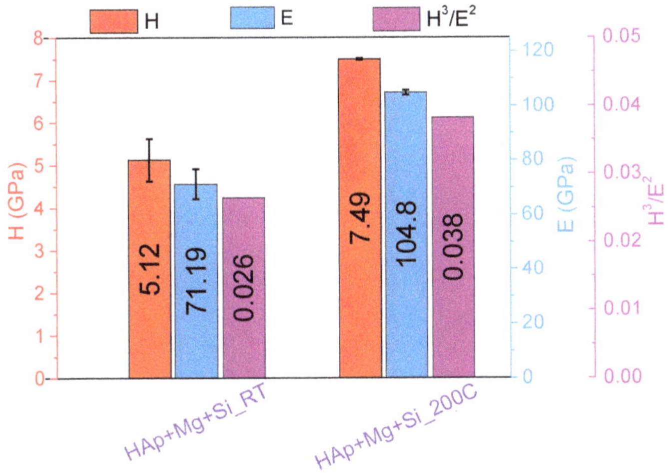

Figure 4. Derived hardness and reduced Young's modulus values of samples based on the Olive–Pharr algorithm.

The results indicated that the mechanical properties of the coatings differ significantly depending on the deposition temperature increment from RT to 200 °C. It was shown that both the hardness and reduced modulus of the coatings increase by approximately 47% as the deposition temperature increases. These results are consistent with our previous findings [47]. The selection of appropriate biomaterials as protective coatings for bare metallic implants is crucial and requires a balance between their hardness and friction properties [48]. One common parameter used to evaluate a material's resistance to plastic deformation is the H^3/Er^2 ratio, which has also increased from 0.026 to 0.038. This suggests that the HAp + Mg + Si_200 °C sample may have superior resistance to plastic deformation when compared to the HAp + Mg + Si_RT sample.

3.3. *In Vitro Corrosion Resistance*

The Tafel plots of the investigated coatings obtained at room temperature and at 200 °C in SBF and DMEM testing media are given in Figure 5. The main electrochemical parameters obtained from the electrochemical tests are presented in Table 3. The values of the i_{corr} parameter evolution as a function of the testing media for the investigated specimens are also shown in Figure 5.

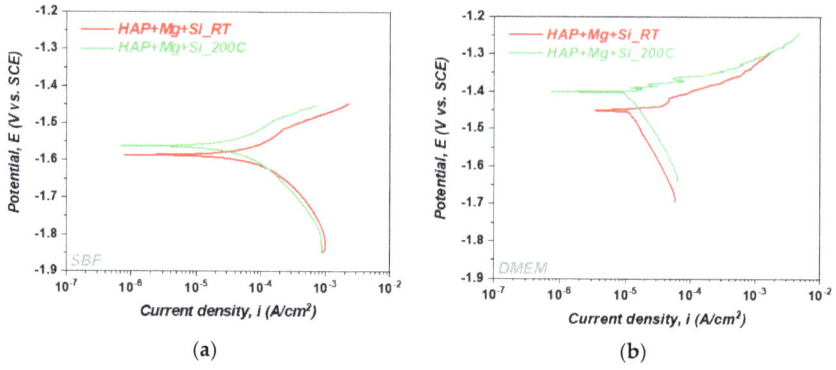

Figure 5. In vitro Tafel results: (**a**) Tests in SBF solution; (**b**) Tests in DMEM solutions.

Table 3. Electrochemical parameters of the coatings obtained in SBF and DMEM solutions.

Coatings	Solutions	E_{oc} (mV)	E_{corr} (mV)	i_{corr} (μA/cm^2)
HAp + Mg + Si_RT	SBF	−1669	−1587	190.73
	DMEM	−1510	−1452	18.05
HAp + Mg + Si_200	SBF	−1668	−1562	77.24
	DMEM	−1511	−1400	12.17

E_{OC}—open circuit potential, E_{corr}—corrosion potential, i_{corr}—current density.

It is well known that a material is more resistant to the hostility of a corrosive medium when the value of the corrosion potential (E_{corr}) has a more electropositive value and the corrosion current density (i_{corr}) has a smaller value. By considering this statement, it can be noticed that better values were reached for the coatings tested in the DMEM solution (Table 3 and Figure 5), indicating that both coatings are more resistant in DMEM. However, in both testing media and irrespective of the deposition temperature used, RT or 200 °C, the electrochemical parameters present a similar evolution.

In SBF, both coatings exhibited almost similar E_{corr} values, while the i_{corr} value decreased considerably (2.5 times) when the deposition temperature increased from RT to 200 °C. This finding shows that the coatings deposited at higher temperatures exhibited higher corrosion resistance in SBF.

In DMEM, the values of the E_{corr} parameter for both coatings are close to each other, while the i_{corr} value was found to be smaller for the coatings prepared at 200 °C, suggesting that an increment of the deposition temperature leads to higher corrosion resistance.

In a previous paper, we reported that the undoped HAp exhibited a value of the i_{corr} from 23.07 μA/cm^2 (RT) to 12.23 μA/cm^2 (200 °C) in DMEM and from 439.18 μA/cm^2 (RT) to 129.16 μA/cm^2 (200 °C) in SBF, indicating that a temperature of 200 °C enhances the electrochemical behaviour of the undoped HAp based coatings. By comparing the parameters obtained for the present coating with those of the undoped HAp, it can be concluded that the addition of Si and Mg into the HAp matrix significantly improves the corrosion resistance in the SBF and DMEM solution of the AZ31B alloy.

3.4. In Vitro EIS Results

Figures 6 and 7 display the Nyquist and Bode plots of the investigated samples immersed in SBF and DMEM electrolytes after the stabilisation of the open circuit potential for 3600 s. Two defined time constants (indicated by arrows in Figures 6 and 7) characterised the HAp + Mg + Si samples, more visible in the case of the SBF test medium. Additionally, for both conditions, the Nyquist plot reveals the presence of an inductive loop in the low-frequency range (emphasised by a red square). According to Flores et al. [49], the time constant present in the medium frequencies range (MF) provides information related to coating–electrolyte interface, while the impedance data present in the low-frequency range (LF) shows the occurrence of a corrosion process at the substrate–solution interface. Thus, the presence of an inductance loop in LF is often associated with the absorption of chloride ions through cracks and defects, causing the initiation of corrosion phenomena on the surface and, consequently, a decrease in substrate resistance [50]. Also, the high rate of Mg ions dissolution caused by low applied frequencies can be a responsible process as well [51].

Even though similar results were shown by the samples immersed in SBF [50,52,53], a slightly different behaviour was promoted by the electrolyte change. Higher semicircle diameters were obtained when DMEM was used, also confirmed by higher impedance modulus revealed by the Bode amplitude plot (Figure 7b). In this case, a more obvious difference was observed among the samples obtained at different deposition conditions, indicating an enhancement of dielectric properties influenced by the temperature increase. According to the literature, larger semicircle diameters are related to higher resistance to charge transfer, leading to a low corrosion rate of specimens under investigation [54].

Based on the impedance data displayed in Nyquist and Bode graphs, the presented electrical equivalent circuit was used to model the physical reactions (inset in Figures 6a and 7b). Thus, processes that occur within the analysed systems, such as fast (i.e., formation of the double layer, ohmic/charge transfer resistance) and slow processes (i.e., adsorbed species or transport phenomena), can be translated into quantitative information [55]. In this case, in addition to the often-used two-time constant equivalent circuit (with a CPE instead of ideal dielectric properties caused by inhomogeneity or current leakage [56]), L was also added, ascribed to corrosion product formation.

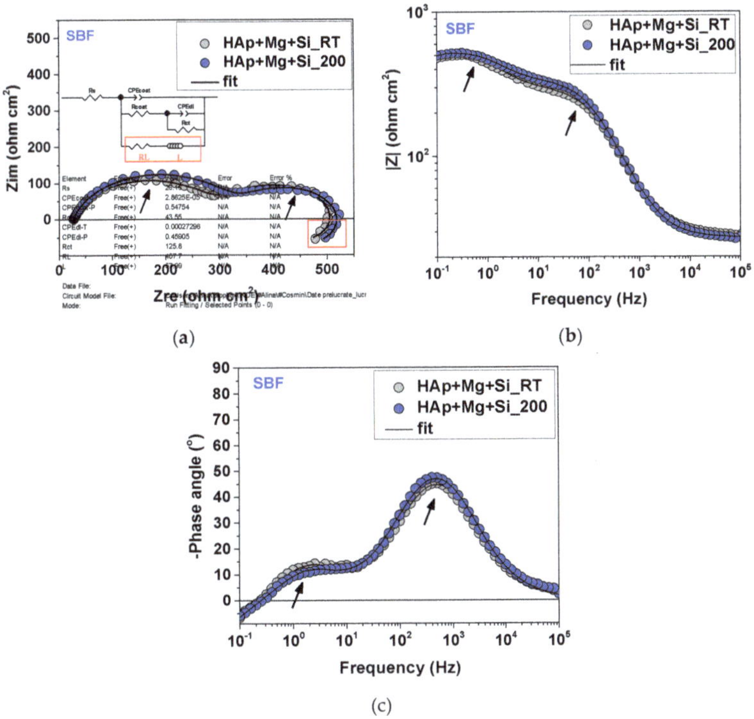

Figure 6. EIS curves of the investigated specimens immersed in SBF: (**a**) Nyquist plots, (**b**) |Z| plots, (**c**) Phase angle plots. (R_s = the solution resistance, CPE_{coat} = coating capacitance, R_{coat} = resistance associated with the current flow, CPE_{dl} = double layer capacitance, R_{ct} = charge transfer resistance, L and RL were used to simulate the inductive behaviour).

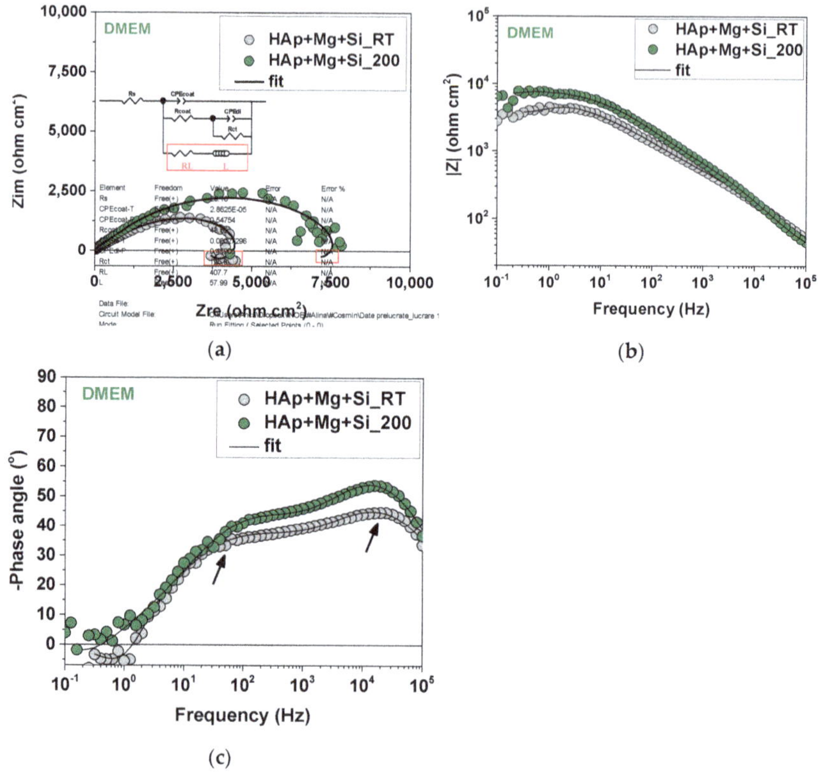

Figure 7. EIS curves of the investigated specimens immersed in DMEM: (**a**) Nyquist plots, (**b**) |Z| plots, (**c**) Phase angle plots. (R_s = the solution resistance, CPE_{coat} = coating capacitance, R_{coat} = resistance associated with the current flow, CPE_{dl} = double layer capacitance, R_{ct} = charge transfer resistance, L and RL were used to simulate the inductive behaviour).

A comparative assessment of the HAp + Mg + Si corrosion protection ability as a function of the used electrolyte (SBF vs. DMEM) was further made based on the electrochemical parameters presented in Table 4. According to Orazem et al. [57–60], the quality of fitting can be evaluated by the χ^2 error parameter, which in this case recorded values of 10^{-4}–10^{-3}. As expected, HAp + Mg + Si showed superior electrochemical behaviour when immersed in a DMEM medium, and even higher dielectric properties were revealed when using an elevated temperature, i.e., 200 °C). Obviously, an increasing tendency of R_{coat} can be observed, according to the coating deposition and immersion conditions, while the lowest value of Q_{coat} was shown by HAp + Mg + Si_200 DMEM. At the electrolyte-substrate interface, similar values of double-layer capacitance (Q_{dl} ~700 μF s(α − 1) cm^{-2}) characterised the samples immersed in SBF, showing no improvement due to temperature in this case. However, the use of DMEM led to much lower values for H-Mg-Si RT (Q_{dl} ~17 μF s(α − 1) cm^{-2}) and HAp + Mg + Si_200 (Q_{dl} ~7 μF s(α − 1) cm^{-2}). The charge transfer reaction parameter was also influenced by the coating deposition and immersion conditions; the highest value obtained, and hence the best corrosion behaviour, was observed in the case of HAp + Mg + Si_200 DMEM (R_{ct} = 42,274 Ω cm^2), proving a hindering effect in case of the mentioned coating, which limited the infiltration of the corrosive medium.

Table 4. EIS fitting results of the coatings obtained in SBF and DMEM solutions.

Sample/Parameters	HAp + Mg + Si			
Temperature	RT	200	RT	200
Test Medium	SBF		DMEM	
R_s (Ω cm^2)	28	27	26	25
Q_{coat} (μF s(α − 1) cm^{-2})	14.396	13.646	2.339	0.813
α_{coat}	0.84	0.84	0.70	0.79
R_{coat} (Ω cm^2)	277	319	558	705
Q_{dl} (μF s(α − 1) cm^{-2})	757.480	749.400	17.533	7.459
α_{dl}	0.82	0.86	0.55	0.62
R_{ct} (Ω cm^2)	235	193	5476	7624
RL	1219	1099	9664	42,274
L	3187	4083	1899	25,337
$\chi 2$	6×10^{-4}	7×10^{-4}	2×10^{-3}	4×10^{-3}

3.5. In Vitro Degradation Tests

Table 5 presents the samples mass loss expressed in mg, while Figure 8 presents the samples mass evolution (mass of the remained material along with the lost mass) expressed in percentages after each time interval (1, 3, 7 and 14 days), accompanied with macroscopic images of the samples.

Table 5. Mass loss of the coatings obtained in SBF and DMEM solutions.

Coatings	Solutions	Day 1	Day 3	Day 7	Day 14
HAp + Mg + Si_RT	SBF	−14.21 ± 0.01	−34.60 ± 0.03	−56.09 ± 0.01	−75.92 ± 0.02
	DMEM	−2.27 ± 0.01	−6.59 ± 0.02	−10.84 ± 0.01	−16.37 ± 0.02
HAp + Mg + Si_200	SBF	−6.,72 ± 0.01	−48.34 ± 0.,02	−68.10 ± 0.01	−82.02 ± 0.01
	DMEM	−1.08 ± 0.02	−5.18 ± 0.01	−13.77 ± 0.02	−17.77 ± 0.01

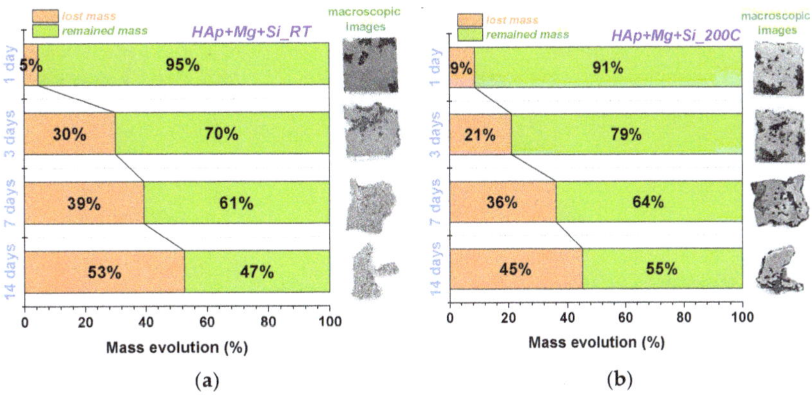

Figure 8. *Cont.*

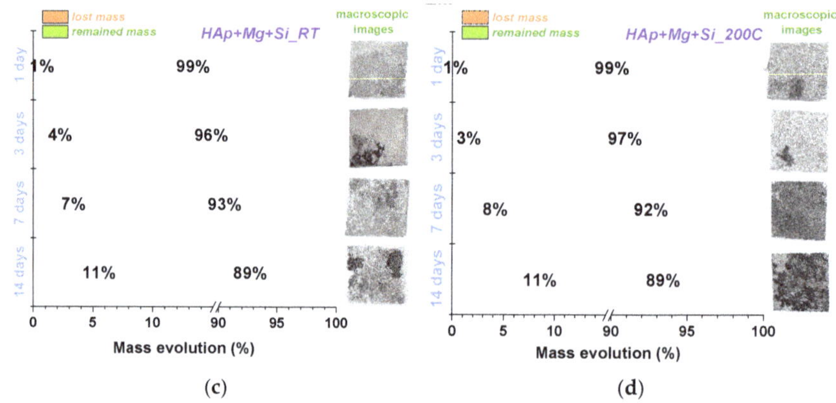

Figure 8. Evolution of mass during in vitro degradation tests: (**a**,**b**) Tests in SBF solution; (**c**,**d**) Tests in DMEM solutions, along with the optical images of the samples after the tests.

Figure 8 shows the results of the test in SBF and DMEM solutions presented in percentages for each immersion period to do a proper correlation of the results. The percentages were obtained by the ratio between the mass before and after tests.

Apparently, the highest degradation rate was recorded for the HAp + Mg + Si deposited at RT in the SBF medium, with a less aggressive degradation in DMEM.

After one day of immersion in SBF, the highest mass loss was noted for the coating deposited at 200 °C. After days 3, 7 and 14 of immersion in SBF, the degradation rate increased for the coatings deposited at RT, indicating a poor resistance in SBF at 37 °C. One may see in Figure 8 that a large surface of the samples deposited at 200 °C and immersed for 14 days in SBF is still there, compared with the surface of the coating prepared at RT. This finding shows that the increase in deposition temperature leads to a decrease in the degradation rate in SBF.

The immersion time in DMEM does not affect the degradation rate of investigated coatings, irrespective of the deposition temperature or immersion time. Both coatings exhibited similar degradation rates in DMEM solution (less than 2% differences in mass loss). Thus, it can be concluded that the investigated coatings are more resistant in DMEM solution compared with SBF.

Figure 9 presents the SEM images with the material's surface morphology after performing the immersion assays in both testing media, SBF and DMEM, respectively. According to the SEM images, it can be noted that in SBF, both coatings begin to present cracks after three days of immersion. The cracks were visible, irrespective of the immersion period. Starting the seventh day of immersion, on the material's surface, some small deposits start to appear, which after 14 days of immersion tend to agglomerate. This can be explained through the competition between biomineralisation (precipitation of some apatite) and biodegradation, a process known as bioactivity. This hypothesis corroborated with the results presented in Table 5 and Figure 8 highlights that, in this case, the degradation process is accelerated by the presence of Mg in the coating but also in the substrate. Moreover, as can be observed in Figure 9a, the presence of the cracks allows the SBF to reach the substrate, accelerating so the degradation of the experimental samples.

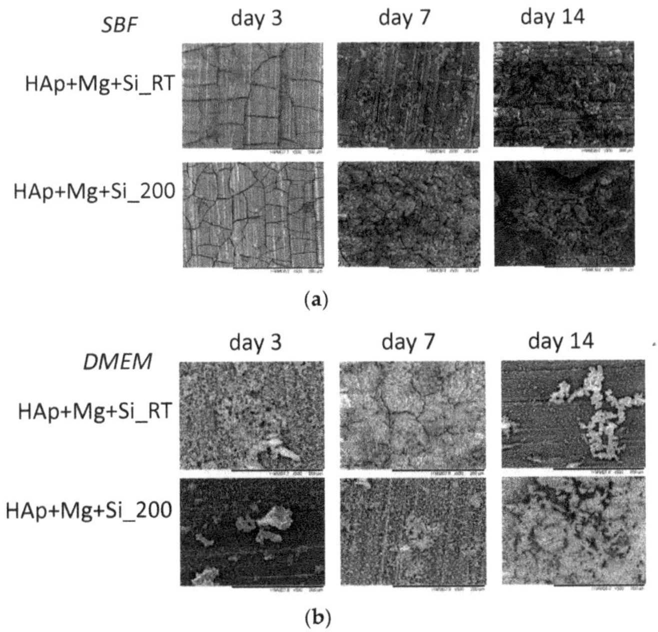

Figure 9. SEM images of degradation of surfaces after (**a**) Tests in SBF solution; (**b**) Tests in DMEM solutions.

In DMEM medium, which is less aggressive than SBF, the material surface does not generally present cracks and/or fissures, with one exception in the case of HAp + Mg + Si_RT, which after seven days of immersion, presented some surface alteration (some thin fissures can be observed along the material surface). These results are in agreement with the mass evolution ones presented in Figure 8, which have shown a lower degradation rate compared to the masses evolution in SBF.

3.6. *Surface Binding of Mussel-Derived Peptide to Coated AZ31B*

A biotin-based ELISA-like assay was performed to investigate the influence of the HAp-based coating as well as the deposition temperature on the binding property of the mussel-derived peptide (Figure 10). Mussle-derived peptides can be used to induce bio-integration, be further modified with cell recruiting or cell-binding segments.

Neither the coating composition nor the deposition temperature did affect the binding affinity of the peptide to the surface. Also, no significant difference between the control and the HAp + Mg + Si-coated samples has been detected.

Figure 10c,d displays 2D SEM images acquired for the mussel-derived peptide bound to the AZ31B substrates with HAp + Mg + Si_RT and HAp + Mg + Si_200 coatings deposited prior to peptide coating. In both cases, the SEM images show a similar coating of the material surface with the mussel-derived peptide, which is in line with the results obtained from the ELISA-like assay. Samples without previous HAp + Mg + Si coating (designated as uncoated) showed less peptide bound on the surface in both ELISA-like assay and SEM imaging (Figure 10b).

In conclusion, future bio-inspired surface-coating approaches do not limit the choice of prior coating compositions and temperatures.

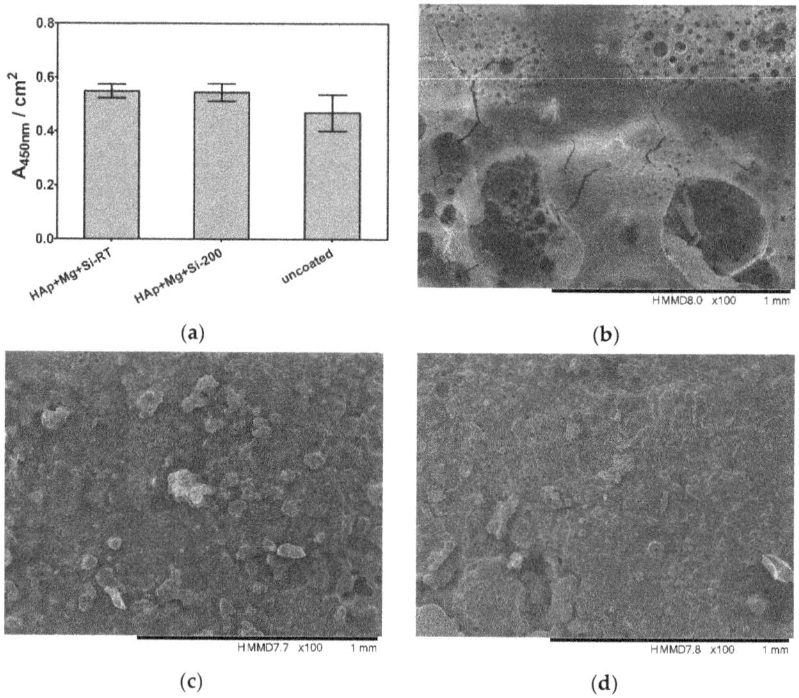

Figure 10. (**a**) Surface binding of the mussel-derived peptide to coated AZ31B. Samples were glued into wells of a 12-well plate prior to coating. Surface binding to uncoated AZ31B served as a control. Data are shown as mean ± standard error of the mean; $n = 2$. (**b–d**) 2D SEM images of (**b**) no, (**c**) HAp + Mg + Si_RT and (**d**) HAp + Mg + Si_200 coatings deposited on AZ31B prior to coating with 1 μM mussel-derived peptide.

4. Conclusions

AZ31B alloys were coated with Mg and Si-doped HAp coatings at RT and 200 °C by RF magnetron sputtering method. The obtained results showed that incrementing the deposition temperature led to the formation of stoichiometric HAp structure irrespective of Mg and Si additions. As expected, the coatings prepared at 200 °C exhibited a denser structure with finer grains compared with ones deposited at RT, which reflected on the hardness and reduced modulus (~47% increase). Further on, a comparative assessment of corrosion resistance was made as a function of the immersion medium used, the coatings deposited at elevated temperature showing a superior corrosion resistance in both SBF and DMEM. However, the lowest i_{corr} value (12 μA/cm^2) and the highest dielectric properties (R_{coat} = 705 Ω cm^2, R_{ct} = 7624 Ω cm^2) were obtained after immersion in DMEM of HAp + Mg + Si_200 coating, proving its protection ability against corrosion. Following the in vitro degradation tests, one can note that the investigated coatings presented a slower degradation rate in DMEM solution compared with SBF during 1, 3, 7 and 14 days of immersion at 37 °C. Moreover, an increment of the deposition temperature increases the resistance to degradation in SBF, while for the DMEM solution, the values are almost identical (less than 2% differences in mass loss). The mussel-derived peptide adhesion, as well as the SEM evaluation, indicated that the investigated coatings are suitable for further bio-inspired tests using mussel-derived peptides, regardless of the deposition temperature.

Importantly, based on the presented results, both coatings have improved the degradation of AZ31B alloy. More efficient protection was found for the coatings prepared at 200 °C, indicating that a higher deposition temperature is desirable when the degradation rate should be slower. If a higher dissolution rate is required, the coatings prepared at RT should be used. For the biodegradable materials used in medical applications, proper control of the degradation rate is very important, and the selection of the coating is dependent on the desired effect, namely a slower or accelerated degradation rate. To conclude, both investigated coatings are proper to control the degradation rate of AZ31B biodegradable materials, but the use of one or another is dependent on the final application of the implant.

Author Contributions: The authors of the present paper had the following individual contributions: C.M.C., D.M.V., C.V.: Methodology, Validation, Writing—Review Editing; M.D.: XRD Investigation; A.V. Conceptualization; Methodology—Coatings preparation; Writing—Original Draft & Review Editing; Project administration; Funding acquisition; A.C.P.: Methodology; Data analysis; Writing—Review & Editing; M.D., D.M.V.: Writing—Original Draft; I.P.: Nanoindentation, FT-IR investigation; Data analysis; Writing—Original Draft; D.A.B.: Peptide synthesis and immobilisation, methodology and data analysis; A.G.B.-S.: Data analysis and Editing; G.S., G.A.: Alloy preparation, Writing—Review Editing. L.R.C. and I.M.M.: Investigation. All authors have read and agreed to the published version of the manuscript.

Funding: This work was funded by the joint research project "ISIDE—Innovative Strategies for bIoactive/antibacterial/advanceD prosthEses", a grant of (i) the Romanian National Authority for Scientific Research and Innovation, CCCDI—UEFISCDI, project number ERANET-M-ISIDE-1: 171/2020 (INOE) and ERANET-M-ISIDE-2: 172/2020 (UPB), within PNCDI III; (ii) the European Regional Development Fund, Calabria Region Grant J28I17000120005, M-ERA.Net 2—Call 2019, Italy, 2020–2023 and (iii) the Free State of Saxony mERA-NET ISIDE (ULEI) 100406843.

Acknowledgments: A.V.D. thanks to Tomsk Polytechnic University within the framework of the Tomsk Polytechnic University-Competitiveness Enhancement Program grant, as well to the Core Program within the National Research Development and Innovation Plan 2022–2027, carried out with the support of MCID, project no. PN 23 05 (id: PN11N-03-01-2023).

Conflicts of Interest: The authors declare no conflict of interest.

References

1. Lemoine, P.; Acheson, J.; McKillop, S.; Beucken, J.J.V.D.; Ward, J.; Boyd, A.; Meenan, B.J. Nanoindentation and nano-scratching of hydroxyapatite coatings for resorbable magnesium alloy bone implant applications. *J. Mech. Behav. Biomed. Mater.* **2022**, *133*, 105306. [CrossRef]
2. Rahman, M.; Li, Y.; Wen, C. HA coating on Mg alloys for biomedical applications: A review. *J. Magnes. Alloy.* **2020**, *8*, 929–943. [CrossRef]
3. Zhuo, X.; Wu, Y.; Ju, J.; Liu, H.; Jiang, J.; Hu, Z.; Bai, J.; Xue, F. Recent progress of novel biodegradable zinc alloys: From the perspective of strengthening and toughening. *J. Mater. Res. Technol.* **2022**, *17*, 244–269. [CrossRef]
4. Agarwal, S.; Curtin, J.; Duffy, B.; Jaiswal, S. Biodegradable magnesium alloys for orthopaedic applications: A review on corrosion, biocompatibility and surface modifications. *Mater. Sci. Eng. C* **2016**, *68*, 948–963. [CrossRef]
5. Zhao, D.; Witte, F.; Lu, F.; Wang, J.; Li, J.; Qin, L. Current status on clinical applications of magnesium-based orthopaedic implants: A review from clinical translational perspective. *Biomaterials* **2017**, *112*, 287–302. [CrossRef]
6. Liu, Y.; Zhang, Y.; Wang, Y.-L.; Tian, Y.-Q.; Chen, L.-S. Research progress on surface protective coatings of biomedical degradable magnesium alloys. *J. Alloy. Compd.* **2021**, *885*, 161001. [CrossRef]
7. Chen, Y.; Xu, Z.; Smith, C.; Sankar, J. Recent advances on the development of magnesium alloys for biodegradable implants. *Acta Biomater.* **2014**, *10*, 4561–4573. [CrossRef]
8. Chandra, G.; Pandey, A. Preparation Strategies for Mg-Alloys for Biodegradable Orthopaedic Implants and Other Biomedical Applications: A Review. *IRBM J.* **2020**, *43*, 229–249. [CrossRef]
9. Song, J.; Chen, J.; Xiong, X.; Peng, X.; Chen, D.; Pan, F. Research advances of magnesium and magnesium alloys worldwide in 2021. *J. Magnes. Alloy.* **2022**, *10*, 863–898. [CrossRef]
10. Dong, J.; Lin, T.; Shao, H.; Wang, H.; Wang, X.; Song, K.; Li, Q. Advances in degradation behavior of biomedical magnesium alloys: A review. *J. Alloy. Compd.* **2022**, *908*, 164600. [CrossRef]
11. Song, J.; She, J.; Chen, D.; Pan, F. Latest research advances on magnesium and magnesium alloys worldwide. *J. Magnes. Alloy.* **2020**, *8*, 1–41. [CrossRef]
12. Singh, N.; Batra, U.; Kumar, K.; Ahuja, N.; Mahapatro, A. Progress in bioactive surface coatings on biodegradable Mg alloys: A critical review towards clinical translation. *Bioact. Mater.* **2023**, *19*, 717–757. [CrossRef]
13. Sahu, M.R.; Kumar, T.S.S.; Chakkingal, U. A review on recent advancements in biodegradable Mg-Ca alloys. *J. Magnes. Alloy.* **2022**, *10*, 2094–2117. [CrossRef]

14. Ding, Z.; Yuan, Q.; Wang, H.; Tang, Y.; Tan, Y.; He, Q. Microstructure and properties of Nb2O5/Mg gradient coating on AZ31 magnesium alloy by magnetron sputtering. *Ceram. Int.* **2023**, *49*, 154–167. [CrossRef]
15. Dorozhkin, S.V. Calcium orthophosphate coatings on magnesium and its biodegradable alloys. *Acta Biomater.* **2014**, *10*, 2919–2934. [CrossRef]
16. Hikku, G.; Arthi, C.; Robert, R.J.; Jeyasubramanian, K.; Murugesan, R. Calcium phosphate conversion technique: A versatile route to develop corrosion resistant hydroxyapatite coating over Mg/Mg alloys based implants. *J. Magnes. Alloy.* **2022**, *10*, 1821–1845. [CrossRef]
17. Zeng, Z.; Salehi, M.; Kopp, A.; Xu, S.; Esmaily, M.; Birbilis, N. Recent progress and perspectives in additive manufacturing of magnesium alloys. *J. Magnes. Alloy.* **2022**, *10*, 1511–1541. [CrossRef]
18. Tsakiris, V.; Tardei, C.; Clicinschi, F.M. Biodegradable Mg alloys for orthopedic implants—A review. *J. Magnes. Alloy.* **2021**, *9*, 1884–1905. [CrossRef]
19. Sanchez, A.H.M.; Luthringer, B.J.; Feyerabend, F.; Willumeit, R. Mg and Mg alloys: How comparable are in vitro and in vivo corrosion rates? A review. *Acta Biomater.* **2015**, *13*, 16–31. [CrossRef]
20. Yin, Z.-Z.; Qi, W.-C.; Zeng, R.-C.; Chen, X.-B.; Gu, C.-D.; Guan, S.-K.; Zheng, Y.-F. Advances in coatings on biodegradable magnesium alloys. *J. Magnes. Alloy.* **2020**, *8*, 42–65. [CrossRef]
21. Bahmani, A.; Arthanari, S.; Shin, K.S. Formulation of corrosion rate of magnesium alloys using microstructural parameters. *J. Magnes. Alloy.* **2020**, *8*, 134–149. [CrossRef]
22. Zhang, A.-M.; Lenin, P.; Zeng, R.-C.; Kannan, M.B. Advances in hydroxyapatite coatings on biodegradable magnesium and its alloys. *J. Magnes. Alloy.* **2022**, *10*, 1154–1170. [CrossRef]
23. Li, B.; Zhang, Z.; Liu, T.; Qiu, Z.; Su, Y.; Zhang, J.; Lin, C.; Wang, L. Recent Progress in Functionalized Coatings for Corrosion Protection of Magnesium Alloys—A Review. *Materials* **2022**, *15*, 3912. [CrossRef]
24. Sharma, G.; Kumar, K.; Satsangi, P.; Sharma, N. Surface Modification of Biodegradable Mg-4Zn Alloy Using PMEDM: An Experimental Investigation, Optimization and Corrosion Analysis. *IRBM J.* **2021**, *43*, 456–469. [CrossRef]
25. Surmeneva, M.; Surmenev, R. Microstructure characterization and corrosion behaviour of a nano-hydroxyapatite coating deposited on AZ31 magnesium alloy using radio frequency magnetron sputtering. *Vacuum* **2015**, *117*, 60–62. [CrossRef]
26. Tong, P.; Sheng, Y.; Hou, R.; Iqbal, M.; Chen, L.; Li, J. Recent progress on coatings of biomedical magnesium alloy. *Smart Mater. Med.* **2021**, *3*, 104–116. [CrossRef]
27. Chowdhury, M.A.; Hossain, H.; Hossain, N.; Hossen, Z.; Kowser, A.; Rana, M. Advances in coatings on Mg alloys and their anti-microbial activity for implant applications. *Arab. J. Chem.* **2022**, *15*, 104214. [CrossRef]
28. Dunne, C.F.; Levy, G.K.; Hakimi, O.; Aghion, E.; Twomey, B.; Stanton, K.T. Corrosion behaviour of biodegradable magnesium alloys with hydroxyapatite coatings. *Surf. Coat. Technol.* **2016**, *289*, 37–44. [CrossRef]
29. Wu, G.; Ibrahim, J.M.; Chu, P.K. Surface design of biodegradable magnesium alloys—A review. *Surf. Coat. Technol.* **2012**, *233*, 2–12. [CrossRef]
30. Yang, H.; Xia, K.; Wang, T.; Niu, J.; Song, Y.; Xiong, Z.; Zheng, K.; Wei, S.; Lu, W. Growth, in vitro biodegradation and cytocompatibility properties of nano-hydroxyapatite coatings on biodegradable magnesium alloys. *J. Alloy. Compd.* **2016**, *672*, 366–373. [CrossRef]
31. Chaharmahali, R.; Fattah-Alhosseini, A.; Babaei, K. Surface characterization and corrosion behavior of calcium phosphate (Ca-P) base composite layer on Mg and its alloys using plasma electrolytic oxidation (PEO): A review. *J. Magnes. Alloy.* **2020**, *9*, 21–40. [CrossRef]
32. Xu, J.; Cai, Q.; Lian, Z.; Yu, Z.; Ren, W.; Yu, H. Research Progress on Corrosion Resistance of Magnesium Alloys with Bio-inspired Water-repellent Properties: A Review. *J. Bionic Eng.* **2021**, *18*, 735–763. [CrossRef]
33. Yadav, V.; Kumar, A.; Das, A.; Pamu, D.; Pandey, L.; Sankar, M. Degradation kinetics and surface properties of bioceramic hydroxyapatite coated AZ31 magnesium alloys for biomedical applications. *Mater. Lett.* **2020**, *270*, 127732. [CrossRef]
34. Parau, A.C.; Cotrut, C.M.; Guglielmi, P.; Cusanno, A.; Palumbo, G.; Dinu, M.; Serratore, G.; Ambrogio, G.; Vranceanu, D.M.; Vladescu, A. Deposition temperature effect on sputtered hydroxyapatite coatings prepared on AZ31B alloy substrate. *Ceram. Int.* **2021**, *48*, 10486–10497. [CrossRef]
35. Wu, G.-S.; Wang, A.-Y.; Ding, K.-J.; Xu, C.-Y.; Dai, W.; Xu, A.-J. Fabrication of Cr coating on AZ31 magnesium alloy by magnetron sputtering. *Trans. Nonferrous Met. Soc. China* **2008**, *18*, s329–s333. [CrossRef]
36. Bakhsheshi-Rad, H.; Hamzah, E.; Daroonparvar, M.; Yajid, M.; Medraj, M. Fabrication and corrosion behavior of Si/HA nano-composite coatings on biodegradable Mg–Zn–Mn–Ca alloy. *Surf. Coat. Technol.* **2014**, *258*, 1090–1099. [CrossRef]
37. Šupová, M. Substituted hydroxyapatites for biomedical applications: A review. *Ceram. Int.* **2015**, *41*, 9203–9231. [CrossRef]
38. Makkar, P.; Kang, H.J.; Padalhin, A.R.; Faruq, O.; Lee, B. In-vitro and in-vivo evaluation of strontium doped calcium phosphate coatings on biodegradable magnesium alloy for bone applications. *Appl. Surf. Sci.* **2020**, *510*, 145333. [CrossRef]
39. McAnulty, T.; Jeswiet, J.; Doolan, M. Formability in single point incremental forming: A comparative analysis of the state of the art. *CIRP J. Manuf. Sci. Technol.* **2017**, *16*, 43–54. [CrossRef]
40. Ambrogio, G.; Gagliardi, F. Temperature variation during high speed incremental forming on different lightweight alloys. *Int. J. Adv. Manuf. Technol.* **2014**, *76*, 1819–1825. [CrossRef]
41. Vranceanu, D.M.; Parau, A.C.; Cotrut, C.M.; Kiss, A.E.; Constantin, L.R.; Braic, V.; Vladescu, A. In vitro evaluation of Ag doped hydroxyapatite coatings in acellular media. *Ceram. Int.* **2019**, *45*, 11050–11061. [CrossRef]

42. *ASTM G59-97(2020)*; Standard Test Method for Conducting Potentiodynamic Polarization Resistance Measurements, (n.d.). ASTM International: Montgomery, PA, USA, 2020. Available online: https://www.astm.org/g0059-97r20.html (accessed on 10 December 2022).
43. Bechir, F.; Bataga, S.M.; Tohati, A.; Ungureanu, E.; Cotrut, C.M.; Bechir, E.S.; Suciu, M.; Vranceanu, D.M. Evaluation of the Behavior of Two CAD/CAM Fiber-Reinforced Composite Dental Materials by Immersion Tests. *Materials* **2021**, *14*, 7185. [CrossRef] [PubMed]
44. Clauder, F.; Czerniak, A.S.; Friebe, S.; Mayr, S.G.; Scheinert, D.; Beck-Sickinger, A.G. Endothelialization of Titanium Surfaces by Bioinspired Cell Adhesion Peptide Coatings. *Bioconjugate Chem.* **2019**, *30*, 2664–2674. [CrossRef] [PubMed]
45. Clauder, F.; Zitzmann, F.D.; Friebe, S.; Mayr, S.G.; Robitzki, A.A.; Beck-Sickinger, A.G. Multifunctional coatings combining bioactive peptides and affinity-based cytokine delivery for enhanced integration of degradable vascular grafts. *Biomater. Sci.* **2020**, *8*, 1734–1747. [CrossRef] [PubMed]
46. Martyniuk, M.; Antoszewski, J.; Walmsley, B.A.; Musca, C.A.; Dell, J.M.; Jung, Y.-G.; Lawn, B.R.; Huang, H.; Faraone, L. Determination of mechanical properties of silicon nitride thin films using nanoindentation. In *Proceedings Spaceborne Sensors II*; SPIE: Toulouse, France, 2005; Volume 5798, pp. 216–226. [CrossRef]
47. Vladescu, A.; Cotrut, C.M.; Azem, F.A.; Bramowicz, M.; Pana, I.; Braic, V.; Birlik, I.; Kiss, A.; Braic, M.; Abdulgader, R.; et al. Sputtered Si and Mg doped hydroxyapatite for biomedical applications. *Biomed. Mater.* **2018**, *13*, 025011. [CrossRef]
48. Dearnaley, G.; Arps, J.H. Biomedical applications of diamond-like carbon (DLC) coatings: A review. *Surf. Coat. Technol.* **2005**, *200*, 2518–2524. [CrossRef]
49. Flores, J.F.; Olaya, J.J.; Colás, R.; Rodil, S.E.; Valdez-Salas, B.; Fuente, I.G. Corrosion behaviour of TaN thin PVD films on steels. *Corros. Eng. Sci. Technol.* **2006**, *41*, 168–176. [CrossRef]
50. Lotfpour, M.; Dehghanian, C.; Emamy, M.; Bahmani, A.; Malekan, M.; Saadati, A.; Taghizadeh, M.; Shokouhimehr, M. In vitro corrosion behavior of the cast and extruded biodegradable Mg-Zn-Cu alloys in simulated body fluid (SBF). *J. Magnes. Alloy.* **2021**, *9*, 2078–2096. [CrossRef]
51. Shkirskiy, V.; King, A.D.; Gharbi, O.; Volovitch, P.; Scully, J.R.; Ogle, K.; Birbilis, N. Revisiting the Electrochemical Impedance Spectroscopy of Magnesium with Online Inductively Coupled Plasma Atomic Emission Spectroscopy. *Chem. Phys. Chem.* **2014**, *16*, 536–539. [CrossRef]
52. Singh, N.; Batra, U.; Kumar, K.; Siddiquee, A.N. Evaluating the Electrochemical and In Vitro Degradation of an HA-Titania Nano-Channeled Coating for Effective Corrosion Resistance of Biodegradable Mg Alloy. *Coatings* **2022**, *13*, 30. [CrossRef]
53. Hu, H.; Wang, X.; Huang, Y.; Yang, Z.; Jia, B.; Sun, K.; Hao, D.; Guo, Y. Electrochemical techniques for monitoring the biodegradability of nanocomposite Mg-alloy/HA for repairing bone fracture. *J. Mater. Res. Technol.* **2022**, *18*, 1669–1681. [CrossRef]
54. Feliu, S., Jr. Electrochemical Impedance Spectroscopy for the Measurement of the Corrosion Rate of Magnesium Alloys: Brief Review and Challenges. *Metals* **2020**, *10*, 775. [CrossRef]
55. Cesiulis, H.; Tsyntsaru, N.; Ramanavicius, A.; Ragoisha, G. *The Study of Thin Films by Electrochemical Impedance Spectroscopy*; Springer: Berlin/Heidelberg, Germany, 2016; pp. 3–42. [CrossRef]
56. Jamesh, M.; Kumar, S.; Narayanan, T.S.N.S. Electrodeposition of hydroxyapatite coating on magnesium for biomedical applications. *J. Coatings Technol. Res.* **2011**, *9*, 495–502. [CrossRef]
57. Watson, W.; Orazem, M.E. IS: Measurement Model Program. Available online: www.ecsarxiv.org/kze9x (accessed on 10 December 2022).
58. Orazem, M.E.; Tribollet, B. A tutorial on electrochemical impedance spectroscopy. *Chemtexts* **2020**, *6*, 41. [CrossRef]
59. Orazem, M.E. Electrochemical impedance spectroscopy: The journey to physical understanding. *J. Solid State Electrochem.* **2020**, *24*, 2151–2153. [CrossRef]
60. Orazem, M.E.; Tribollet, B. (Invited) On Teaching Electrochemical Impedance Spectroscopy. *ECS Meet. Abstr.* **2019**, *28*, 2170. [CrossRef]

coatings

MDPI

Article

Investigation into Effect of Natural Shellac on the Bonding Strength of Magnesium Substituted Hydroxyapatite Coatings Developed on Ti6Al4V Substrates

Ritwik Aravindakshan *, Kaiprappady Kunchu Saju and Reghuraj Aruvathottil Rajan

Department of Mechanical Engineering, School of Engineering, Cochin University of Science & Technology, Kochi 682022, India; kksaju@cusat.ac.in (K.K.S.); reghuraj@alameen.edu.in (R.A.R.)

* Correspondence: ritwik@tistcochin.edu.in

Abstract: The bioactive and biocompatible properties of hydroxyapatite (HA) promote the osseointegration process. To enhance other bio-functions of HA such as improving the antibacterial property of the implant, increasing the rate of cell proliferation, or improving tissue generation capability, HA is substituted with many elements such as Zn, Cl, Ba, Fe, Cu, Ag, Sr, F, Na, etc. This study reports development of Magnesium substituted HA (Mg-HA) coatings on Ti_6Al_4V substrates using the dip coating technique. To improve the adhesion and stability of the coating, an intermediate layer of shellac was applied between the coating and Ti6Al4V substrate. The dip coating process parameters were optimized using the Taguchi technique and it was found that dipping time of 35 s and 13% w/w of shellac concentration provided the maximum bonding strength of 12.5 MPa. The biocompatibility, dissolution, and corrosion study of the developed coating using the optimal parameters obtained were carried out in this study. An improvement in cell growth and cell proliferation was observed in the extract medium prepared from coated substrates. Release of Ca, P and Mg ions from the surface of the coated substrate into the simulated body fluid (SBF) was found to be almost constant which shows the stability of the thin film coating formed. The Mg-HA coated substrate also exhibited better corrosion resistance than the uncoated substrate.

Keywords: Mg-HA; dip-coating; shellac; Taguchi technique

Citation: Aravindakshan, R.; Saju, K.K.; Aruvathottil Rajan, R. Investigation into Effect of Natural Shellac on the Bonding Strength of Magnesium Substituted Hydroxyapatite Coatings Developed on Ti6Al4V Substrates. *Coatings* **2021**, *11*, 933. https://doi.org/10.3390/coatings11080933

Academic Editor: Liviu Duta

Received: 30 June 2021
Accepted: 29 July 2021
Published: 4 August 2021

1. Introduction

Hydroxyapatite ($Ca_{10}(PO_4)_6(OH)_2$) (HA) is one of the prominent biomaterials among all other calcium phosphate ceramics used. It has been commercially used since the 1980s because of its biocompatibility and osteoconductive properties [1]. HA exhibited a composition similar to the bone minerals phase and it has been widely used for orthopedic and dental applications [2]. Researchers tried to modify the structural composition of HA to improve its bio-functionality. The naturally formed bio-apatite has trace amounts of fluoride, magnesium, sodium, potassium, carbonate, chloride, zinc, barium, iron, strontium, copper, etc [3,4]. Studies showed that the incorporation of different foreign cations and anions into synthetic HA structures enhanced its osteoblastic response [5]. HA in general, is brittle and is not suitable for load-bearing applications. To overcome this disadvantage, HA coating is developed on metallic substrates [2]. Commercially pure titanium (CpTi), titanium alloys, stainless steel, and cobalt-chromium (Co–Cr) alloys are mainly used as substrates because of their bio-compatibility and superior mechanical properties [6].

In recent years, various types of substituted HA were developed by researchers. Many of them processed it in form of scaffolds and some others deposited it as a thin film coating on metal substrates. A considerable amount of literature showed the presence of zinc (Zn) in the human body. It plays a vital role in enzyme activity, nucleic acid metabolism, hormonal activity, and bio-mineralization. Zn is one of the important minerals found in human teeth. Studies also showed that Zn incorporation into metal implants promotes bone formation [7]. Zn substituted HA (Zn-HA) was used as an alternative material for autograft and allograft bone replacements. Studies reported contradictory findings on the crystalline nature of HA structure when substituted with Zn cations [8]. Few other studies showed that silicon substituted HA (Si-HA) promotes osteoblast cell activity and apatite formation in simulated body fluid (SBF) compared to pure HA. Silicon is essential for bone mineralization and soft tissue development [2].

Bacterial infection post-implant placement is very serious and many studies showed that the success of the implant over a while depends on its antibacterial property. Silver (Ag) is a key element with antibacterial properties. Ag substituted HA (Ag-HA) related studies showed that substituting Ag^+ ions into the crystal lattice structure of HA can control bacterial infection significantly [9]. Another researcher showed that adding titanium dioxide (TiO_2) into Ag-HA can kill almost 98% of S. aureus and 99% of E. coli bacteria [10]. The potential cytotoxic effect of Ag in the Ag-HA compound was reduced by doping strontium (Sr). Ag-HA/Sr was seen to augment the cell proliferation in a better way compared to the Ag-HA [1]. Strontium substituted HA (Sr-HA) was also studied extensively. It was shown that Sr-HA can enhance pre-osteoblastic cell replication. SR-HA also inhibits the activity of osteoclasts [5].

Chitosan ((1,4)-2-amino-2-deoxy-β-D-glucan) is a partially deacetylated form of chitin, which is abundantly produced from marine copepods [11]. Chitosan substituted HA (CS-HA) was used to enhance tissue regenerative efficacy and osteoconductivity. To improve mechanical strength, HA was incorporated into the chitosan polymer [12]. Some researchers showed that CS-HA composite scaffolds have better cell proliferation compared to chitosan scaffolds [13]. Studies also showed that a chitosan polymer offers the advantage of low-temperature processing of composite materials [14]. Researchers also developed novel HA-silica-chitosan composites (HA-CS-SiO_2). This composite showed lower porosity which can be used for developing laminates [15].

Lithium (Li) is a bioelectric material and very few studies have been carried out to analyze its biocompatibility. Researchers demonstrated that adding Li decreases the solubility of HA and it is a good indication of biocompatibility. Lithium substituted HA (Li-HA) showed good adherence to osteoblasts [16]. Carbonate substituted HA (CO_3-HA) has been explored by many. CO_3-HA mimics the chemical composition of hard tissue. It is also reported to be thermally stable as it would not decompose into undesirable secondary phases at high calcination temperatures [17].

Magnesium (Mg) substituted HA (Mg-HA) was investigated in detail for the last few decades. Studies showed that Mg-HA stimulates cell proliferation [18]. Mg-HA, due to its ion exchanging properties is known to be highly toxic to microbes and is used to control bacterial growth [19]. It also can enhance the osteoblast bonding strength. Researchers demonstrated that Mg-HA is more biocompatible than phase pure HA [20].

Apart from being a biocompatible material, Mg-HA has wide clinical applications. It has the potential to improve angiogenesis which is the fundamental requirement for bone regeneration. Bone neo-formation was also observed during in vivo studies conducted with Mg-HA [21]. Clinical studies pertaining to the use of Mg-HA in dental application received much attention lately. Because of decay or damage, a tooth might have to be removed from its position and an extraction socket is formed there. Immediately placing an implant directly to the extraction socket might not be practically viable and compatible. Studies showed that filling these sockets with Mg-HA prior to implant placement provides better results [22].

Digital dentistry techniques such as coherent-scatter computed tomography (CSCT) and cone-beam computed tomography (CBCT) have been used to measure the concentration of HA [23,24]. Recently 3D printing of substituted HA is evolving as a forefront technology [25,26]. Robust Mg-based scaffolds have also been 3D printed for promoting the osteointegration process [27].

Various methods have been adopted for developing a substituted-HA coating on metal substrates. Coating methods such as the sputtering process, sol-gel technique, electrophoretic method, plasma spray technique, etc., have been used. This study reports development of a thin film of Mg-HA on titanium alloy (Ti6Al4V) using a dip coating technique. It was observed that the coating thus developed has poor bonding strength. To enhance the bonding strength, an intermediate shellac layer (natural resin) was applied between the coating film and substrate. The Taguchi optimization technique was employed to obtain optimal parameters for developing Mg-HA Coating on Ti6Al4V materials. The dissolution rate, cell viability, and corrosion behavior of the developed coating were also studied as part of this research work.

2. Materials and Methodology

2.1. Mg-HA Preparation

Analytical grade of diammonium hydrogen phosphate ($(NH_4)_2HPO_4$, Nice Chemicals Ltd., Kochi, India), calcium nitrate tetrahydrate ($Ca(NO_3)_2 \cdot 4H_2O$, Nice Chemicals Ltd., Kochi, India), magnesium chloride hexahydrate ($MgCl_2 \cdot 6H_2O$, Nice Chemicals Ltd., Kochi, India), and ammonium hydroxide (NH_4OH, Nice Chemicals Ltd., Kochi, India) were used for the preparation of nano-sized Mg-HA. It was prepared as per the methodology reported in reference [18]. Aqueous diammonium hydrogen phosphate was added to the solution having magnesium chloride hexahydrate and calcium nitrate tetrahydrate in a drop-wise manner. The molar ratio of Mg/Ca was 0.18 and the pH value was maintained at 11. The mixture was stirred for 24 hours continuously using a magnetic stirrer. The precipitate thus formed was further centrifuged and then washed with distilled water. The substrates thus formed were dried at 70 °C overnight and then sintered at 900 °C for one hour. The elemental composition of as-prepared Mg-HA is shown in Table 1.

Table 1. Elemental Composition.

Material	Elemental Composition (%)				
Ti6Al4V	89.7% Ti	6.1% Al	3.9% V	0.3% Fe	-
Mg-HA	43.31% O	7.79% Mg	15. 08% P	0.25% Cl	33.57% Ca

2.2. Characterization of Mg-HA

The surface morphology of the as-prepared Mg-HA nanopowders was analyzed using scanning electron microscopy (SEM, JEOL JSM-6390LV, JEOL Ltd., Tokyo, Japan). High-resolution transmission electron microscopy (HR-TEM, JEOL JEM 2100, JEOL Ltd., Tokyo, Japan) was used to analyze the crystalline morphology of the nanopowders prepared. Energy-dispersive X-ray spectroscopy (EDS, Oxford XMX N, Wiesbaden, Germany) was used to quantify elemental composition of Mg-HA. The crystallinity of the powders was characterized using the X-ray powder diffraction (technique (XRD, Bruker AXS D8, Bruker AXS, Berlin, Germany)).

2.3. Experiment Methodology

Ti6Al4V (grade 5) substrates of dimensions 10 mm × 20 mm × 2 mm thickness were used as base substrates. The composition of the substrate used is shown in Table 1. Initially, the substrates were scoured using emery papers of grit size ranging from 80 to 400. The substrates were then sandblasted to remove any impurities on the surface [28]. Further, they was soaked in the acetone solution, sonicated for 45 min, and then they were air-dried. The coating procedure was carried out on a custom-fabricated dip coating apparatus. The substrate after being dipped in shellac solution was then dipped further in Mg-HA solution.

Natural shellac is produced from lac insects (*Laccifer Lacca*) and it acts as a bio-resin [29]. The shellac layer imparts adhesion to the coating. The strength of the coating formed on the substrates is dependent on various coating parameters. Parameters such as dipping duration, dipping and withdrawal speed, the concentration of the dipping solution, and temperature. The shellac solution was prepared by varying the concentration of shellac in ethanol. Three different concentrations of shellac were considered in this study. Concentrations of 12.5% (*w*/*w*), 13.0% (*w*/*w*) and 13.5% (*w*/*w*) were fixed after conducting numerous trial experiments. It was seen that a concentration less than 12.5% does not help in producing a stable coating. Similarly, a concentration of more than 13.5% affected the uniformity of the coating. The dipping time in the shellac solution varied from 15 to 35 s. Experiment trials showed that a duration less than 15 s in shellac solution did not yield any coating and more than 35 s produced thicker coating which was undesirable.

Mg-HA solution for the dip coating process was prepared using as-developed Mg-HA as the solute and ethanol (99.99% purity) as the solvent. Glycerin and polyethylene glycol (PEG 600, Nice Chemical Ltd., Kochi, India) were used from a stable solution [30]. Magnetic stirring was performed for 8 hours at room temperature to obtain a homogenous solution. The concentration of Mg-HA in the Mg-HA solution was maintained at 20% (*w*/*w*). The dipping time for the HA solution was fixed at 25 s. The dipping and withdrawal speed for shellac and Mg-HA solution was fixed as 15 mm/min. The schematic diagram of the coating process and layer by layer coating is illustrated in Figure 1. In this study, all the experiments were carried out at room temperature.

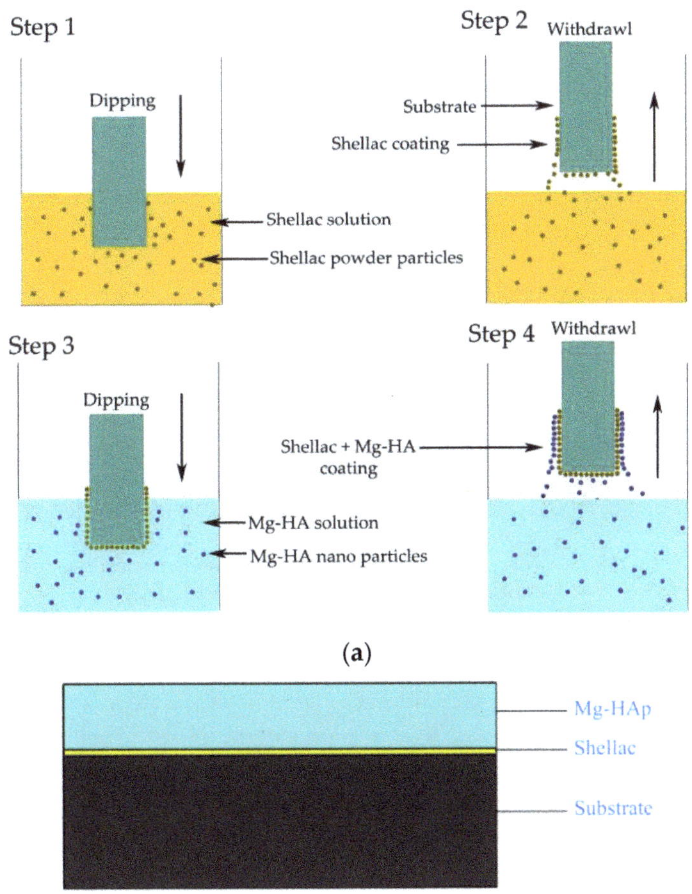

(a)

(b)

Figure 1. Schematic diagram of the coating process and coated substrate. (**a**) Schematic diagram of the coating process; (**b**) coated substrate.

2.4. Bonding Strength and Thickness

The bonding strength of the coating was measured using the standard pull-out test equipment following ASTM D 7234 standard [31] and the thickness of the coating was measured following the IS- 3203-2001 standards [32].

2.5. Optimization

For optimization of coating parameters, the Taguchi technique was used. The parameters considered in this study are dipping time and shellac concentration (Table 2). The required response was the bonding strength. Statistical software package Minitab was employed to perform signal-to-noise (S/N) ratio calculations, analysis of variance (ANOVA), and to obtain a regression equation. S/N ratio gives the ratio of mean to standard deviation. ANOVA provides details of the percentage contribution of each parameter towards the bonding strength. Likewise, the relationship between the coating parameters and the bonding strength is put into a mathematical model using regression analysis. The strength of the obtained model is validated by defining the value of the coefficient of determination (R^2).

Table 2. Levels of parameters.

Parameters	Experiment Levels		
	1	2	3
Dipping time (s) (w_1)	15	25	35
Shellac concentration (w/w %) (w_2)	12.5	13.0	13.5

2.6. In Vitro Studies

The substrate coated with optimal parameters was further investigated for its dissolution, in vitro, and corrosion behavior. The dissolution studies of Mg-HA were conducted by soaking substrates in simulated body fluid (SBF) for 4 weeks. Preparation of SBF was carried out using Kokubo methodology [33] where the pH value was maintained between 7.3 and 7.4. SBF ionic concentration is quantified using inductively coupled plasma-mass spectrometry (ICP-MS). The coated substrates were soaked in 20 mL of SBF without shaking at 37 ± 0.5 °C following ISO 10993-14 standards [34]. The rate at which the dissolution of calcium (Ca), phosphorous (P), and magnesium (Mg) ions occurred was recorded for 4 weeks.

MG-63 osteosarcoma human cell line was utilized to perform cell viability studies following ISO 10993-12 [35]. A 96-well plate was used in this study and each cell was seeded with 10,000 cells. A complete medium (CM) was prepared by adding a minimum essential medium (MEM) and 10% fetal bovine solution (FBS). It was incubated at 37 °C in 5% CO_2 for a 24 hour duration. To prepare the cell culture medium (CCM), cells were seeded to the CM. As the thin layer of a coated substrate cannot be peeled off, an extract medium is to be prepared. The extract medium preparation is carried out by soaking the coated substrates in CM for 24 hours. Before soaking, it is ensured that all the substrates were ethylene oxide (ETO) sterilized. After carefully removing the substrate from CM, it was further was diluted to 12.5%, 25% and 50% CCM. For quantification of cell viability, assay was utilized. The absorbance value from the assay was recorded using a TECAN microplate reader at 595 nm.

2.7. Corrosion

Usually, substrates are susceptible to various types of corrosion attacks. Pitting corrosion, crevice corrosion, galvanic corrosion, fretting corrosion and corrosion fatigue are some common types of corrosion that are normally found in substrates [36]. In this study, the potentiodynamic polarization test (Metrohm Autolab, Chennai, India) was performed following ASTM G5-94 [37]. The substrates were exposed to a simulated body fluid environment in a three-electrode electrochemical cell. Here, platinum was the counter electrode, saturated calomel electrode (SCE) was the reference electrode, and the substrates were kept as the working electrode.

The potentiodynamic polarization curve was obtained for the coated and uncoated substrates with a stabilization period of 3600 s and the electrode potential range was varied from −1000 to 1000 mV. The scanning rate was 0.005 mVs^{-1}. The corrosion potential (E_{corr}), polarization resistance (R_p), anodic Tafel slope(β_a), and cathodic Tafel slope (β_c) were measured during this study to analyze the corrosion effect.

3. Results and Discussions

3.1. Surface Morphology and Characterization

The as-prepared Mg-HA powders are spherical in appearance in Figure 2. It is observed that the powder particles are tightly packed with very little porosity. The tighter packing might be a result of Mg^{2+} ions occupying the vacant Ca^{2+} sites. Previous studies showed that the spherical-sized particles transform themselves into needle-like structures during apatite formation [38].

Figure 2. SEM micrograph of as-prepared Mg-HA.

Figure 3 shows high-resolution TEM images of the Mg-HA nanopowders. The nanopowder size varied from 50.28 to 98.20 nm. The average size of the powders was observed to be ~65 nm. Each nanoparticle has a definite shape and can be seen agglomerating. The nano size of the particles makes it easier to form a stable and uniformly suspended dip coating solution.

Figure 3. HR-TEM image of as-prepared Mg-HA.

EDS analysis was carried out to identify different elements in the as-prepared Mg-HA nanopowders. Characteristic peaks of O, Mg, P, Cl and Ca can be observed in Figure 4 and its corresponding atomic percentage is also presented. Presence of Mg, Ca and P indicate that Mg has entered into the lattice structure of HA, which corroborates studies previously conducted [39].

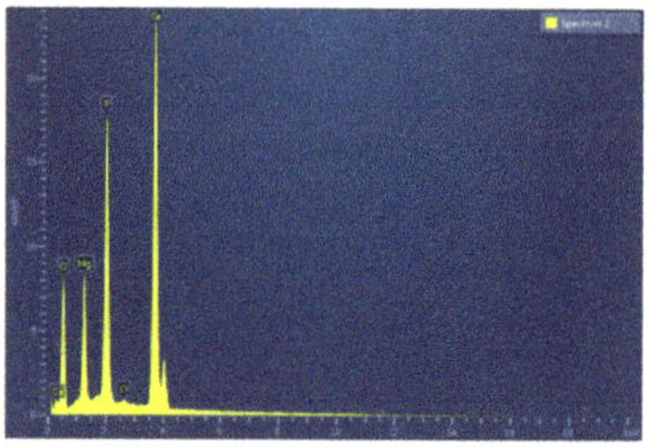

Element	Atomic %
O	63.11
Mg	7.59
P	11.15
Cl	0.13
Ca	18.02

Figure 4. EDS spectra of as-prepared Mg-HA.

X-ray diffraction characterization of as-prepared Mg-HA is shown in Figure 5. Crystalline phases were identified by comparing them with JCPDS standards. A HA crystalline structure is observed at a 2θ value of 27.808° and 31.113° which is in good agreement with JCPDS No. 09-432. The $Mg(OH)_2$ crystalline phase was observed at a 2θ value of 34.463°, confirmed by JCPDS No. 07-0239. The MgO phases are also seen in the XRD profile between the 2θ value of 48° and 55°. In the XRD profile, an amorphous phase is seen at the beginning. It is due to the incorporation of Mg^{2+} ions into the HA structure. Studies showed that an increase in magnesium content can lead to an amorphous phase [40].

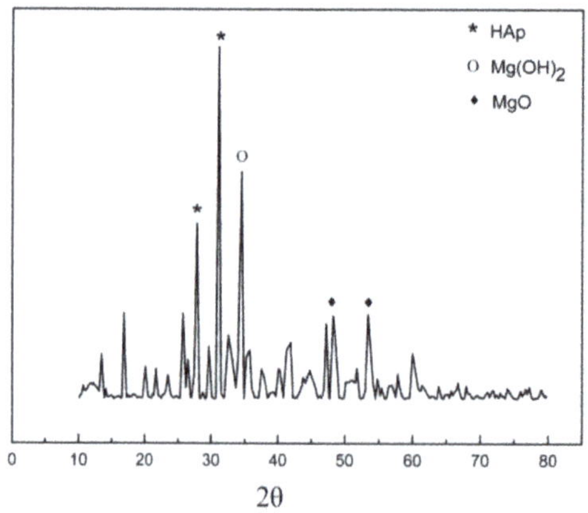

Figure 5. XRD of as-prepared Mg-HA.

3.2. Bonding Strength and Thickness Measurement

A thin film of Mg-HA coating was developed on Ti6Al4V substrates using the as-prepared Mg-HA precursor powder. A thin layer of shellac applied between the substrate and HA coating provided stability and improved bonding strength of Mg-HA coatings. The thickness of shellac coating applied in between was measured as 2 μm and the total Mg-HA thickness was measured as 12 μm. Mg-HA coating developed on the substrate without shellac has a bonding strength of 1.19 MPa. A tremendous increase in the bonding strength could be observed because of the application of natural shellac.

3.3. SBF Composition

Table 3 shows the ionic concentration of the prepared SBF measured using inductively coupled plasma mass spectrometry (ICP-MS, Thermo Fisher Scientific, Mumbai, India). It was observed that SBF had a similar ionic concentration as that of human blood [28]. It mimics the real human body environment.

Table 3. Composition of human blood and SBF prepared.

Ionic Concentrations	Human Blood (ppm)	SBF Prepared (ppm)
Na^+	142.0	136.0
Cl^-	103.0	107.0
HCO_3^-	27.0	21.0
K^+	5.0	8.0
Mg^{2+}	1.5	2.0
Ca^{2+}	2.5	3.0
HPO_4^{2-}	1.0	0.5
SO_4^{2-}	0.5	0.3

3.4. Optimization Results

The values of bonding strength and corresponding S/N ratios for various combinations of experiments are shown in Table 4. The mean effect plot is shown in Figure 6. It can be seen that substrate with code H (dipping time 35 s and shellac concentration of 13.0% w/w) provides the maximum bonding strength.

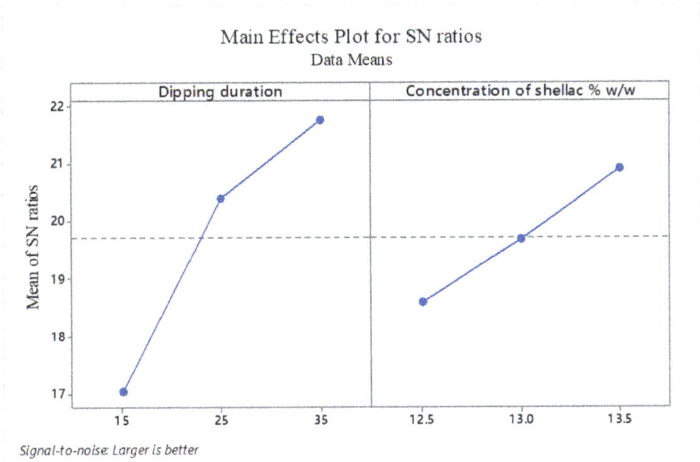

Figure 6. Coating parameters effect on bonding strength.

Table 4. Measured experimental values with S/N ratio.

Orthogonal Array with Parameters				
Substrate Code	**Dipping Time (S)(w_1)**	**Concentration of Shellac % (w/w)(w_2)**	**Bonding Strength (MPa)**	**Bonding Strength(S/N Ratio)**
A	1	1	5.9	15.417
B	1	2	6.7	16.521
C	1	3	9.1	19.181
D	2	1	8.8	18.890
E	2	2	10.7	20.588
F	2	3	12.1	21.656
G	3	1	11.8	21.438
H	3	2	12.5	21.938
I	3	3	12.4	21.868

The ANOVA results for bonding strength are shown in Table 5. The p value and individual parameter contribution percentage indicate that dipping time has a greater influence on the coating process compared to shellac concentration.

Table 5. ANOVA.

Parameters	DF	SS	MS	F-Value	p-Value	% Contribution
Dipping time	2	38.78	19.39	26.62	0.005	77.41
Shellac concentration% w/w	2	8.47	4.203	5.77	0.066	16.91
Error	4	2.93	0.7283			
Total	8	50.1				

A linear mathematical model was developed to study the effect of parameters on bonding strength using regression analysis and is expressed in Equation (1). The significance of the mathematical model is interpreted using the R-sq value. Usually, this value varies from zero to one and if the value obtained is nearing one, it indicates a good fit and can be used for predicting the bonding strength values. In the present study models developed for bonding strength have an R-sq value of 70.56%.

$$\text{Bonding strength} = -27.02 + 0.25w_1 + 2.367w_2 \quad (1)$$

Figure 7 shows the residual plots for bonding strength. The straight-line residual plot indicates that the residual errors are following a normal distribution and the coefficients in the model are significant.

To validate the developed mathematical model random values, parameters well within the range were selected and substituted. Experimental and predicted results showed slight deviations. Confirmation tests for a random experiment combination with 19 s and 13.2% of dipping time and shellac concentration respectively were conducted and the results are tabulated in Table 6.

Table 6. Confirmation test.

Bonding Strength	
Experimental result	Predicted results
9.231	8.974

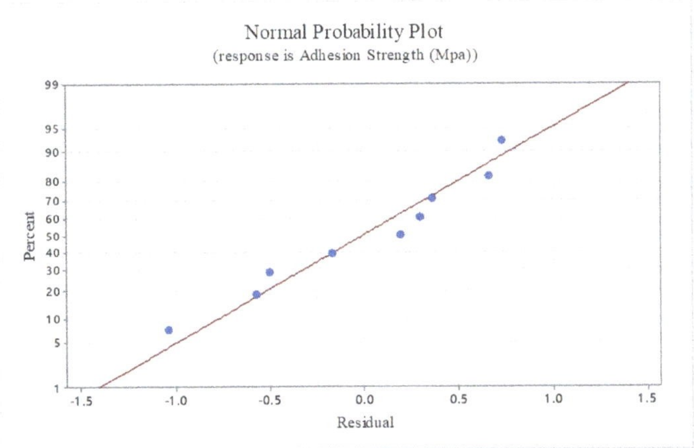

Figure 7. Normal probability plot.

3.5. Dissolution and Cell Viability

The dissolution behavior of the coating developed is evaluated by immersing it in the prepared SBF. The results are shown in Figure 8. It can be seen that the dissolution of Ca and P ions are almost constant indicating their stability in the body fluid environment. The dissolution rate of Mg ions is seen to increase slightly after the second week and thereafter it shows a constant value. The intermediately applied shellac layer is seen to provide stability for the developed coating.

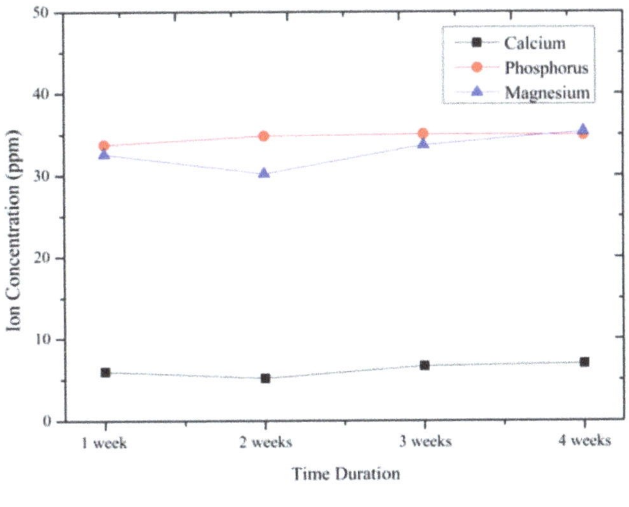

Figure 8. Change in ionic concentration.

The cell viability percentage of the Mg-HA coated substrate was calculated using the following Equation (2).

$$\%\text{ Cell Viability} = \frac{\text{A(Substrate)} - \text{A(Blank)}}{\text{A(Control)} - \text{A(blank)}} \quad (2)$$

Here A (Substrate) is the absorbance value of the extract medium seeded with cells, A (Blank) is the absorbance value of CM, and A(control) is the CCM. The calculated value of cell viability percentage of control and Mg-HA coated substrates are illustrated in Figure 9. The extracted medium as mentioned was diluted to 50% (50% extract medium + 50% CM), 25% (25% extract medium + 75% CM) and 12.5% (12.5% extract medium + 87.5% CM). The cell viability was computed for every single dilution percentage. With an increasing percentage of extract medium in the CM, the cell viability was seen reducing. In every dilution proportion, Mg-HA coated substrates are exhibiting better cell viability compared to the control. At 50% dilution percentage cell viability was approximately recorded to be 92% which is admissible as per ISO 10993-5 standards [41]. From the results obtained it is observed that the Mg-HA coatings developed are non-toxic to the human body.

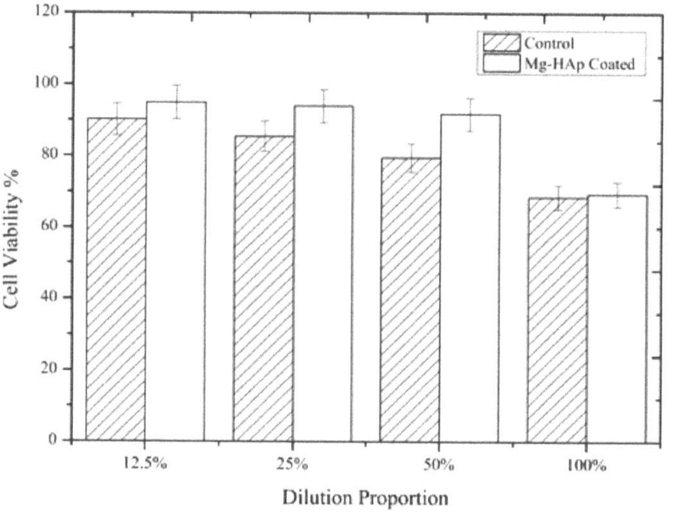

Figure 9. Cell viability at different dilution ratios.

The cell proliferation of MG-63 cells on extract medium was observed for 24, 48, and 72 hours. From Figure 10 it can be seen that the number of cells is proliferating in the extract medium at 12.5% and 25% dilution. At 72 hours duration for 50% dilution proportion, a decline in cell proliferation is noticed. More importantly, at 100% dilution proportion, a significant increase in cell deaths is seen. The reduction in cell viability percentage at 100% dilution proportion can be justified by this cell death process. One probable reason for the cell death may be because the lack of space for the cells to grow. After a thrust point, the cells need more surface area for further multiplication. In this study, the surface area is constrained and it leads to overcrowding which results in cell death.

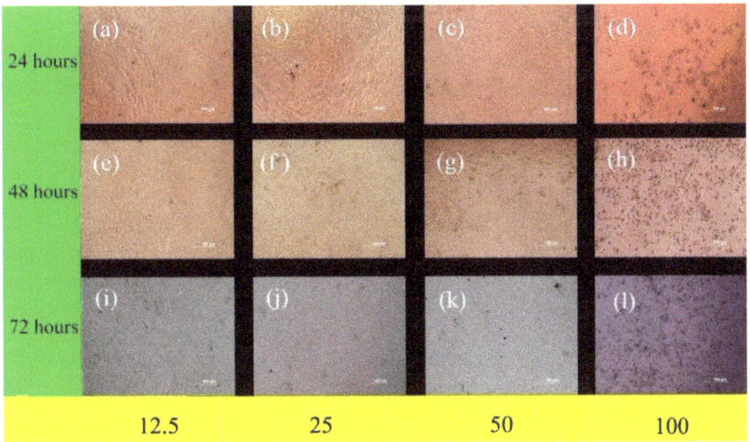

Figure 10. Microscopic images of MG-63 human cells seeded to different proportions of extract medium.

3.6. Corrosion Studies

The corrosion behavior of the coated substrates is confirmed by electrochemical studies in a simulated body fluid environment. The potentiodynamic polarization curves of the uncoated and Mg-HA coated substrates are plotted and shown in Figure 11. The I_{corr} value of uncoated and coated substrates was measured as 1.10×10^{-6}A and 2.93×10^{-8}A respectively. Corrosion resistance is proportional to the value of I_{corr}. Here the coated substrate has a lower I_{corr} Value and so it exhibits better corrosion resistance. The corrosion resistance of the coated substrate improves because the thin film coating acts as a barrier to the substrate. If the thin film obtained on the substrate is non-uniform and porous, the substrate is exposed to the SBF environment. A further ionic release takes place and pitting corrosion starts to appear. Even a small amount of released debris may cause local inflammation, irritation, and tissue damage [42,43]. The thickness of the coating also plays an important part in corrosion studies. Thicker coatings tend to develop cracks at a rapid rate and therefore thin coating is suitable as a surface modification technique for metal implants [44].

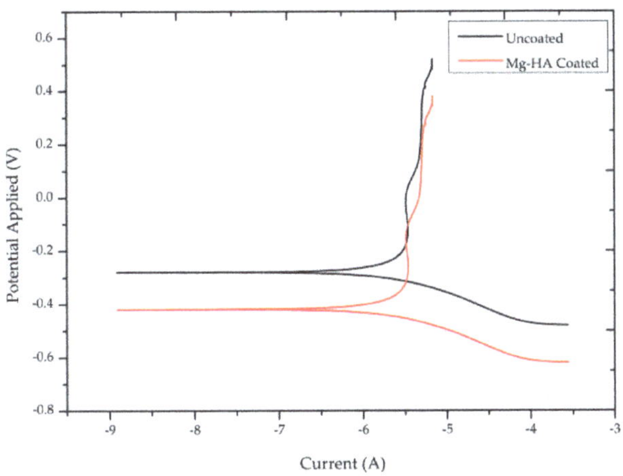

Figure 11. Potentiodynamic polarization curve of uncoated and Mg-HA coated substrates.

4. Conclusion

The Mg-HA powder used for coating purposes was synthesized using a wet chemical method and was characterized using SEM, HR-TEM, EDS, and XRD analysis. The surface morphology of the powder using SEM analysis showed spherical-shaped particles. Similarly, nano-scale crystalline morphology of the prepared powders was observed using HR-TEM and it was noticed that the nano size of the powders enabled the preparation of a uniformly suspended colloidal solution. EDS analysis confirmed the incorporation of Mg in the HA lattice structure. A dip coating technique was used for coating the substrates. A thin film of Mg-HA was developed on Ti6Al4V substrate. The shellac layer applied in between the Mg-HA coating and the substrate improved the bonding strength and provided stability for the developed coating. The coating parameter values of the dip coating process were optimized using the Taguchi optimization technique. Regression models were developed to predict the bonding strength value.

Dissolution studies conducted for 4 weeks in SBF showed that the release of Ca, P, and Mg ions into the body fluid was not that significant and it further indicated the adhesiveness of the developed coating onto the metal substrate. The cell viability studies performed using MG-63 osteosarcoma cell lines confirmed the non-toxic behavior of the developed coating and make it suitable for application in orthopedic and dental implants. Corrosion studies showed that Mg-HA coated substrate has better corrosion resistance properties than the uncoated substrate.

These results suggest the use of Mg-HA coatings developed on Ti6Al4V for biocompatible applications. A specific application such as the combination of bone marrow mesenchymal stem cells with substituted HA [45,46] for promoting the osseointegration process and for reducing the bone healing period can be studied further. Initial studies reported that using Mg-HA combined with stem cells yields good results [47]. Such composite coatings can also be applied to abutments and double conical connections, but further investigations on the effect of mechanical loading and torque required to loosen the screws must be carried out in a detailed manner. Implant stability also must be studied using some non-invasive techniques such as periotest, resonance frequency analysis, and implant insertion torque measurement [48] before suggesting it for clinical applications.

Author Contributions: R.A. has contributed to the experimental and analytical work of the research, K.K.S. has involved himself in the validation and evaluation of the findings reported in the paper, and R.A.R. contributed towards the optimization technique. All authors have read and agreed to the published version of the manuscript.

Funding: We did not receive any specific grant from funding agencies in the public, commercial, or not-for-profit sectors.

Institutional Review Board Statement: Not applicable.

Informed Consent Statement: Not applicable.

Data Availability Statement: The authors declare that data supporting the findings of this study are available within the article.

Acknowledgments: We would like to thank DST-SAIF, Cochin for providing all necessary support towards this research work.

Conflicts of Interest: The authors declare that they have no competing interests.

Abbreviations

Mg-HA	Magnesium substituted Hydroxyapatite
XRD	X-ray powder diffraction method
SEM	Scanning Electron Microscopy
SBF	Simulated body fluid

ICP-MS	Inductively Coupled Plasma–Mass Spectrometry
CM	complete medium
ETO	Ethylene oxide
MEM	Minimum essential medium
FBS	Fetal bovine solution
CCM	Cell culture medium
SCE	Saturated calomel electrode
DF	Degrees of freedom
SS	Sum of squares
MS	Mean square

References

1. Fielding, G.A.; Roy, M.; Bandyopadhyay, A.; Bose, S. Antibacterial and biological characteristics of silver containing and strontium doped plasma sprayed hydroxyapatite coatings. *Acta Biomater.* **2012**, *8*, 3144–3152. [CrossRef] [PubMed]
2. Thian, E.S.; Huang, J.; Best, S.M.; Barber, Z.H.; Bonfield, W. Magnetron co-sputtered silicon-containing hydroxyapatite thin films—an in vitro study. *Biomaterials* **2005**, *26*, 2947–2956. [CrossRef] [PubMed]
3. Tite, T.; Popa, A.-C.; Balescu, L.; Bogdan, I.; Pasuk, I.; Ferreira, J.; Stan, G. Cationic substitutions in hydroxyapatite: Current status of the derived biofunctional effects and their in vitro interrogation methods. *Materials* **2018**, *11*, 2081. [CrossRef] [PubMed]
4. Šupová, M. Substituted hydroxyapatites for biomedical applications: A review. *Ceram. Int.* **2015**, *41*, 9203–9231. [CrossRef]
5. Boyd, A.R.; Rutledge, L.; Randolph, L.D.; Meenan, B.J. Strontium-substituted hydroxyapatite coatings deposited via a co-deposition sputter technique. *Mater. Sci. Eng. C* **2015**, *46*, 290–300. [CrossRef] [PubMed]
6. Saini, M. Implant biomaterials: A comprehensive review. *World J. Clin. Cases* **2015**, *3*, 52–57. [CrossRef]
7. Tang, Y.; CHApell, H.F.; Dove, M.T.; Reeder, R.J.; Lee, Y.J. Zinc incorporation into hydroxylapatite. *Biomaterials* **2009**, *30*, 2864–2872. [CrossRef] [PubMed]
8. Ren, F.; Xin, R.; Ge, X.; Leng, Y. Characterization and structural analysis of zinc-substituted hydroxyapatites. *Acta Biomater.* **2009**, *5*, 3141–3149. [CrossRef]
9. Chen, W.; Liu, Y.; Courtney, H.; Bettenga, M.; Agrawal, C.M.; Bumgardner, J.D.; Ong, J.L. In vitro anti-bacterial and biological properties of magnetron co-sputtered silver-containing hydroxyapatite coating. *Biomaterials* **2006**, *27*, 5512–5517. [CrossRef]
10. Mo, A.; Liao, J.; Xu, W.; Xian, S.; Li, Y.; Bai, S. Preparation and antibacterial effect of silver–hydroxyapatite/titania nanocomposite thin film on titanium. *Appl. Surf. Sci.* **2008**, *255*, 435–438. [CrossRef]
11. Viala, S.; Freche, M.; Lacout, J. Preparation of a new organic-mineral composite: Chitosan-hydroxyapatite. *Ann. Chim. Sci. Des. Mater.* **1998**, *23*, 69–72. [CrossRef]
12. Li, H.; Zhou, C.R.; Zhu, M.Y.; Tian, J.H.; Rong, J.H. Preparation and characterization of homogeneous hydroxyapatite/chitosan composite scaffolds via in-situ hydration. *J. Biomater. Nanobiotechnol.* **2010**, *1*, 42–49. [CrossRef]
13. Kong, L.; Gao, Y.; Cao, W.; Gong, Y.; Zhao, N.; Zhang, X. Preparation and characterization of nano-hydroxyapatite/chitosan composite scaffolds. *J. Biomed. Mater. Res. Part A* **2005**, *75A*, 275–282. [CrossRef]
14. Pang, X.; Zhitomirsky, I. Electrodeposition of composite hydroxyapatite–chitosan films. *Mater. Chem.Phys.* **2005**, *94*, 245–251. [CrossRef]
15. Grandfield, K.; Zhitomirsky, I. Electrophoretic deposition of composite hydroxyapatite–silica–chitosan coatings. *Mater.Charact.* **2008**, *59*, 61–67. [CrossRef]
16. Shainberg, A.P.M.; Valério, P.; Zonari, A.; Oktar, F.N.; Ozyegin, L.S.; Graça, M.P.F.; Leite, M.F.; Goes, A.M. Attachment and proliferation of osteoblasts on lithium-hydroxyapatite composites. *Adv. Mater. Sci. Eng.* **2012**, *2012*, 1–10. [CrossRef]
17. Gibson, I.R.; Bonfield, W. Novel synthesis and characterization of an AB-type carbonate-substituted hydroxyapatite. *J. Biomed. Mater. Res.* **2001**, *59*, 697–708. [CrossRef]
18. Farzadi, A.; Bakhshi, F.; Solati-Hashjin, M.; Asadi-Eydivand, M.; abu Osman, N.A. Magnesium incorporated hydroxyapatite: Synthesis and structural properties characterization. *Ceram. Int.* **2014**, *40*, 6021–6029. [CrossRef]
19. Gayathri, B.; Muthukumarasamy, N.; VelautHAillai, D.; Santhosh, S.B.; asokan, V. Magnesium incorporated hydroxyapatite nanoparticles: Preparation, characterization, antibacterial and larvicidal activity. *Arab. J. Chem.* **2018**, *11*, 645–654. [CrossRef]
20. Mishra, V.K.; Bhattacharjee, B.N.; Parkash, O.; Kumar, D.; Rai, S.B. Mg-doped hydroxyapatite nanoplates for biomedical applications: A surfactant assisted microwave synthesis and spectroscopic investigations. *J. Alloys Compd.* **2014**, *614*, 283–288. [CrossRef]
21. Santos, G.G.; Nunes, V.L.C.; Marinho, S.M.O.C.; Santos, S.R.A.; Rossi, A.M.; Miguel, F.B. Biological behavior of magnesium-substituted hydroxyapatite during bone repair. *Braz. J. Biol.* **2021**, *81*, 53–61. [CrossRef]
22. Crespi, R.; Capparè, P.; Gherlone, E. Osteotome sinus floor elevation and simultaneous implant placement in grafted biomaterial sockets: 3 years of follow-up. *J. Periodontol.* **2010**, *81*, 344–349. [CrossRef] [PubMed]
23. Batchelar, D.L.; Davidson, M.T.M.; Dabrowski, W.; Cunningham, I.A. Bone-composition imaging using coherent-scatter computed tomography: Assessing bone health beyond bone mineral density. *Med. Phys.* **2006**, *33*, 904–915. [CrossRef]

24. Pauwels, R.; Jacobs, R.; Singer, S.R.; Mupparapu, M. CBCT-based bone quality assessment: Are Hounsfield units applicable? *Dentomaxillofacial Radiol.* **2015**, *44*, 20140238. [CrossRef] [PubMed]
25. Han, Y.; Wei, Q.; Chang, P.; Hu, K.; Okoro, O.V.; Shavandi, A.; Nie, L. Three-dimensional printing of hydroxyapatite composites for biomedical application. *Crystals* **2021**, *11*, 353. [CrossRef]
26. Luo, Y.; Chen, S.; Shi, Y.; Ma, J. 3D printing of strontium-doped hydroxyapatite based composite scaffolds for repairing critical-sized rabbit calvarial defects. *Biomed. Mater.* **2018**, *13*, 065004. [CrossRef]
27. Golafshan, N.; Vorndran, E.; Zaharievski, S.; Brommer, H.; Kadumudi, F.B.; Dolatshahi-Pirouz, A.; Malda, J. Tough magnesium phosphate-based 3D-printed implants induce bone regeneration in an equine defect model. *Biomaterials* **2020**, *261*, 120302. [CrossRef]
28. Jemat, A.; Ghazali, M.J.; Razali, M.; Otsuka, Y. Surface Modifications and Their Effects on Titanium Dental Implants. *BioMed Res. Int.* **2015**, *2015*, 1–11. [CrossRef]
29. Farag, Y. Characterization of Different Shellac Types and Development of Shellac-Coated Dosage Forms. Ph.D. Thesis, The Faculty of Mathematics, Computer Science Andnatural Sciences, University of Hamburg, Hamburg, Germany, 2010.
30. Mavis, B.; Taş, A. Dip coating of calcium hydroxyapatite on Ti-6Al-4V substrates. *J. Am. Ceram. Soc.* **2004**, *83*, 989–991. [CrossRef]
31. *ASTM D7234-19 Standard Test Method for Pull-Off Adhesion Strength of Coatings on Concrete Using Portable Pull-off Adhesion Testers*; ASTM International: West Conshohocken, PA, USA, 2019.
32. *IS 3203: 1982 Methods of Testing Local Thickness of Electroplated Coatings*; Bureau of Indian Standards: Old Delhi, India, 2016.
33. Kokubo, T.; Takadama, H. How useful is SBF in predicting in vivo bone bioactivity? *Biomaterials* **2006**, *27*, 2907–2915. [CrossRef] [PubMed]
34. *ISO 10993-14 Biological Evaluation of Medical Devices—Part 14: Identification and Quantification of Degradation Products from Ceramics*; International Organization for standardization: Geneva, Switzerland, 2019.
35. *ISO 10993-12 Biological Evaluation of Medical Devices—Part 12: Sample preparation and Reference Materials*; International Organization for standardization: Geneva, Switzerland, 2021.
36. Kamachimudali, U.; Sridhar, T.M.; Raj, B. Corrosion of bio implants. *Sadhana* **2003**, *28*, 601–637. [CrossRef]
37. *ASTM G5-94(2011)e1 Standard Reference Test Method for Making Potentiostatic and Potentiodynamic Anodic Polarization Measurements*; ASTM International: West Conshohocken, PA, USA, 2011.
38. Lotsari, A.; Rajasekharan, A.K.; Halvarsson, M.; Andersson, M. Transformation of amorphous calcium phosphate to bone-like apatite. *Nat.Commun.* **2018**, *9*, 4170. [CrossRef]
39. Mo, N.; Zhu, Z.; Zhu, Y.; Liu, Y.; Wang, X.; Yang, H.; Zhao, N. Purification behavior of Zn(II) in water by magnesium hydroxyapatite: Surface complexation, and dissolution–precipitation. *Int. J. Environ. Res. Public Health* **2020**, *17*, 3804. [CrossRef]
40. Suchanek, W.L.; Byrappa, K.; Shuk, P.; Riman, R.E.; Janas, V.F.; TenHuisen, K.S. Preparation of magnesium-substituted hydroxyapatite powders by the mechanochemical–hydrothermal method. *Biomaterials* **2004**, *25*, 4647–4657. [CrossRef]
41. *ISO 10993-5 Biological Evaluation of Medical Devices—Part 5: Tests for in Vitro Cytotoxicity*; International Organization for standardization: Geneva, Switzerland, 2017.
42. Eliaz, N. Corrosion of metallic biomaterials: A review. *Materials* **2019**, *12*, 407. [CrossRef]
43. Munir, S.; Oliver, R.A.; Zicat, B.; Walter, W.L.; Walter, W.K.; Walsh, W.R. The histological and elemental characterisation of corrosion particles from taper junctions. *Bone Jt. Res.* **2016**, *5*, 370–378. [CrossRef] [PubMed]
44. Li, M.; Jin, Z.-X.; Zhang, W.; Bai, Y.-H.; Cao, Y.-Q.; Li, W.-M.; Li, A.-D. Comparison of chemical stability and corrosion resistance of group IV metal oxide films formed by thermal and plasma-enhanced atomic layer deposition. *Sci. Rep.* **2019**, *9*, 10438. [CrossRef] [PubMed]
45. Yang, H.W.; Lin, M.H.; Xu, Y.Z.; Shang, G.W.; Wang, R.R.; Chen, K. Osteogenesis of bone marrow mesenchymal stem cells on strontium-substituted nano-hydroxyapatite coated roughened titanium surfaces. *Int. J. Clin. Exp. Med.* **2015**, *8*, 257–264. [PubMed]
46. Michel, J.; Penna, M.; Kochen, J.; Cheung, H. Recent advances in hydroxyapatite scaffolds containing mesenchymal stem cells. *Stem Cells Int.* **2015**, *2015*, 1–13. [CrossRef] [PubMed]
47. Kim, D.-H.; Shin, K.-K.; Jung, J.S.; Chun, H.H.; Park, S.S.; Lee, J.K.; Yoon, S.-Y. The role of magnesium ion substituted biphasic calcium phosphate spherical micro-scaffolds in osteogenic differentiation of human adipose tissue-derived mesenchymal stem cells. *J. Nanosci. Nanotechnol.* **2015**, *15*, 5520–5523. [CrossRef] [PubMed]
48. Baldi, D.; Lombardi, T.; Colombo, J.; Cervino, G.; Perinetti, G.; Di Lenarda, R.; Stacchi, C. Correlation between Insertion Torque and Implant Stability Quotient in Tapered Implants with Knife-Edge Thread Design. *BioMed Res. Int.* **2018**, *2018*, 7201093. [CrossRef] [PubMed]

coatings

MDPI

Article

Fabrication of Silver- and Zinc-Doped Hydroxyapatite Coatings for Enhancing Antimicrobial Effect

Daniela Predoi [1], Simona Liliana Iconaru [1] and Mihai Valentin Predoi [2,*]

[1] National Institute of Materials Physics, Atomistilor Street, No. 405A, P.O. Box MG 07, 077125 Magurele, Romania; dpredoi@gmail.com (D.P.); simonaiconaru@gmail.com (S.L.I.)
[2] Department of Mechanics, University Politehnica of Bucharest, BN 002, 313 Splaiul Independentei, Sector 6, 060042 Bucharest, Romania
* Correspondence: predoi@gmail.com or mihai.predoi@upb.ro

Received: 23 August 2020; Accepted: 18 September 2020; Published: 20 September 2020

Abstract: This study develops, for the first time, composite coatings based on silver and zinc doped hydroxyapatite in chitosan matrix (AgZnHApCs). The AgZnHApCs composite coatings were prepared by dip coating method. The hydroxyapatite (HAp), biocompatible material for regenerating and strengthening damaged bones were doped with silver and zinc ions and coated with chitosan in order to produce a uniform and homogenous coating with biocompatibility and antimicrobial properties. The stability of AgZnHApCs suspensions was evaluated by ultrasound measurements. The value of stability parameters of AgZnHApCs suspension is in good agreement with the value of bidistilled water used as reference fluid. Homogeneously dispersed solutions of AgZnHApCs were synthesized to endeavor to optimize the physico-chemical and biological characteristics of the coatings obtained at room temperature. The AgZnHApCs composite suspension and coatings were analyzed using various investigation techniques, such as X-ray diffraction (XRD), Fourier transformed infrared spectroscopy (FTIR), MTT (3-[4,5-dimethylthiazol-2-yl]-2,5-diphenylte-2H-tetrazolium bromide) assay and antimicrobial studies. The optical spectroscopy, atomic force microscopy (AFM), metallographic examination and X-ray photoelectron spectroscopy (XPS) on AgZnHApCs composite coatings were also conducted. Cell culture and MTT assays demonstrate that AgZnHApCs composite suspension and coatings have no negative effect on the cell viability and proliferation. The cell morphology was not affected in presence of AgZnHApCs composite suspension and coatings. The antimicrobial assays conducted against *Staphylococcus aureus* ATCC 25923, *Escherichia coli* ATCC 25922, and *Candida albicans* ATCC 90029 microbial strains revealed that both the AgZnHApCs composite suspension and coatings exhibited great antimicrobial properties.

Keywords: hydroxyapatite; composite coatings; silver; zinc; chitosan; antimicrobial activity

---

1. Introduction

The development of new materials with enhanced antimicrobial properties is one of the priority research areas in the last years. An important property of these materials is represented by the possibility of their use in the biomedical field [1]. Several studies reported in the literature have shown on the one hand that the occurrence of infections leads to a significant increase in medication costs (hospitalization), and on the other hand, it has been shown that the use of antibiotics before surgery does not bring real benefit to patients; they can even lead (when misused) to the appearance of antibiotic-resistant bacteria, these infections being much more difficult to treat [2]. One of the most studied materials with application in various domain such as dentistry and implantology is hydroxyapatite (HAp, $Ca_{10}(PO_4)_6(OH)_2$, Ca/P ratio of 1.67) [1,3].

HAp is the major inorganic component of bone tissue and has a structure with a special affinity that allows it to be substituted with various ions, such as: Zinc (Zn^{2+}), Magnesium (Mg^{2+}), Silver (Ag^{+}), carbonate (CO_3^{2-}), etc. [2]. Therefore, once a substitution occurs, there is a significant change in the physicochemical and biological properties of hydroxyapatite [4]. This behavior allows one to improve the properties of HAp, including antibacterial ones, through substitutions [4–8]. Among the properties of synthetic HAp we mention: excellent biocompatibility, good bioactivity, osteoconductivity, chemical and crystallographic similarity with natural hydroxyapatite [1]. Due to the fact that hydroxyapatite does not have antimicrobial properties, and yet it is used in various biomedical applications, one way to improve these properties is to dope HAp with antimicrobial agents. Therefore, to make it suitable to fight with antibiotic resistant bacteria, we chose to dope HAp with Ag^+ and Zn^{2+} ions. Materials based on HAp doped with Ag and/or zinc ions can be obtained by several methods, among which we mention: hydrothermal method [2], microwave [9] wet-chemical method [1], sol-gel process [10,11], co-precipitation [12–14], chemical solution deposition [5], etc.

Silver is one of the most used and studied antimicrobial agents due to its broad spectrum of action and low toxicity at low concentrations [15,16]. Compared to Silver, Zinc is a chemical element that is found naturally in bone tissue and is essential for biological processes such as: enzyme activity, DNA synthesis, mitosis, cell proliferation, etc. [5,9]. Recently, in their study, van Hengel et al. [4] have shown that surfaces functionalized with zinc and silver ions exhibit antimicrobial activity, even against bacterial strains of methicillin-resistant *Staphylococcus aureus* (MRSA). The use of metallic ions (Silver and Zinc) as antimicrobial agents give to hydroxyapatite the ability to fight against gram-positive (*Staphylococcus aureus, Streptococcus mutans, Bacillus cereus*) [1,7,17,18], gram-negative (*Escherichia coli, Aggregatibacter actinomycetemcomitans, Fusobacterium nucleatum*) [7,10] bacterial strains and fungi (*Candida albicans*) [1].

On the other hand, studies reported in the literature have shown that the antimicrobial properties of compounds based on hydroxyapatite doped with silver or zinc ions are strongly dependent on the concentration of antimicrobial agent in the samples [8,13]. Taking into account the aspects mentioned above, the mechanism of antimicrobial action of both Ag and Zn compounds are not yet fully understood [1,19]. Furthermore, a widely used material due to its remarkable biodegradability and biocompatibility in recent years, mostly in dental applications, is chitosan. Chitosan is a biopolymer isolated from shellfish, crab and shrimp which was reported as having antimicrobial properties and antioxidant activity [20–22].

The aim of this study was to develop a novel AgZnHApCs composite with superior biocompatibility and significant antibacterial activity. A mixture suspension of silver and zinc doped hydroxyapatite and chitosan with a good stability was analyzed by ultrasound measurement and XRD, FTIR, MTT and antimicrobial activity in order to be used in obtaining uniform and homogeneous coatings that preserve the biological properties of the suspension. The novel AgZnHApCs composite coatings obtained were evaluated by XRD, FTIR, optical spectroscopy, atomic force microscopy (AFM), metallographic examination, XPS, MTT and antimicrobial activity.

2. Materials and Methods

2.1. Materials

The reagents used in the synthesis of the materials such as chitosan (Cs) powder (low molecular weight, 75–85% deacetylated), calcium nitrate tetrahydrate, zinc nitrate hexahydrate, silver nitrate, ammonium hydrogen phosphate, ethanol absolute and acetic acid were purchased from Sigma Aldrich, St. Louis, MO, USA, and all were of analytical grade. The double distilled water and deionized water were also used in the experiment.

2.2. Silver and Zinc Doped Hydroxyapatite in Chitosan Matrix (AgZnHApCs)

The synthesis of silver- and zinc-doped hydroxyapatite in chitosan matrix with $x_{Ag} = 0.1$ and $x_{Zn} = 0.1$ was effectuated using an adapted coprecipitation method [23] with (Ca + Ag + Zn)/P was fixed to 1.67 [24,25]. In the first step, 0.1 g of chitosan (Cs) was added to 100 mL deionized water containing 1% and stirred until a clear solution is obtained (about 2 h). In the second step, $(NH_4)_2HPO_4$ was dissolved in 50 mL ethanol and added in 100 mL chitosan solution and stirred for 2 h at 40 °C. At the same time, $Ca(NO_3)_2{\cdot}4H_2O$, $(AgNO_3)$ and $Zn(NO_3)_2{\cdot}6H_2O$ were dissolved in ethanol and mixed for 2 h at 40 °C. The solution containing Ca, Ag and Zn was added dropwise in the solution containing P and Cs. The final solution was stirred for 2 h at 100 °C. The resulting solution of AgZnHApCs was centrifuged and redispersed in solution of chitosan and stirred at room temperature for 4 h. The resulting suspension was analyzed from a stability point of view and used to prepare the coatings.

2.3. *Preparation of AgZnHApCs Coatings*

The AgZnHApCs coatings were obtained using a dip coating method which was previously described in [26]. AgZnHApCs coatings were deposited on the glass substrate. The obtained AgZnHApCs layer was dried at 100 °C for 24 h.

2.4. *Characterization Methods*

Ultrasonic measurements were performed on 100 mL of concentrated suspension of 5Sm-HAp [6,27,28]. The digitalized ultrasonic signals were recorded on the digital oscilloscope at a very precise interval of 5.00 s. In order to have an accurate evaluation of the stability of the 5Sm-HAp suspension, the double distilled water (the most stable suspension) was chosen as the reference fluid, under the same experimental conditions.

The synthesized silver and zinc doped hydroxyapatite in chitosan matrix were characterized by X-ray diffraction (XRD, Bruker D8 Advance diffractometer, Bruker, Billerica, MA, USA, CuKα radiation) technique in the range of 2θ values from 10–60° at a step of 0.02° and 34 s measuring time per step.

The molecular structure of the samples was also studied by FTIR-ATR spectroscopy in the 400–4000 cm^{-1} spectral range using a Perkin Elmer instrument (Waltham, MS, USA).

The X-ray photoelectron spectroscopy measurements (XPS) were recorded with a multimethod SPECS surface analysis system, as previously described in [29,30].

Furthermore, the surface morphology of the coatings was investigated by atomic force microscopy (AFM) with the aid of an NT-MDT NTEGRA Probe Nano Laboratory instrument (NT-MDT, Moscow, Russia), as previously reported in [31]. The data analysis of the 2D surface topographies, as well as the 3D representation of the AFM images, were performed with the aid of Gwyddion 2.55 software [32]

In addition, the morphology of the surface coatings was studied by optical and metallographic microscopy as previously described in [33]. The image processing of the recorded data was performed using Image J software (ImageJ 1.51j8, National Institutes of Health, Bethesda, MD, USA) [32].

The cytotoxicity of the AgZnHApCs suspensions and coatings was assessed by MTT (3-4,5-dimethylthiazol 2,5-diphenyltetrazolium bromide) assay, in agreement with previous studies [34]. For this purpose, human fetal osteoblasts (hFOB 1.19) from the American Type Culture Collection (ATCC) were used. The cells were grown in Dulbecco's modified Eagle's medium (DMEM), enriched using L-glutamine (2.5 mM), sodium pyruvate (0.5 mM), non-essential amino acids (0.1 mM), baking soda (1.2 g/L) and fetal bovine serum. Afterwards, the cultures were incubated in the atmosphere with 5% CO_2 at 37 °C. All subsequent passages were performed at confluence (approximately 2×10^6 cells/plate), on average at 5–6 days, the ratio being 1/3. After that, the cells were seeded in a Petri dish, including in the control sample (Petri dish without sample), at a concentration of 3×10^6 cells/flask of 75 cm^2. The cultures were incubated with the samples in the atmosphere with 5% CO_2, at 37 °C, and after 1, 7 and 14 days, the cells are visualized under an inverted microscope and pictures were taken in the visible (VIS) with a digital camera. For the quantitative cytotoxic assay, the cells were treated with MTT solution [3-(4,5-dimethylthiazolyl)-2,5-diphenyltetrazolium bromide] and incubate in an atmosphere with 5% CO_2 at 37 °C and the optical density of formazane solubilized at 595 nm using a TECAN spectrophotometer (Tecan GENios, Grödic, Germany) was quantified in order to obtain the cell viability of the cells. The experiments were performed 4 times and the data were presented as mean ± SD.

The antimicrobial properties of the AgZnHAp suspensions and coatings were studied against reference microbial strains *Staphylococcus aureus* ATCC 25923, *Escherichia coli* ATCC 25922, and *Candida albicans* ATCC 90029. The antimicrobial assays were performed by preparing 0.5 McFarland standard microbial cultures, as previously described in [35]. Afterwards, the AgZnHApCs suspensions and coatings were exposed to 2 mL of microbial suspension of 5×10^6 CFU/mL (colony forming units/mL) in phosphate-buffered saline (PBS) in a Petri dish and incubated for 24, 48 and 72 h. As a positive control, a free microbial culture was assessed at the same time intervals. Afterwards, the suspension was collected at different time intervals (24, 48 and 72 h) and incubated on LB agar medium for 24 h at 37 °C. The number of CFU/mL was determined for each of the incubated samples with the microbial suspensions. The experiments were performed 4 times and the data were presented as mean ± SD. Statistical analysis was done using the ANOVA single factor test.

3. Results

The stability of AgZnHApCs suspension was evaluated by ultrasonic measurements performed on 100 mL suspension. Ultrasonic waves propagate in liquids and di-phasic systems (e.g., suspensions), bringing information about the medium in which these waves propagate. One advantage of this method, compared to optical methods, is the possibility of characterizing opaque fluids. The frequencies which are currently used are between 1 MHz and 50 MHz. The digitalized ultrasonic signals were recorded using two pairs of coaxial ultrasonic transducers; one pair with a central frequency of 5 MHz and the other pair with a central frequency of 25 MHz. The interval at which the ultrasonic signals were recorded was 5.00 s. We consider a suspension to be stable if the amplitude of the overall signal has a weak change of amplitude during the monitored sedimentation process. Of course, in pure water, our reference is perfectly stable (zero variation of ultrasonic echo amplitude). We determine the slope of the amplitude during the monitoring period, to determine a quantitative stability parameter, defined in the manuscript. A short echo comprises several superposed signals of different frequencies, as the Fourier series can provide. We determine the variation of amplitudes for a number of frequencies also, during the monitoring period. The evolution of the signals in time gives us important information about the stability of the suspension and the attenuation vs. time. The value of the stability parameter $s = \frac{1}{A_m}\left|\frac{dA}{dt}\right|$ calculated for the AgZnHApCs suspension had the value 1.77×10^{-5} s^{-1} using a pair of coaxial ultrasonic transductors, both of central frequency of 5 MHz and 6.2×10^{-7} s^{-1} using a pair of coaxial ultrasonic transductors, both of central frequency of 25 MHz. The values obtained for the stability parameter indicated a very good stability. Another characteristic regarding the stability of AgZnHApCs suspension is the frequency spectrum of the first transmitted echo (Figure 1). Using transducers of 5 MHz central frequency, the peaks of amplitudes at 4 MHz, for the reference fluid, are also peaks for the measured signals. More relevant information is represented by the spectral amplitudes' variation during the experiment (Figure 1a). From the Fourier spectrum, were selected frequencies 2–8 MHz (Figure 1a). On (Figure 1b) is presented the frequency spectrum for the 10–40 MHz range using transducers of 25 MHz central frequency. The behavior of the measured signals is similar to that of the reference fluid. The peaks of amplitudes at 11.5 and 23 MHz, for the reference fluid, are also peaks for the measured signals (Figure 1b). For AgZnHApCs suspension, the typical decreasing values with increasing frequency components were obtained (Figure 2). Based on the 30 mm distance between the coaxial transducers, the attenuation vs. frequency in nepper/m can be obtained, averaged over the 5000 s recording period (Figure 2). Compared to distilled water, the ultrasonic signals attenuation is larger in the tested AgZnHApCs suspension. The ultrasonic signals attenuation has classical, continuously increasing values vs. frequency, over the selected frequency range. The experiments were resumed using a pair of coaxial transducers of 25 MHz central frequency. The most relevant supplementary information was presented in Figure 2b. In intervals of 10–25 MHz, the absorbed energy translates into signal attenuation, visible in the local peak at 20 MHz (Figure 2b). The other peak at 30 MHz, is obtained in a frequency interval of weak signals, and has to be verified in future experiments centered at these higher frequencies. In the range of 25–40 MHz, the behavior of attenuation vs.

frequency for the first transmitted echo at 25 MHz central frequency for the AgZnHApCs suspension has a behavior similar to that of bi-distilled water as a reference fluid (Figure 2b). Based on the results obtained, we can say that the suspension is very stable. The information brought by analyzing the first transmitted echo provides relevant information for a characterization from the ultrasonic point of view, in terms of stability, overall and spectral attenuation.

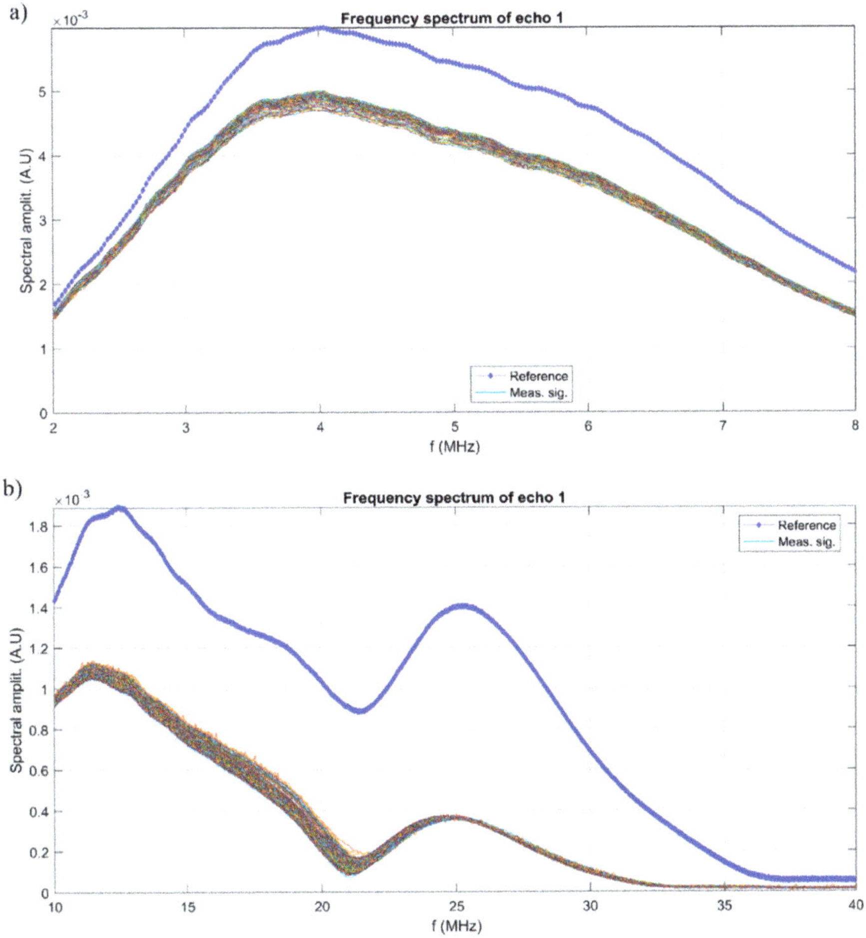

Figure 1. Frequency spectrum of the first transmitted echo at 5 MHz (**a**) and 25 MHz (**b**) central frequency for the AgZnHApCs suspension.

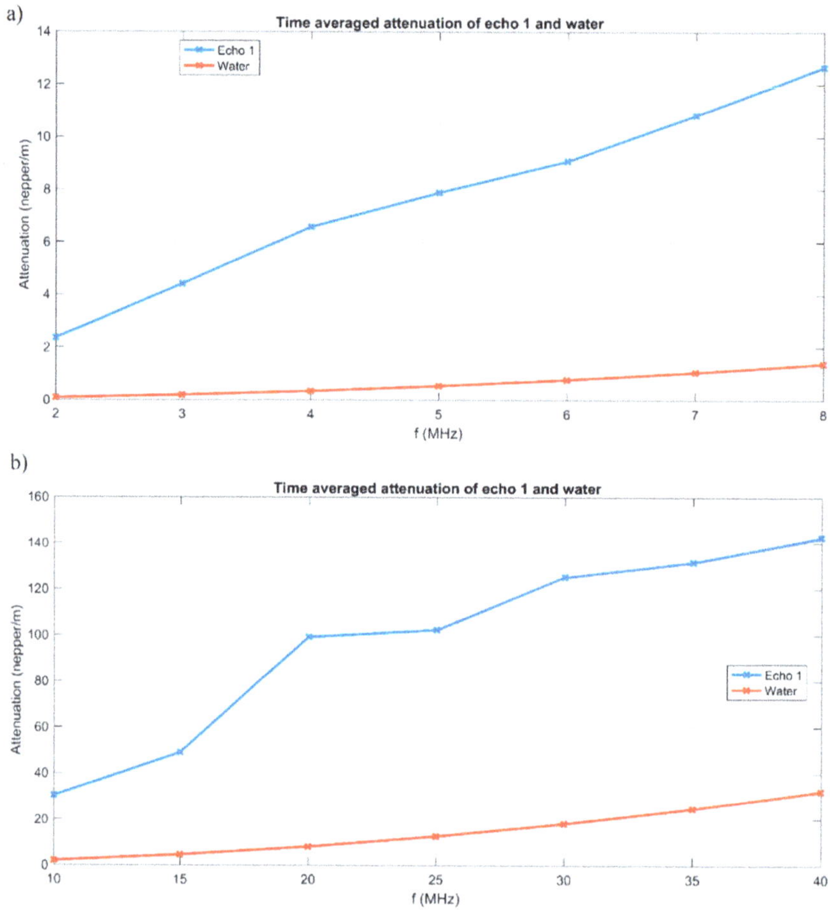

Figure 2. Attenuation vs. frequency for the first transmitted echo at 5 MHz (**a**) and 25 MHz (**b**) central frequency for the AgZnHApCs suspension.

The XRD studies were conducted in order to evaluate the crystalline structure of biocomposites suspension and coatings based on silver and zinc doped hydroxyapatite in chitosan matrix prepared from the sol-gel process (Figure 3). XRD patterns on the powder obtained from the AgZnHApCs gel are shown in Figure 3a. XRD pattern of the AgZnHApCs coatings (Figure 3b) and chitosan (Figure 3c) and the standard pattern of the HAp (JCPDS card No.09-0432) and chitosan PDF-00-054-1952 are also presented.

The XRD patterns of AgZnHApCs powder (Figure 3a) and coatings (Figure 3b) revealed the main diffraction peaks identified at different 2θ values that were assigned to the position of the planes of pure hexagonal HAp at (002), (211), (112), (300), (202), (310), (222), (213), (004), (323), in agreement with JCPDScardNo.09-0432 [36]. In order to calculate the average size all, the peaks of the XRD pattern were taken into account. The mean crystallite size was calculated using Scherrer's formula

$$D = 0.9\lambda/\beta \times \cos\theta. \quad (1)$$

In the Scherrer's formula, D represents the average crystallite size, β the peak broadening of the diffraction line measured at half of its maximum intensity, λ the wavelength of X-rays, and θ the

Bragg's diffraction angle. The mean crystallite size of AgZnHApCs powders (Figure 3a) was found to be 18 ± 2 nm, while the mean crystallite size of AgZnHApCs coatings (Figure 3b) was found to be 15 ± 3 nm.

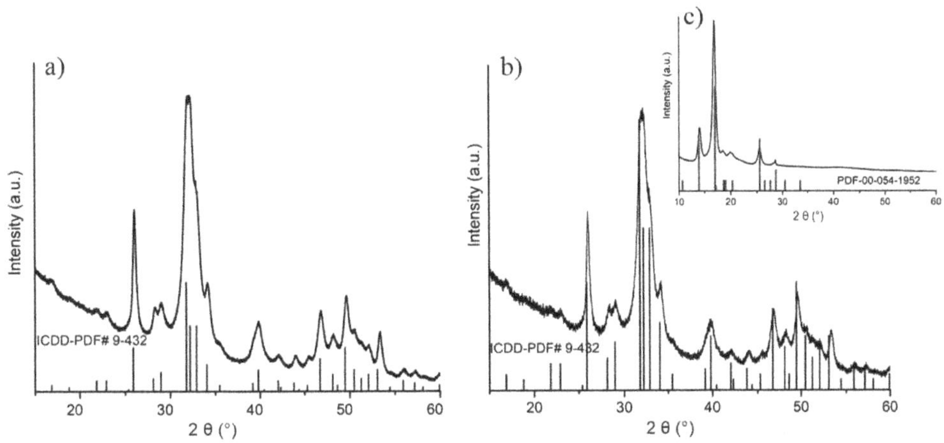

Figure 3. XRD pattern of the AgZnHApCs powder (**a**), AgZnHApCs coatings (**b**) and Chitosan (**c**) and the standard pattern of the HAp (JCPDS card No.09-0432) and Chitosan PDF-00-054-1952.

FTIR analysis revealed the presence of chitosan on the AgZnHApCs suspension and coating. FTIR spectra of AgZnHApCs suspension, AgZnHApCs coating and Cs recorded over the range of 400–4000 cm^{-1} were presented in Figure 4. The spectra of AgZnHApCs suspension and AgZnHApCs coating exhibit characteristic vibrational modes of HAp and chitosan, proving that the AgZnHAp nanoparticles were successfully coated with chitosan. The two samples spectra (AgZnHApCs suspension and AgZnHApCs coating) showed specific phosphate vibrational modes (ν_4) at 571, 604 and 636 cm^{-1}. The peaks at 571 and 604 cm^{-1} correspond to a triply degenerated bending mode of the O–P–O band [34]. The peak at 636 cm^{-1} was also assigned to the hydroxyl group deformation mode [36]. According to previous studies [36,37], the peak at 632 cm^{-1} is considered to be overlapped with an asymmetric P–O stretching vibration of PO_4^{3-}. The bands at around 1088 and 1047 cm^{-1} in the region 900-1200 cm^{-1} were attributed to the C–O–P stretching and phosphorylated hydroxyl group, while the peaks at around 473 and 998 cm^{-1} may be assigned for P–OH groups. The bands assigned to P–O, P=O stretching of phosphate group can be observed at around 1100 to 1250 cm^{-1} in the region 900–1200 cm^{-1}. According to previous studies on the interactions of calcium phosphates with chitosan [38,39], the peaks in the region 1300–1450 cm^{-1} can be attributed to a combination of CN–NH, CH_2–OH and CH_3 bands. The band from 1422 cm^{-1} can be assigned to the bending CH vibrations, while the peak from 1076 cm^{-1} in the region 1008–1150 cm^{-1} can be assigned to groups C–O–C. The absorption bands at 1315, 1372 and 1407 cm^{-1} were assigned to the carbonate bands of acetic acid which were not observed in the spectra. On the other hand, the presence of carbonate bands of acetic acid should be avoided for biocompatible applications [38,39]. The peak at 1642 cm^{-1} can be assigned to the N–H from chitosan. The AgZnHApCs in suspension and AgZnHAp coating showed typical bands characteristic to HAp and chitosan, as can be exhibited in Figure 4b,c.

The morphology of the AgZnHApCs coating's surface was studied by AFM analysis. The 2D AFM topography of the AgZnHApCs coating's surface is presented in Figure 5a and the 3D representation of the surface morphology of AgZnHApCs coating is depicted in Figure 5b. The 2D AFM micrograph and the 3D representation of the AgZnHApCs coating's surface highlighted that the AgZnHApCs coating's surface exhibited the morphology of a uniformly deposited layer. In addition, the AFM topography also suggested that the surface of the coating does not present any cracks or fissures and is composed of equally distributed nanoaggregates. The AFM results emphasized that the surface topography of the AgZnHApCs coating was homogenous, having a roughness (R_{RMS}) value of 43.1125 nm.

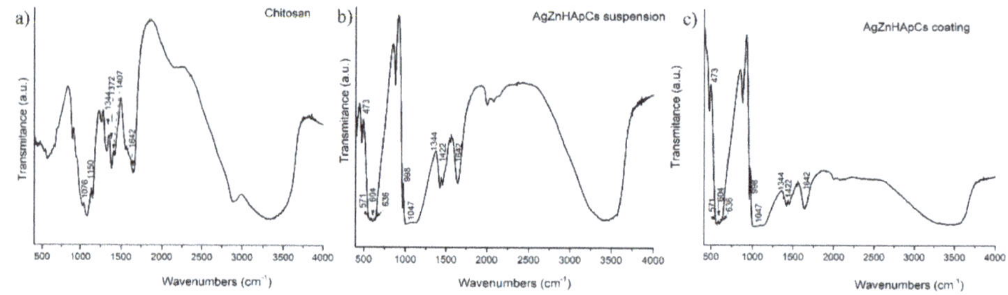

Figure 4. FTIR spectra of chitosan (**a**), AgZnHApCs suspension (**b**) and AgZnHApCs coating (**c**).

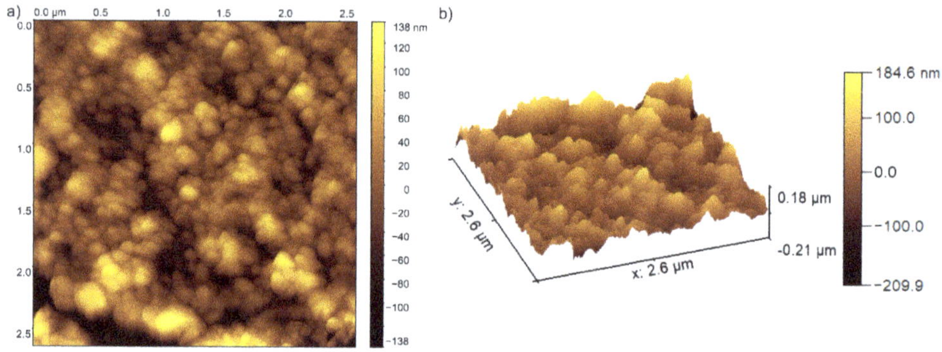

Figure 5. Representation of the atomic force microscopy (AFM) topography image of the AgZnHApCs coating's surface (**a**) and 3D representation of the surface topography of AgZnHApCs coatings (**b**).

Moreover, the surface morphology of the AgZnHApCs coating was also studied by optical and metallographic microscopy. The 2D images with the optical and metallographic microscope were acquired using the 10× objective. The results are presented in Figures 6a and 7a. The optical images of the surface of the AgZnHApCs coating emphasized that the coatings deposited on glass substrate are homogenous and uniform, having no visible discontinuities, fissures or irregularities. In addition, the 3D representation of the optical image of AgZnHApCs coating's surface obtained using ImageJ software [32] is depicted in Figure 6b.

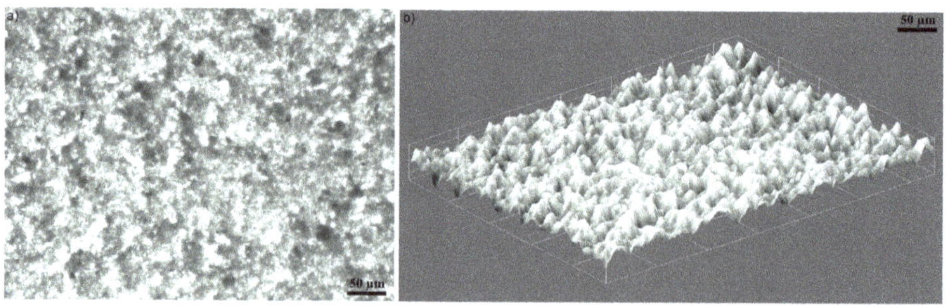

Figure 6. Optical microscopy image (10× magnification) of the AgZnHApCs coating's surface (**a**) and 3D representation of the surface of the AgZnHApCs coating (**b**).

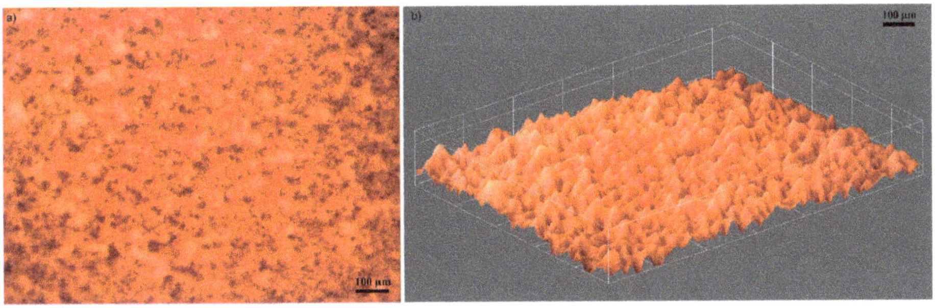

Figure 7. Metallographic microcopy image (10× magnification) of the AgZnHApCs coating's surface (**a**) and 3D representation of the surface of the AgZnHApCs coating (**b**).

Complementary information regarding the morphology of the AgZnHApCs coating deposited on glass substrate was also obtained using a metallographic microscope. The 2D image of the AgZnHApCs coating's surface obtained with the 10x objective of the metallographic microscope is depicted in Figure 7a. Similar to the optical microscopy studies, the metallographic investigations highlighted that the surface of AgZnHApCs coating has an evenly homogenous morphology and do not exhibit any fissures or cracks. Moreover, the 3D representation of the metallographic microscope 2D image of the AgZnHApCs coating's surface depicted in Figure 7b confirmed the results of the optical microscopy studies, and revealed that the AgZnHApCs coatings are continuous, with no cracks or fissures.

The in vitro biocompatibility of the AgZnHApCs suspensions and coatings was assessed using human fetal osteoblasts (hFOB 1.19) cells. For this purpose, the suspension and coatings were incubated with the hFOB 1.19 cell suspension for different time intervals (1, 7 and 14 days) and their viability was measured by MTT assay. The MTT results of the hFOB 1.19 cell viability are depicted in Figure 8. The experiments were repeated four times and the data were presented as mean ± SD. Furthermore, a *t*-test: paired two sample for means was used to perform the statistical analysis, and all the p calculated values were $p < 0.05$. The results revealed that both investigated samples presented a cell viability above 85% for all tested time intervals. Moreover, the results also highlighted that there was no significant difference between the samples in the form of suspension and coatings. However, a slight increase in cell viability was observed in the case of AgZnHApCs suspensions compared to the AgZnHApCs coatings for all incubation periods. Moreover, the data also suggested that the cell viability increased with the increase of the incubation time. The results showed that both samples present very good biocompatible and bioactive properties. The results are in agreement with previous studies performed to date, regarding the hydroxyapatite properties of facilitating the attachment and growth of osteoblast cells due to its high hydrophilic property [40,41].

The morphology of the hFOB 1.19 cells incubated for 1, 7 and 14 days with the AgZnHApCs suspensions and coatings was also assessed using an inverted microscope. The morphology of the hFOB 1.19 cells treated with AgZnHApCs suspensions and coatings after 1, 7 and 14 days of incubation are depicted in Figure 9. The images of the hFOB 1.19 cells incubated with AgZnHApCs suspensions and coatings at different time intervals show the normal morphology of the hFOB 1.19 cells and revealed that the suspensions and coatings did not induce any significant changes in the hFOB 1.19 cells. These results are in agreement with the results of the quantitative MTT in vitro assay, and highlighted that both AgZnHApCs suspensions and coatings have a good biocompatibility and are suitable for being used in biomedical applications.

The antimicrobial activities of the AgZnHApCs suspensions and coatings were also assessed using reference gram positive, gram negative and fungal microbial strains. The antimicrobial properties of AgZnHApCs suspensions and coatings were assessed at three different time intervals of incubation (24, 48 and 72 h) against *S. aureus* ATCC 25923, *E. coli* ATCC 25922, and *C. albicans* ATCC 9002, by assessing the colony forming units (CFU) inhibition for each tested microbial strain. The results of the antimicrobial assay are presented in Figure 10. The antimicrobial assays were repeated four times and the data were represented as mean ± SD. A free microbial suspension was also assessed and used as positive control (C+) for all tested time intervals. The statistical analysis was done using the ANOVA single factor algorithm and the calculated p-values were less than 0.0001 [42,43]. The data revealed that the AgZnHApCs suspensions and coatings exhibited strong antimicrobial activity against all the tested strains, even after 24 h of incubation. The CFU values were significantly reduced after 24 h of incubation with the suspensions and coatings. The results of the antimicrobial assays also suggested that the CFU inhibition was significant for all tested samples and that was correlated with the incubation time. Moreover, the results revealed that the antimicrobial activity of the suspensions was greater than that of the coatings for all tested microbial strains at all tested time intervals. In addition, a fungicidal effect was noticed in the case of AgZnHApCs suspensions after 48 h and 72 h of exposure, and in the case of AgZnHApCs coatings after 72 h of incubation. The results revealed by the antimicrobial assays are in agreement with previously reported data from the literature regarding the antimicrobial properties of silver and zinc ions [5–8,44–53]. Moreover, the results are in accordance with literature data reported on the potential antimicrobial properties of materials and composite layers based on hydroxyapatite enhanced with metal ions in polymeric matrix which possess antimicrobial activity. Furthermore, the statistical analysis emphasized that the data supports that both AgZnHApCs suspensions and coatings present antimicrobial activities on the tested microbial strains and that the AgZnHApCs suspensions exhibited a greater antimicrobial activity than AgZnHApCs coatings.

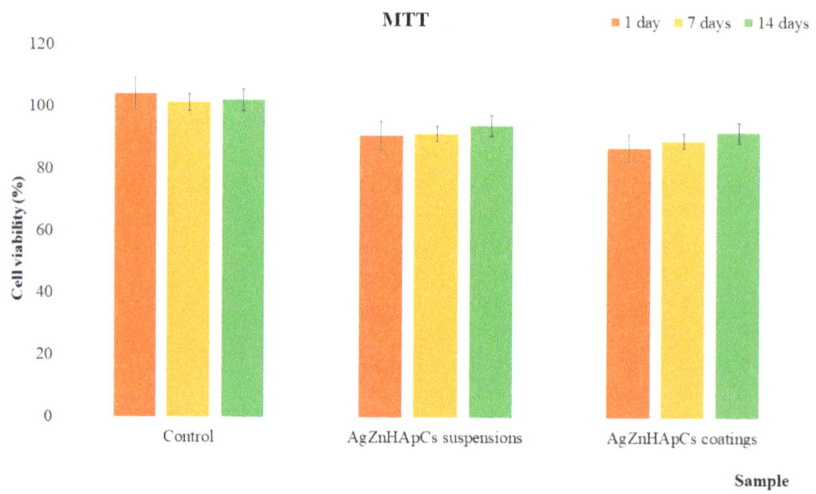

Figure 8. MTT assay for the viability of hFOB 1.19 cells incubated on AgZnHApCs suspensions and coatings for 1 day, 7 days and 14 days. The results are presented as means ± standard error of the mean of four independent experiments. The data were statistically analyzed by using a t-Test: Paired Two Sample for Means and p values, $p \leq 0.05$, were accepted as statistically significant.

Despite the reported studies regarding the antimicrobial activity of metallic ions such as silver and zinc, data regarding the synergies formed between the chemical components of composites based on hydroxyapatite and metal ions in polymeric matrix are still limited. In this case, the antimicrobial properties of the AgZnHApCs suspensions and coatings could be attributed both to their constituent elements and to the synergies that appear in the composite. Therefore, the antimicrobial properties obtained in our studies could be attributed both to the presence of the silver and zinc ions, as well as to the synergy that formed between the zinc, silver and chitosan components with the hydroxyapatite structure and with the glass substrate in the case of AgZnHApCs coatings. Moreover, our results also suggested that the nature of the sample, either suspension or coating, has an important role in their biological activity. Both the cytotoxic assay and the antimicrobial assay revealed that the biological properties of the AgZnHApCs suspensions were better than the ones of the AgZnHApCs coatings.

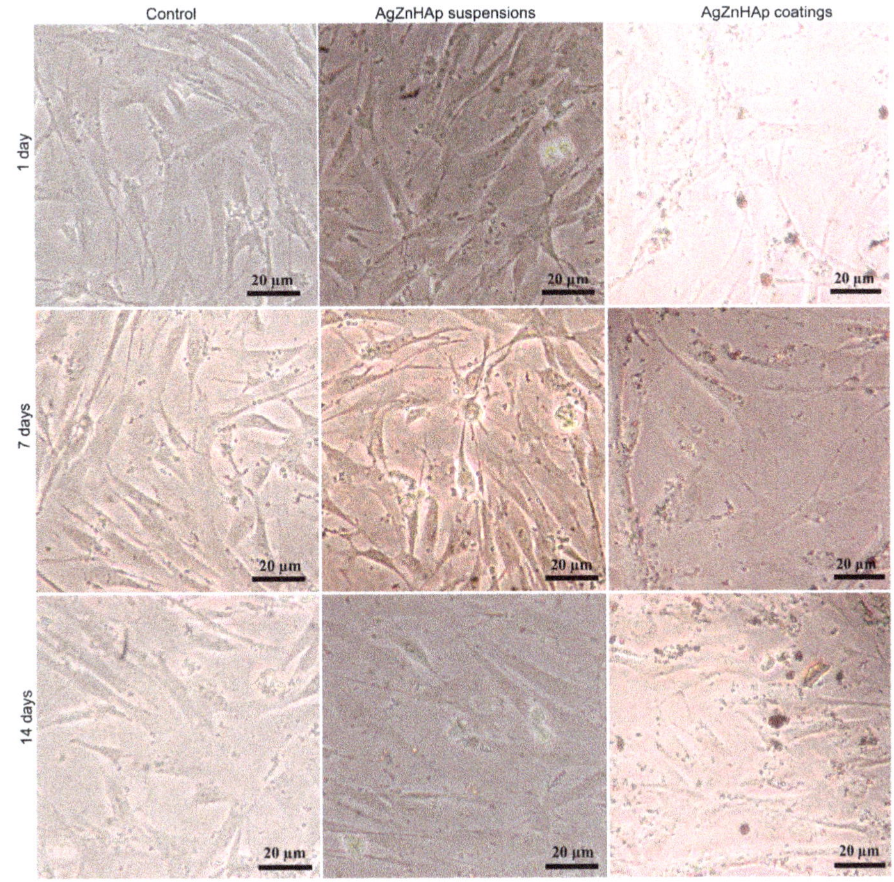

Figure 9. The morphology of the hFOB 1.19 cells incubated with AgZnHApCs suspensions and coatings after 1, 7 and 14 days.

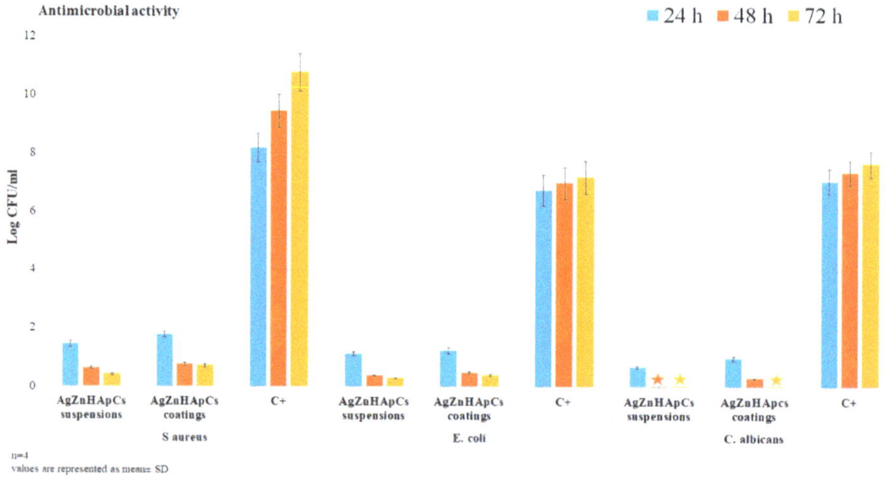

Figure 10. The graphical representation of the Log colony forming units (CFU)/mL of *S. aureus* ATCC 25923, *E. coli* ATCC 25922, and *C. albicans* ATCC 9002 for different incubation times with AgZnHApCs suspensions and coatings. The results are presented as means ± standard error of the mean of 4 independent experiments. The data was statistically analyzed by ANOVA Single Factor algorithm and the calculated *p*-values were less than 0.0001.

4. Conclusions

A novel AgZnHApCs composite suspension with superior biocompatibility and significant antibacterial activity has been synthesized by an adapted coprecipitation method, in order to prepare the AgZnHApCs composite coatings by dip coating method. The FTIR and XRD evaluation showed the bonding interaction between silver and zinc doped hydroxyapatite and chitosan. The homogeneous dispersion and the stability of nanoparticles in the chitosan matrix have been evaluated. The morphology and homogeneity of the surface coatings were investigated by optical spectroscopy, AFM and metallographic examination and XPS. The use of chitosan provides a significant increasing of dispersion of nanoparticles. FTIR analysis revealed the chemical bond interaction between Ca ions and –OH groups of HAp and a combination of CN–NH, CH_2–OH and CH_3 bands assigned of chitosan. MTT test confirms that the AgZnHApCs suspension and coatings composite are cytocompatible. hFOB 1.19 cell culture study proves that the AgZnHApCs suspension and coatings composite do not influence cell morphology and proliferation. The antimicrobial assays conducted against *Staphylococcus aureus* ATCC 25923, *Escherichia coli* ATCC 25922, and *Candida albicans* ATCC 90029 microbial strains revealed that both the AgZnHApCs composite suspension and coatings exhibited great antimicrobial properties. In addition, the results have emphasized that the suspension were more effective against the tested microbial strains. The results of our studies suggest that the novel developed AgZnHApCs nanocomposite coatings can be used in different fields, such as the surface modification of bone implants, and ameliorate the osteoconductivity of bone regeneration.

Author Contributions: Conceptualization, D.P. and M.V.P.; methodology, D.P.; software, M.V.P.; validation, D.P., S.L.I. and M.V.P.; formal analysis, D.P.; investigation, S.L.I.; resources, D.P.; data curation, D.P.; writing—original draft preparation, D.P., S.L.I. and M.V.P.; writing—review and editing, D.P., S.L.I. and M.V.P.; visualization, D.P., S.L.I. and M.V.P.; supervision, D.P.; project administration, M.V.P.; funding acquisition, M.V.P. All authors have read and agreed to the published version of the manuscript.

Funding: This research was funded by Romanian Ministry of Research and Innovation with the grant number PN-III-P1-1.2-PCCDI-2017-0629/contract No. 43PCCDI/2018 and Scientific Research Contract

Nr.1/4.06.2020] and "The APC was funded by [Romanian Ministry of Research and Innovation] project number PN-III-P1-1.2-PCCDI-2017-0629/contract No. 43PCCDI/2018."

Acknowledgments: We would like to thank M.L. Badea for her help with the acquisition of the biological assays data.

Conflicts of Interest: The authors declare no conflict of interest.

References

1. Mocanu, A.; Furtos, G.; Rapuntean, S.; Horowitz, O.; Flore, C.; Garbo, C.; Danisteanu, A.; Rapuntean, G.; Prejmerean, C.; Tomoaia-Cotisel, M. Synthesis; characterization and antimicrobial effects of composites based on multi-substituted hydroxyapatite and silver nanoparticles. *Appl. Surf. Sci.* **2014**, *298*, 225–235. [CrossRef]
2. Radovanovic, Z.; Jokic, B.; Veljovic, D.; Dimitrijevic, S.; Kojic, V.; Petrovic, R.; Janackovic, D. Antimicrobial activity and biocompatibility of Ag^+- and Cu^{2+}-doped biphasic hydroxyapatite/α-tricalcium phosphate obtained from hydrothermally synthesized Ag^+- and Cu^{2+}-doped hydroxyapatite. *Appl. Surf. Sci.* **2014**, *307*, 513–519. [CrossRef]
3. Shanmugam, S.; Gopal, B. Antimicrobial and cytotoxicity evaluation of aliovalent substituted hydroxyapatite. *Appl. Surf. Sci.* **2014**, *303*, 277–281. [CrossRef]
4. Van Hengel, I.A.J.; Putra, N.E.; Tierolf, M.W.A.M.; Minneboo, M.; Fluit, A.C.; Fratila-Apachitei, L.E.; Apachitei, I.; Zadpoor, A.A. Biofunctionalization of selective laser melted porous titanium using silver and zinc nanoparticles to prevent infections by antibiotic-resistant bacteria. *Acta Biomater.* **2020**, *107*, 325–337. [CrossRef] [PubMed]
5. Ohtsu, N.; Kakuchi, Y.; Ohtsuki, T. Antibacterial effect of zinc oxide/hydroxyapatite coatings prepared by chemical solution deposition. *Appl. Surf. Sci.* **2018**, *445*, 596–600. [CrossRef]
6. Predoi, D.; Iconaru, S.L.; Predoi, M.V. Dextran-coated zinc-doped hydroxyapatite for biomedical applications. *Polymers* **2019**, *11*, 886. [CrossRef]
7. Predoi, D.; Iconaru, S.L.; Predoi, M.V. Bioceramic layers with antifungal properties. *Coatings* **2018**, *8*, 276. [CrossRef]
8. Predoi, D.; Iconaru, S.L.; Deniaud, A.; Chevallet, M.; Michaud-Soret, I.; Buton, N.; Prodan, A.M. Textural, structural and biological evaluation of hydroxyapatite doped with zinc at low concentrations. *Materials* **2017**, *10*, 229. [CrossRef]
9. Iqbal, N.; Kadir, M.R.A.; Mahmood, N.H.; Salim, N.; Froemming, G.R.A.; Balaji, H.R.; Kamarul, T. Characterization, antibacterial and in vitro compatibility of zinc–silver doped hydroxyapatite nanoparticles prepared through microwave synthesis. *Ceram. Int.* **2014**, *40*, 4507–4513. [CrossRef]
10. Chen, W.; Oh, S.; Ong, A.P.; Oh, N.; Liu, Y.; Courtney, H.S.; Appleford, M.; Ong, J.L. Antibacterial and osteogenic properties of silver-containing hydroxyapatite coatings produced using a sol–gel process. *J. Biomed. Mater. Res. A* **2007**, *82*, 899–906. [CrossRef]
11. Helen, S.; Ruban Kumar, A. Study of structural, mechanical and dielectrical properties of ions doped apatite for antibacterial activity. *Mater. Chem. Phys.* **2019**, *237*, 121867. [CrossRef]
12. Ciobanu, C.S.; Iconaru, S.L.; Chifiriuc, M.C.; Costescu, A.; Le Coustumer, P.; Predoi, D. Synthesis and antimicrobial activity of silver-doped hydroxyapatite nanoparticles. *Biomed. Res. Int.* **2013**, *2013*, 916218. [CrossRef] [PubMed]
13. Ciobanu, C.S.; Massuyeau, F.; Constantin, L.V.; Predoi, D. Structural and physical properties of antibacterial Ag-doped nano-hydroxyapatite synthesized at 100 °C. *Nanoscale Res. Lett.* **2011**, *6*, 613. [CrossRef] [PubMed]
14. Costescu, A.; Ciobanu, C.S.; Iconaru, S.L.; Ghita, R.V.; Chifiriuc, C.M.; Marutescu, L.G.; Predoi, D. Fabrication, characterization, and antimicrobial activity, evaluation of low silver concentrations in silver-doped hydroxyapatite nanoparticles. *J. Nanomater.* **2013**, *2013*, 194854. [CrossRef]
15. Furko, M.; Havasi, V.; Kónya, Z.; Grünewald, A.; Detsch, R.; Boccaccini, A.R.; Balázsi, C. Development and characterization of multi-element doped hydroxyapatite bioceramic coatings on metallic implants for orthopedic applications. *Boletín de la Sociedad Española de Cerámica y Vidrio* **2018**, *57*, 55–65. [CrossRef]
16. Chou, Y.J.; Ningsih, H.S.; Shih, S.-J. Preparation, characterization and investigation of antibacterial silver-zinc co-doped β-tricalcium phosphate by spray pyrolysis. *Ceram. Int.* **2020**, *46*, 16708–16715. [CrossRef]
17. Ofudje, E.A.; Adeogun, A.I.; Idowu, M.A.; Kareem, S.O. Synthesis and characterization of Zn-doped hydroxyapatite: Scaffold application, antibacterial and bioactivity studies. *Heliyon* **2019**, *5*, e01716. [CrossRef]

18. Chen, X.; Tang, Q.-L.; Zhu, Y.-J.; Zhu, C.-L.; Feng, X.-P. Synthesis and antibacterial property of zinc loaded hydroxyapatite nanorods. *Mater. Lett.* **2012**, *89*, 233–235. [CrossRef]
19. Xie, Y.; He, Y.; Irwin, P.L.; Jin, T.; Shi, X. Antibacterial activity and mechanism of action of zinc oxide nanoparticles against *Campylobacter jejuni*. *Appl. Environ. Microbiol.* **2011**, *77*, 2325–2331. [CrossRef]
20. Mirhashemi, A.; Bahador, A.; Kassaee, M.; Daryakenari, G.; Ahmad-Akhoundi, M.; Sodagar, A. Antimicrobial Effect of nano-zinc oxide and nano-chitosan particles in dental composite used in orthodontics. *J. Med. Bacteriol.* **2013**, *2*, 1–10.
21. Haldorai, Y.; Shim, J.-J. Chitosan-zinc oxide hybrid composite for enhanced dye degradation and antibacterial activity. *Compos. Interfaces* **2013**, *20*, 365–377. [CrossRef]
22. Wu, S.; Ma, S.; Zhang, C.; Cao, G.; Wu, D.; Gao, C.; Lakshmanan, S. Cryogel biocomposite containing chitosan-gelatin/cerium–zinc doped hydroxyapatite for bone tissue engineering. *Saudi J. Biol. Sci.* **2020**, *27*, 2638–2644. [CrossRef]
23. Ciobanu, C.S.; Iconaru, S.L.; Massuyeau, F.; Constantin, L.V.; Costescu, A.; Predoi, D. Synthesis, structure, and luminescent properties of europium-doped hydroxyapatite nanocrystalline powders. *J. Nanomater.* **2012**, *2012*, 942801. [CrossRef]
24. Iconaru, S.L.; Motelica-Heino, M.; Guegan, R.; Predoi, M.V.; Prodan, A.M.; Predoi, D. Removal of zinc ions using hydroxyapatite and study of ultrasound behavior of aqueous media. *Materials* **2018**, *11*, 1350. [CrossRef]
25. Ciobanu, C.S.; Iconaru, S.L.; Popa, C.L.; Motelica-Heino, M.; Predoi, D. Evaluation of samarium doped hydroxyapatite, ceramics for medical application: Antimicrobial activity. *J. Nanomater.* **2015**, *2015*, 849216. [CrossRef]
26. Rodriguez, L.; Matoušek, J. Preparation of TiO_2 sol-gel layers on glass. *Ceram. Silik.* **2003**, *47*, 28–31.
27. Predoi, D.; Iconaru, S.L.; Predoi, M.V.; Motelica-Heino, M.; Guegan, R.; Buton, N. Evaluation of antibacterial activity of zinc-doped hydroxyapatite colloids and dispersion stability using ultrasounds. *Nanomaterials* **2019**, *9*, 515. [CrossRef]
28. Predoi, D.; Predoi, M.V.; Iconaru, S.L.; Ech Cherif El Kettani, M.; Leduc, D.; Prodan, A.M. Ultrasonic Measurements on β Cyclodextrin/Hydroxyapatite Composites for Potential Water Depollution. *Materials* **2017**, *10*, 681. [CrossRef]
29. Predoi, D.; Iconaru, S.L.; Predoi, M.V.; Motelica-Heino, M. Removal and oxidation of As(III) from water using iron oxide coated CTAB as adsorbent. *Polymers* **2020**, *12*, 1687. [CrossRef]
30. Yamaguchi, M. Role of zinc in bone formation and bone resorption. *J. Trace Elem. Exp. Med.* **1998**, *11*, 119–135. [CrossRef]
31. Predoi, D.; Iconaru, S.L.; Predoi, M.V.; Motelica-Heino, M.; Buton, N.; Megier, C. Obtaining and characterizing thin layers of magnesium doped hydroxyapatite by dip coating procedure. *Coatings* **2020**, *10*, 510. [CrossRef]
32. Gwyddion. Available online: http://gwyddion.net/ (accessed on 20 January 2020).
33. ImageJ. Available online: http://imagej.nih.gov/ij (accessed on 10 January 2018).
34. Predoi, D.; Vatasescu-Balcan, R.A. Osteoblast interaction with iron oxide nanoparticles coated with dextrin in cell culture. *J. Optoelectron. Adv. Mat.* **2008**, *10*, 152–157.
35. Predoi, D.; Iconaru, S.L.; Buton, N.; Badea, M.L.; Marutescu, L. Antimicrobial activity of new materials based on lavender and basil essential oils and hydroxyapatite. *Nanomaterials* **2018**, *8*, 291. [CrossRef] [PubMed]
36. Ciobanu, C.S.; Iconaru, S.L.; Le Coustumer, P.; Predoi, D. Vibrational investigations of silver-doped hydroxyapatite with antibacterial properties. *J. Spectrosc.* **2013**, *2013*, 1–5. [CrossRef]
37. Negrila, C.C.; Predoi, M.V.; Iconaru, S.L.; Predoi, D. Development of zinc-doped hydroxyapatite by sol-gel method for medical applications. *Molecules* **2018**, *23*, 2986. [CrossRef]
38. Ruphuy, G.; Saralegi, A.; Lopes, J.C.; Dias, M.M.; Barreiro, M.F. Spray drying as a viable process to produce nano-hydroxyapatite/chitosan (n-HAp/CS) hybrid microparticles mimicking bone composition. *Adv. Powder Technol.* **2016**, *27*, 575–583. [CrossRef]
39. Fadeeva, I.V.; Barinov, S.M.; Fedotov, A.Y.; Komlev, V.S. Interactions of calcium phosphates with chitosan. *Dokl. Chem.* **2011**, *441*, 387–390. [CrossRef]
40. Silverstein, R.M.; Webster, F.X.; Kiemle, D.; Bryce, D.L. *Spectrometric Identification of Organic Compounds*, 8th ed.; John Wiley & Sons: Hoboken, NJ, USA, 2014.

41. Thian, E.S.; Ahmad, Z.; Huang, J.; Edirisinghe, M.J.; Jayasinghe, S.N.; Ireland, D.C.; Brooks, R.A.; Rushton, N.; Bonfield, W.; Best, S.M. The role of surface wettability and surface charge of electrosprayed nanoapatites on the behaviour of osteoblasts. *Acta Biomater.* **2010**, *6*, 750–755. [CrossRef]
42. Cabral, D.J.; Penumutchu, S.; Norris, C.; Morones-Ramirez, J.R.; Belenky, P. Microbial competition between Escherichia coli and Candida albicans reveals a soluble fungicidal factor. *Microb. Cell* **2018**, *5*, 249–255. [CrossRef]
43. Lorowitz, W.; Saxton, E.; Sondossi, M.; Nakaoka, K. Integrating statistics with a microbiology laboratory activity. *J. Microbiol. Biol. Educ.* **2005**, *6*, 14–19. [CrossRef]
44. Kilpadi, K.L.; Chang, P.-L.; Bellis, S.L. Hydroxylapatite binds more serum proteins, purified integrins, and osteoblast precursor cells than titanium or steel. *J. Biomed. Mater. Res.* **2001**, *57*, 258–267. [CrossRef]
45. Zhang, D.; Lepparanta, O.; Munukka, E.; Ylanen, H.; Viljanen, M.K.; Eerola, E.; Hupa, M.; Hupa, L. Antibacterial effects and dissolution behavior of six bioactive glasses. *J. Biomed. Mater. Res. A* **2010**, *93*, 475–483. [CrossRef] [PubMed]
46. Gopi, D.; Sathishkumar, S.; Karthika, A.; Kavitha, L. Development of Ce^{3+}/Eu^{3+} Dual-Substituted Hydroxyapatite Coating on Surgical Grade Stainless Steel for Improved Antimicrobial and Bioactive Properties. *Ind. Eng. Chem. Res.* **2014**, *53*, 20145–20153. [CrossRef]
47. Kolmas, J.; Groszyk, E.; Kwiatkowska-Różycka, D. Substituted Hydroxyapatites with Antibacterial Properties. *BioMed Res. Int.* **2014**, *2014*, 1–15. [CrossRef] [PubMed]
48. Wang, Y.Z.; Xue, X.X.; Yang, H. Preparation and characterization of zinc and cerium co-doped titania nano-materials with antibacterial activity. *J. Inorg. Mater.* **2013**, *28*, 117–122. [CrossRef]
49. Bhadra, P.; Mitra, M.K.; Das, G.C.; De, R.; Mukherjee, S. Interaction of chitosan capped ZnO nanorods with Escherichia coli. *Mater. Sci. Eng. C* **2011**, *31*, 929–937. [CrossRef]
50. Mujeeb Rahman, P.; Abdul Mujeeb, V.M.; Muraleedharan, K.; Steni, K.T. Chitosan/nano ZnO composite films: Enhanced mechanical, antimicrobial and dielectric properties. *Arab. J. Chem.* **2018**, *11*, 120–127. [CrossRef]
51. Goy, R.C.; De Britto, D.; Assis, O.B.G. A review of the antimicrobial activity of chitosan. *Polímeros* **2009**, *19*, 241–247. [CrossRef]
52. Zheng, L.-Y.; Zhu, J.-F. Study on antimicrobial activity of chitosan with different molecular weights. *Carbohydr. Polym.* **2003**, *54*, 527–530. [CrossRef]
53. Devlieghere, F.; Vermeulen, A.; Debevere, J. Chitosan: Antimicrobial activity, interactions with food components and applicability as a coating on fruit and vegetables. *Food Microbiol.* **2004**, *21*, 703–714. [CrossRef]

coatings

Article

Development of Arabinoxylan-Reinforced Apple Pectin/Graphene Oxide/Nano-Hydroxyapatite Based Nanocomposite Scaffolds with Controlled Release of Drug for Bone Tissue Engineering: In-Vitro Evaluation of Biocompatibility and Cytotoxicity against *MC3T3-E1*

Wafa Shamsan Al-Arjan [1], Muhammad Umar Aslam Khan [2,3,4,*], Samina Nazir [1], Saiful Izwan Abd Razak [3,5] and Mohammed Rafiq Abdul Kadir [3]

1 Department of Chemistry, College of Science, King Faisal University, P.O. Box 400, Al-Ahsa 31982, Saudi Arabia; walarjan@kfu.edu.sa (W.S.A.-A.); ssamina@kfu.edu.sa (S.N.)
2 Department of Polymer Engineering and Technology, University of the Punjab, Lahore 54590, Pakistan
3 School of Biomedical Engineering and Health Sciences, Faculty of Engineering, Universiti Teknologi Malaysia, Skudai 81300, Johor, Malaysia; saifulizwan@utm.my (S.I.A.R.); rafiq@biomedical.utm.my (M.R.A.K.)
4 School of Biomedical Engineering, Med-X Research Institute, Shanghai Jiao Tong University (SJTU), 1954 Huashan Road, Shanghai 200030, China
5 Centre of Advanced Composite Materials, Faculty of Engineering, Universiti Teknologi Malaysia, Skudai 81300, Johor, Malaysia
* Correspondence: umar007khan@gmail.com or umar-786@sjtu.edu.cn

Received: 28 October 2020; Accepted: 18 November 2020; Published: 20 November 2020

Abstract: Fabrication of reinforced scaffolds to repair and regenerate defected bone is still a major challenge. Bone tissue engineering is an advanced medical strategy to restore or regenerate damaged bone. The excellent biocompatibility and osteogenesis behavior of porous scaffolds play a critical role in bone regeneration. In current studies, we synthesized polymeric nanocomposite material through free-radical polymerization to fabricate porous nanocomposite scaffolds by freeze drying. Functional group, surface morphology, porosity, pore size, and mechanical strength were examined through Fourier Transform Infrared Spectroscopy (FTIR), Single-Electron Microscopy (SEM), Brunauer-Emmet-Teller (BET), and Universal Testing Machine (UTM), respectively. These nanocomposites exhibit enhanced compressive strength (from 4.1 to 16.90 MPa), Young's modulus (from 13.27 to 29.65 MPa) with well appropriate porosity and pore size (from 63.72 ± 1.9 to 45.75 ± 6.7 μm), and a foam-like morphology. The increasing amount of graphene oxide (GO) regulates the porosity and mechanical behavior of the nanocomposite scaffolds. The loading and sustained release of silver-sulfadiazine was observed to be 90.6% after 260 min. The in-vitro analysis was performed using mouse pre-osteoblast (*MC3T3-E1*) cell lines. The developed nanocomposite scaffolds exhibited excellent biocompatibility. Based on the results, we propose these novel nanocomposites can serve as potential future biomaterials to repair defected bone with the load-bearing application, and in bone tissue engineering.

Keywords: arabinoxylan; apple pectin; biocompatibility; material science; nanotechnology; bone tissue engineering

1. Introduction

Bone tissue engineering (BTE) is an advanced and possible alternative solution for current surgical bone grafting approaches using porous biomaterials. BTE aims to support, facilitate, and regenerate the damaged or fractured bones using porous scaffolds. It is an effective and quick solution to heal defects of bones, to eliminate the donor deficiency problem; source limitation, and problems with conventional tissue implants (allograft and autograph) [1,2]. The design and composition of the artificial bone is a major objective in bone regeneration, and choosing appropriate materials to fabricate nanocomposite scaffolds can meet this challenge. The nanocomposite scaffold should resemble the extracellular matrix (ECM) to facilitate the osteogenesis and must regenerate the host bone by restoring its functionality [3]. The porous scaffolds should facilitate the growth of osteogenic cells, which is an ideal property of the porous scaffolds. The porous scaffolds with appropriate structure, interconnected porosity, and greater surface area have a positive impact on bone regeneration that facilitates the tissue growth response [4]. The nanocomposite materials furnished with tailorable properties and desired characteristics have become potential biomaterials. The nanocomposite scaffolds can be fabricated with high porosity, interconnected surface area to mimic the properties of the natural bone, and high tensile strength for load-bearing applications.

Porous nanocomposite scaffolds are the possible and potential channel to address the defective bone issue through bone tissue engineering conventions. These enable the synthesis of structural properties to meet the requirements for pertinent bone regeneration and growth [5]. Freeze drying is a well-known method to fabricate porous scaffolds from polymeric nanocomposite with desired pore size and porosity. The freeze-drying technique is famous among researchers and scientists within the last decade to obtain porous biomaterials [6,7]. Nanocomposite scaffolds can be fabricated using different nanocomposite materials to get the desired shape, pore size, % porosity, and other physicochemical characteristics. Several researchers have been focused on the synthesis of porous nanocomposite scaffolds using nanocomposite materials, using freeze drying methodology, to develop the multi-functional properties [8]. The self-bone regeneration mechanism can be activated by adding certain growth factors that encourage bone regeneration by securing neovascularization. These nanocomposite scaffolds provide a microenvironment due to their microstructure and polymeric matrix that resembles with extracellular matrix (ECM). The microenvironment and polymeric matrix support the bone mechanically by facilitating cell proliferation, cell differentiation, and migration. Hydroxyapatite (HAp) is a well-recognized biomaterial that is biocompatible, effectively bioavailable to the ECM, and can regulate the biomechanical properties of the host bone [9,10].

Natural polymers have been frequently utilized as potential biomaterials in medical sciences due to their biocompatibility, biodegradability, and non-immunological response. The researchers are focused on natural polymers to develop biomaterials for addressing different medical issues. These natural polymers have popularity momentum due to their capability in fabricating effectively engaged porous nanocomposite scaffolds with controlled biocompatible and biodegradable behavior [11,12]. Arabinoxylan (AX) is a natural polymer. AX and its derivatives have potential applications in tissue engineering, and other medical applications, due to its biocompatible, antioxidant, and anti-inflammatory properties. The backbone of AX is similar to that of glycosaminoglycans and it is a major ECM component of the cartilage bone. Henceforth, AX-based scaffolds can facilitate and support defective bone regeneration [8,13]. Pectin is a well-known natural polymer and is obtained from apple pulp, beetroot, and citrus fruits, etc. Pectin is an anionic polysaccharide and it possesses different chemical characteristics with several biomedical applications [14]. It is hydrophilic by nature and retains substantial water that facilitates cell proliferation and migration. Pectin has a similar structure to ECM and can induce different cellular functionalities to enhance the growth of cells [14]. The polymeric matrix of pectin has various carboxylate groups that help the crystallization of calcium and phosphate ions to regenerate calcified bone tissue [15]. The natural polymers based composite material are lacking in mechanical strength, which can be addressed by adding different fillers. Graphene oxide (GO) is a well-known filler that has been used in biomedical applications [16].

It has also gained recognition among researchers because of its high surface area, extremely versatile structure, flow, and movement of electrons with extraordinary mechanical strength [17,18]. In recent years, GO-based composite materials have been introduced in tissue engineering. GO-based composite materials have relatively low toxicity and high osteoinductive efficiency that facilitates rapid cell growth with loading application in bone tissue engineering [19,20]. Different composite materials have been introduced to regenerate and repair defective bone in tissue engineering.

In this research, polymeric nanocomposite materials have been synthesized using AX, AP, and nHAp with a variable amount of GO through free-radical polymerization. These polymeric nanocomposites were used to fabricate nanocomposite scaffolds (NCSs) using freeze drying methods. According to the best of our knowledge, these compositions have never been reported before in tissue engineering. The structural analysis and morphology have been analyzed using FTIR, XRD, and SEM. The swelling, biodegradation, drug loading, and release analysis were performed. The in-vitro analysis of developed scaffold materials was performed using mouse pre-osteoblast (*MC3T3-E1*) cells to determine biocompatibility. The obtained results demonstrated that these nanocomposite scaffolds have excellent morphology, physicochemical behavior, and biocompatibility. These nanocomposite scaffolds may be promising future biomaterial to repair and regenerate fracture bone in bone tissue engineering.

2. Method and Material

2.1. Material

Arabinoxylan was extracted from the husk of *Plantago ovata* using a well-reported method by Saeed et al. [21]. Apple pectin (Cas No. 76828-100G), Acrylic acid ($C_3H_4O_2$) AAc), *N,N'*-methylene-bis-acrylamide (*N,N'* MBA ($C_7H_{10}N_2O_2$)), silver nitrate ($AgNO_3$) and potassium persulfate ($K_2S_2O_8$), n-HAp (<100 nm particle size, ≥95%), GO (763713-1G), Aqua ammonia, phosphate buffer saline (PBS) solution, and hydrochloric acid (HCl) was received from Sigma-Aldrich, Malaysia. All chemicals are analytical grades and utilized without any purification.

Fetal bovine serum (FBS), L-glutamine penicillin/streptomycin, and Alpha-MEM (α-MEM) were supplied by ThermoFisher Scientific and Hyclone Laboratories Inc. The mouse pre-osteoblast (*MC3T3-E1*) cell-lines were purchased from the American Type Culture Collection (ATCC-USA).

2.2. Synthesis Polymeric Nanocomposite

The polymeric nanocomposites materials were synthesized through the free-radical polymerization process using arabinoxylan (AX) and apple pectin (AP) were grafted and doping of nHAp and GO into the polymeric matrix of AX-g-AP. Briefly, AX (1.0 g) and AP (1.0 g) was dispersed in deionized water separately and transferred into a three-necked round bottom flask. Then, 0.45 mL of AAc as monomer and *N,N'*-MBA as crosslinker (0.045% by weight of AAc) was added into the solution. Nano-hydroxyapatite (2 g) as a primary component and GO (0.1, 0.2, 0.3, and 0.4 mg) as an auxiliary component was added slowly into three-neck round bottom flask. The reaction mixture was stirred for 2 h at 60 °C. Later, 0.05 g of the initiator ($K_2S_2O_8$) was added and stirred under an inert environment for another 3 h at 60 °C. After that, the reaction was stopped by removing flow of nitrogen gas and reaction media was allowed to cool. The reaction media was vacuum filtered and the residue was washed several times with deionized water to remove unreacted chemicals. The residue was dried overnight at 50 °C in an oven and ground to get fine powder of polymeric nanocomposite materials. The nHAp and GO were doped into the polymeric matrix of AX-g-AP to produce polymeric nanocomposites as shown in the proposed chemical reaction (Figure 1 (Left-side)).

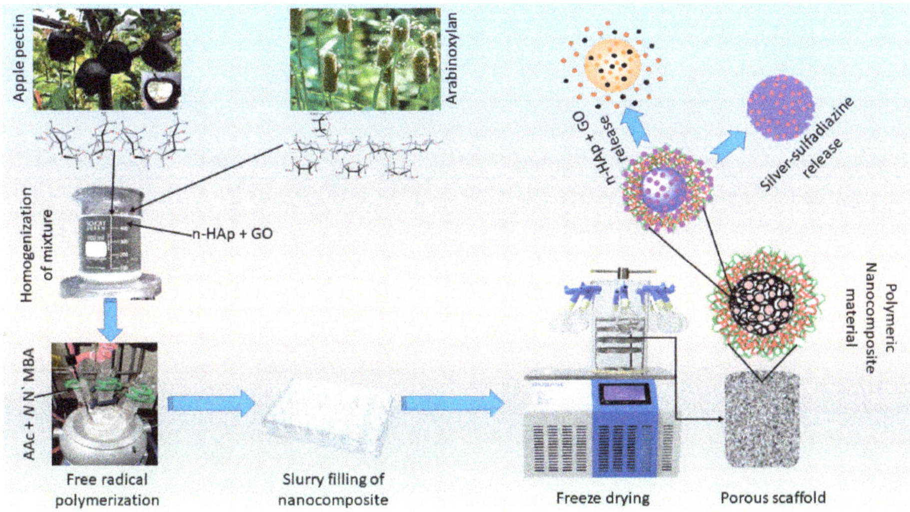

Figure 1. Experimental design. Polymeric nanocomposite scaffolds were synthesized by free radical polymerization to fabricate porous scaffolds via the freeze-drying method. Briefly, biopolymers (arabinoxylan, apple pectin), acrylic acid (monomer), n-HAp, and GO were stirred to have a homogenized mixture and crosslinked by using *N,N′*-methylene-bis-acrylamide to form the hybrid nanocomposite. These hybrid nanocomposites were then freeze-dried to have porous scaffolds. Finally, the mouse pre-osteoblast (*MC3T3-E1*) cell line was used to evaluate in vitro behavior of these scaffolds.

2.3. Fabrication of Nanocomposite Scaffolds

The porous nanocomposite scaffolds were fabricated through the freeze-drying technique. Concisely, polymeric nanocomposite powder (5 g) was dispersed into deionized water and a homogenized slurry was made. The slurry was filled into a 24-well plate (Figure 2 (Right side)) and allow to freeze at −80 °C for 24 h. The frozen molds were freeze-dried to get porous nanocomposite scaffolds without any crack. Different codes NCS-1, NCS-2, NCS-3, and NCS-4 were assigned to nanocomposite scaffolds after different amounts of GO (0.1, 0.2, 0.3, and 0.4 mg). The proposed chemical reaction has been presented in the following schematic diagram (Figure 1).

2.4. Loading of Silver-Sulfadiazine

Silver-sulfadiazine (25 mg) was dissolved carefully into ethanol (5 mL) and added into deionized water (45 mL), and it was stirred to have a homogenized solution. Then, nanocomposite scaffold (NCS-4) was immersed into a beaker containing a silver-sulfadiazine solution for 1 h at room temperature. The beaker was kept in the oven (50 °C) under vacuum, until complete evaporation of the solvent, and to deposit the silver-sulfadiazine into the porous structure of the scaffold as shown in schematic diagram-I (Lower right-side). The drug release was determined into a PBS buffer solution (pH 7.4 at 37 °C). A 5 mL solution was taken after every 20 min and drug release was measured using a double beam UV–VIS spectrophotometer. The PBS buffer solution is taken as standard reference and the cumulative release of the drug was presented through the graph (Figure 7).

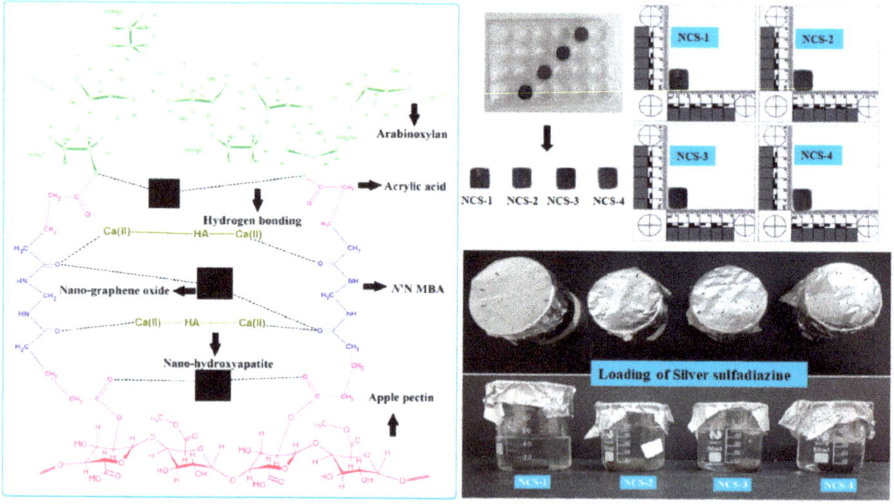

Figure 2. The scheme diagram presenting a proposed chemical reaction among the arabinoxylan (biopolymer), apple pectin (biopolymer), nano-hydroxyapatite (ceramic), nano-graphene, acrylic acid (monomer), and *N,N′*-methylene-bis-acrylamide (crosslinker) for the synthesis of bioactive nanocomposite scaffolds through free radical polymerization method. These nanocomposites were freeze-dried to obtain porous scaffolds for bone tissue engineering and silver-sulfadiazine was loaded simple dipping technique.

3. Characterizations

3.1. Fourier Transformation Infrared (FTIR)

The functional group analysis of nanocomposite scaffolds was analyzed through Fourier transformation infrared (Shimadzu FTIR-8100A, Tokyo, Japan) with the range of 4000–400 cm^{-1} and 96 scans/s.

3.2. Scanning Electron Microscopy (SEM)

The surface morphological behavior of a well-dried and gold-sputtered nanocomposite scaffold was determined via scanning electron microscopy (JEOL-JSM 6480, Peabody, MA, USA). The porosity and pore sizes of nanocomposite scaffolds were analyzed using Brunauer–Emmett-Teller (Micromeritics Gemini II 2370, Norcross, GA, USA).

3.3. Water Contact-Angle

The homogenized slurry of the corresponding nanocomposite was placed into Petri dishes and allow drying in the oven at 45 °C to measure the accurate contact angle. The wetting behavior of nanocomposite scaffolds was determined through contact angle measurement using contact angle meter (XCA-50) (VCA-Optima, AST Inc., Tacoma, WA, USA) and snaps were captured at different time intervals (1 and 10 min) after dropping water drop (4 μL) over the surface of nanocomposite film and tests were performed in triplicates.

3.4. Mechanical Testing

The mechanical testing strength of dried nanocomposite scaffolds was determined by a universal testing machine (ASTM D638) [22], (Testometrics, UK). The mechanical testing was performed at 10 mm/min speed and the dimension of all scaffolds (height 1.7 cm, diameter 1.5 cm) was identical. The testing was performed in triplicate.

3.5. Biodegradation

The nanocomposite scaffolds were cut into 1 × 1 cm^2 pieces, the weight was measured, and they were incubated in the PBS solution (pH 7.4) at 37 °C for 35 days to observe the degradation rate. These nanocomposite scaffolds recovered from the PBS solution were rinsed with deionized water and dried in the oven for 1 h at 55 °C. The percentage of biodegradation was determined by following Equation (1).

$$\text{Biodegradation } (\%) = \frac{W_o - W_t}{W_o} \times 100, \quad (1)$$

whereas: W_t = weight at a specific time, W_o = initial weight

3.6. Swelling Analysis

The swelling behavior of nanocomposite scaffolds was determined by PBS solution and deionized water at different temperatures (35 and 37 °C). All scaffold samples were dried well and weighed before analysis. After a specific interval of time, the samples were taken out and surface water was removed carefully using tissue paper to record gained weight. The percentage swelling of the nanocomposite scaffolds was determined using Equation (2).

$$\text{Swelling } (\%) = \frac{W_f - W_i}{W_i} \times 100, \quad (2)$$

whereas: W_f = final wet weight and W_i = initial dry weight.

3.7. Release of Silver-Sulphadiazine from Nanocomposite Scaffold

The silver release profile was studied through atomic absorption spectrophotometer (PerkinElmer, Inc., Waltham, MA, USA). The nanocomposite scaffolds were soaked into deionized water under standard in-vitro conditions (37 °C and 5% CO_2). The cumulative release of silver was measured after every 20 min to analyze releasing.

3.8. In-Vitro Biological Activities

3.8.1. Cell Viability and Optical Density

Different concentrations of biomaterials (0.50–2.00 mg/mL) were prepared to study the cell viability. The *MC3T3-E1* cell-lines were cultured by taking 0.1% gelatin (+ive control) and incubated under standard in-vitro conditions for different time intervals (24, 48, and 72 h). A well-reported method by Repetto et al. [23] was used to treat cells by incubating into the neutral red medium (40 μg/mL) for 2 h and washed PBS solution for another 2 h to remove the unnecessary neutral red stain. The dye-staining solution (50% distilled water, 49% absolute ethanol, and 1% glacial acetic acid) was used at 37 °C for 10 min. The optical density of cultured *MC3T3-E1* cell lines was observed at 570 nm using an absorbance microplate reader (ELx-800) (Bio-Tek, Winooski, VT, USA). The experiment was performed in triplicate and the percentage of cell viability was calculated by Equation (3).

$$\text{Cell viability } (\%) = \frac{OD_S}{OD_C} \times 100, \quad (3)$$

whereas: OD_S = sample concentration, OD_C = positive control

3.8.2. Cell Morphology

The pre-osteoblast cell-lines were maintained in *α-MEM*, 10% FBS, 1% (2 mM) L-glutamine, 1% penicillin/streptomycin. The density of pre-osteoblast cell-lines was assumed to be 5000 cells/cm^2 in a 100 μm cell culture plate (96 well plate). The well-plates were coated using a gelatin solution

(0.1%) as a coating agent. The well-plates were incubated with scaffolds (e.g., NCS-1, NCS-2, NCS-3, and NCS-4) at 37 °C with 5% CO_2 and 90% humidity, which are standard in-vitro conditions.

3.8.3. Cell Culture SEM Morphological Analysis

The *MC3T3-E1* cell was cultured over the nanocomposite scaffolds for 7 days and cell adherence was observed by SEM (JEOL-JSM-6480, Peabody, MA, USA). PBS solution was used to carefully wash the attached cells and at room temperature. The cell was fixed with absolute ethanol for 5 min. These scaffolds samples were dried and gold-sputtered. The SEM analysis was performed at an accelerated voltage of 1 kV under an operating pressure of 7×10^{-2} bar with the deposition current of 20/2.0 mA/min.

3.9. Statistical Analysis

Experimental data was performed in triplicate and presented as mean standard errors (S.E). The statistical software (IBM, SPSS Statistics 21) was used to perform statistical analysis. The means and standard errors of means (mean ± S.E) were analyzed for each analysis shown as *Y*-error bars in figures. ($p < 0.05$; $n = 3$).

4. Results and Discussions

4.1. FTIR

The FTIR spectral analysis has been conducted to determine different functional groups and chemical interactions of nanocomposite materials as shown in Figure 3. The broadband at 3600–3200 cm^{-1} presents the stretching vibrations of the hydroxyl (–OH) group; these vibrations' peaks present the hydrogen bonding. The increasing peak intensity causes more hydrogen bonding due to the increasing amount of GO [24,25]. The stretching vibration at 2930 cm^{-1} is due to a saturated aliphatic C–H functional group. Furthermore, the broadband 3600–3200 cm^{-1} vibration may explain hydrogen bonding AX, AP, and GO [8]. The stretching vibration peaks at 1729, 1643, and 1430 cm^{-1} were attributed to C=O, C=C, and C–H, respectively, and these are basic functional groups of GO [26,27]. The vibration peak at 1056 cm^{-1} is a characteristic peak of saccharine that confirms the presence of polysaccharide (Arabinoxylan and apple pectin) [8,28]. The vibration bands at 1062 and 965 cm^{-1} attributes the triply degenerated P–O stretching and PO_4^{3-} bending of HAp, respectively [8]. The asymmetric bending vibration of P–O was attributed to 703–654 cm^{-1} that resembles HAp materials [26]. The successful synthesis of nanocomposite scaffolds was confirmed by these available spectral functional groups.

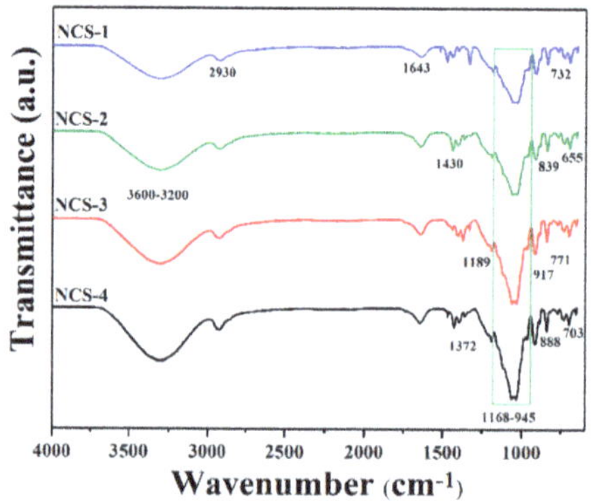

Figure 3. FTIR spectral profile of nanocomposite scaffolds to determine the different functional groups.

4.2. SEM Analysis

Porosity allows the transportation of nutrients, the exchange of gases, cell infiltration, and migration into scaffolds to help growing new blood vessels. The surface morphologies of nanocomposite scaffolds have been shown at different magnifications (Figure 4). The increasing amount of the GO was found to be more interconnected with homogeneous pore distribution. The highly interconnected porous structure of nanocomposite scaffolds is necessary. The morphological properties are important to exchange gases, nutrient permeability wastes disposal, and retain maximum tissue fluid that facilitates the cell growth activities. These nanocomposite scaffolds can be suitable for cell proliferation and migration in bone tissue engineering applications [29,30]. The porous and interconnected morphology of nanocomposite scaffolds also confirms the successful scaffold fabrication and interaction of GO through H-bonding. Therefore, these nanocomposite scaffolds play a key role in the release of growth factor, drug delivery, cell migration for fracture bone regeneration [31]. Therefore, the optimized amount of GO can produce the desired porosity, as an increasing amount of GO regulates porosity and pore size. The addition of GO as filler imparts additional porosity. These nanocomposite scaffolds demonstrated the complete interconnected porosity with foam-like morphology (Figure 4). These characteristics are considered to be very significant for tissue and pore size 50–250 μm was found as best for cell adherence, growth, and proliferation [32,33]. No crack or major defect has been seen over the scaffolds and scaffolds displayed interconnected porosity and pore sizes that indicated strong control over the manufacturing process. The pore size and personage of porosity have been described in Table 1. The successful fabrication of porous nanocomposite scaffolds plays a crucial role in the fabrication of porous hybrid nanocomposite scaffolds with rough morphology [34].

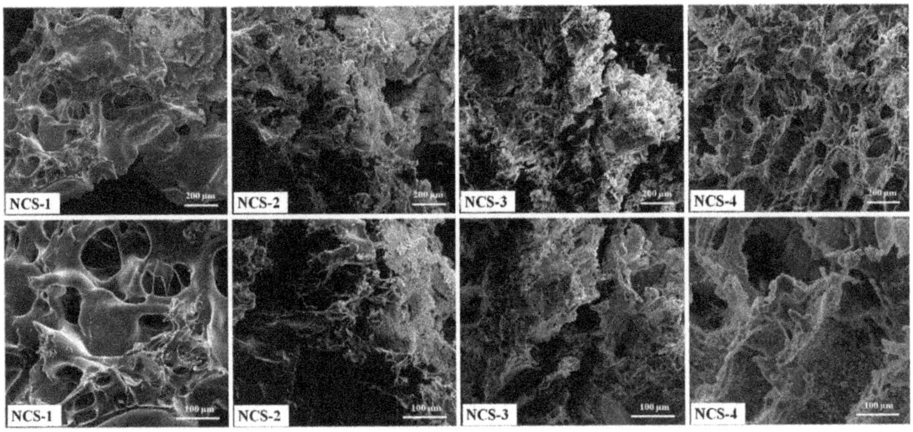

Figure 4. Presents the porous morphologies of the nanocomposite scaffolds at different magnifications.

Table 1. Present the mechanical properties of nanocomposite scaffolds.

Scaffold	Strain (%)	Stress (MPa)	Young's Modulus (MPa)	Porosity (%)	Pore Area (μm^2)
NCS-1	57.01	16.90	29.65	63.72 ± 1.9	0.64×10^3
NCS-2	49.80	12.9	25.90	56.23 ± 3.4	0.55×10^3
NCS-3	36.71	8.20	22.34	50.94 ± 8.3	0.46×10^3
NCS-4	30.92	4.1	13.27	45.75 ± 6.7	0.38×10^3

4.3. Water Contact-Angle

The wetting behavior of nanocomposite scaffolds was determined using water contact angles after a different interval of time (1 and 5 min) as shown in Figure 5. The sample NCS-4 was found to be more hydrophilic as compared to NCS-1, due to the increasing amount of GO, which shifts the wetting behavior from hydrophobicity toward hydrophilicity. The increasing time of water contact to the surface also increases the hydrophilicity. Hence, an increasing amount of GO in the nanocomposite scaffolds can lower the water contact angle and shift the wetting behavior from hydrophobic to hydrophilic. More hydrophilic character is an indication of hydrogen bonding that is fundamental for cell adhesion and cell proliferation [35,36]. It is also assumed that the increasing quantity of GO induced the nanocomposite scaffold increases the hydrogen bonding and the high biological activity was observed NCS-4.

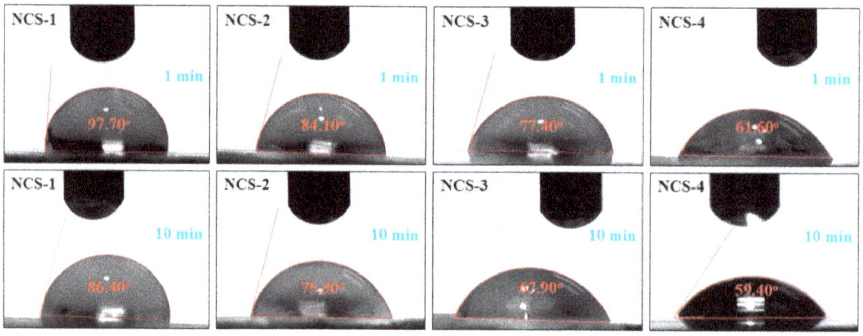

Figure 5. Wetting behavior of the nanocomposite scaffolds at different time intervals.

4.4. Mechanical Testing

The mechanical behavior of nanocomposite scaffolds was analyzed through the strain-stress curve (Figure 6A). The mechanical performance of the scaffold is an important character to support the fractured bone and helps the regeneration of defected bone by enhancing cellular physiology [37]. The mechanical strength of trabecular bone ranging from 1 to 10 MPa as reported by different groups [38,39]. The mechanical strength of the composite scaffold should mimic the mechanical strength of the natural bone. A detailed mechanical behavior (compressive strength, strain %, and Young's modulus) of all nanocomposite scaffolds have been summarized in the Table 1. The increasing amount of reinforcement (GO) into the polymeric composite increases the strength and modulus nanocomposite scaffolds. Therefore, all nanocomposite scaffolds began to be reinforced progressively as the amount of reinforcement agents increased, which could be attributed to the enhancement of stress transfer from NCS-1 scaffold to NCS-4 scaffold [40].

The value of Young's modulus is different for all scaffold samples due to the variable amount of GO. Collectively, the incorporation of GO on the mechanical properties of scaffolds imparted chemical interaction (as mentioned in Figure 2 (Left side)) that increased the mechanical strength of the nanocomposite scaffolds. It was worth noting, however, that HAp and GO were homogeneously mixed with a polymeric solution; this implies that HAp and GO were uniformly hybridized with the polymer phase at the molecular stage [41]. GO may also generate additional physical crosslinking due to possible H-bonding among polymeric matrix and nHAp, and can serve as an efficient physical cross-linking agent. This might have contributed to microstructural improvements, and provided a close packaging to strengthen the nanocomposite scaffolds. Nonetheless, such nanocomposite scaffolds can be a potential biomaterial and might provide a similar mechanical strength as of human trabecular bone [38]. Henceforth, we could conclude that the mechanical properties of the porous composite scaffolds were optimal and can be applied in tissue engineering to treat bones with different mechanical strengths.

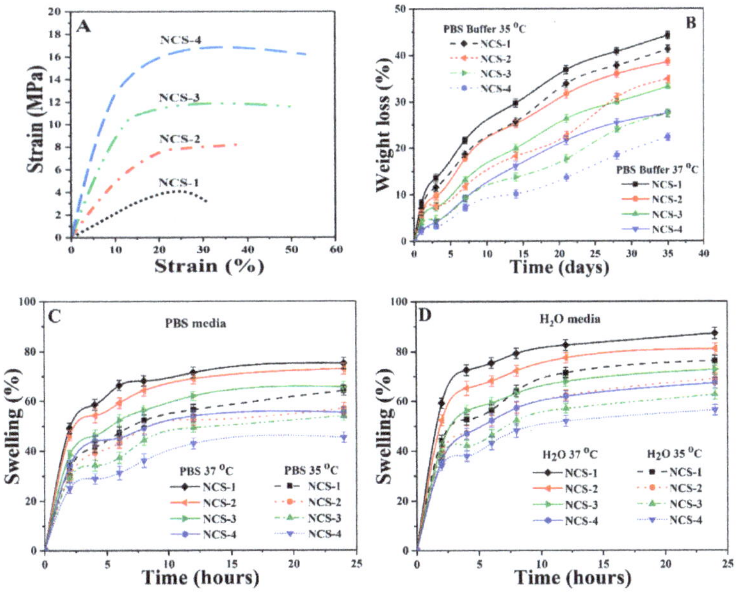

Figure 6. Presents (**A**) mechanical behavior of the nanocomposite scaffolds at room temperature, (**B**) the biodegradation under standard in-vitro conditions at different temperatures, and (**C**,**D**) the swelling of nanocomposite scaffolds at different temperatures and in different media.

4.5. In-Vitro Biodegradation

An ideal biomaterial for bone tissue corresponds to mechanical strength for host bone tissue with controllable degradation and resorption rate that creates space to facilitate bone growth [42]. The biodegradation of nanocomposite scaffolds was conducted under standard in-vitro condition using PBS solution for 35 days (Figure 6B). The nanocomposite scaffolds exhibited different degradation behavior due to different incorporated amounts of GO, and the mass loss profile was found to be linear as the time increases. The sample NCS-1 demonstrated maximum degradation at elevated temperature and NCS-4 presented the least degradation after 30 days. The degradation behavior of the nanocomposite scaffolds is dependent upon the hydrophobic behavior of polymeric matrix and crystallinity characteristics due to nHAp. The degradation of NCS-1 for 30 days was considerably greater than NCS-4, which confirms that the lower GO content leads to a faster scaffold degradation. Since GO has several oxygen-based functional groups that facilitated more H-bonding, and therefore the behavior of GO could be considered as a crosslinker. Therefore, we can conclude that GO content controlled the degradation of the scaffolds by improving matrix properties [43,44]. Biodegradation of polymeric-based scaffold or implants is an important issue in tissue engineering. The quick in-vivo degradation was reported due to the strong autocatalytic effect that produced acidic degradation products. These products are accumulated by the medium surrounding the implants. By changing frequent PBS media for in-vitro degradation can minimize the autocatalytic effect and the absence of certain enzymes helps to estimate in-vivo degradation behavior [45]. Monireh and colleagues have reported 10–12 weeks as the normal period for bone tissue integration [46]. The spinal fusion takes 9 months and cranio-maxillofacial takes 3–6 months as reported by Susmita and coworkers [1]. Therefore, the scaffolds with different degradation rates can be applied to different defective bone sites. Hereafter, we conclude that bone degradation can be controlled by regulating the amount of GO, which corresponds to the rate of formation of new bone tissue. After implantation of a porous scaffold, scaffold material should support the host bone mechanically by retaining structural integrity on degradation into a complex in-vivo environment.

4.6. Swelling Analysis

The swelling behavior of well-dried nanocomposite scaffolds was performed in phosphate-buffered saline (PBS) and aqueous media at room temperature (Figure 6C,D). An increase in fast swelling was found at the beginning of 5–7 h, then became stable as it attained an equilibrium state after 12 h. It was observed that an increasing amount of GO caused the swelling due to its water-loving nature, which facilitated H-bonding with water molecules. Swelling is an important property of scaffolds that helps to regulate the exchange of gases and waste, and nutrients transported during metabolism [47]. The interconnected porosity and pore size increased, and surface area were observed to be increased during swelling, which facilitated cell adhesion, proliferation, and migration to form new tissues. The optimized swelling plays a vital role to enhance cell performance by controlled degradation of scaffolds, which are essential for bone regeneration [48]. Swelling phenomena is different at different temperatures (35 and 37 °C) and in different media (PBS solution and deionized water) due to the polymeric matrix. It was found that the maximum degree of hydration was found in water media at 37 °C for NCS-4. Hence, an increasing amount of GO increases swelling at all temperatures and media [8]. Therefore, swelling of nanocomposite scaffolds confirmed the successful crosslinking and role of GO.

4.7. Release of Silver Nanoparticles from Nanocomposite Scaffolds

The release of silver sulfadiazine under in-vitro conditions has been presented in the PBS (Figure 6.). Silver-based nanocomposite scaffolds are significant in biomaterials, which release the silver over time to clear the microorganisms at the implant site. The sustained and prolonged release of silver is effective for antimicrobial activity [29]. The gradual release profile of silver from nanocomposite scaffolds fulfills the antimicrobial behavior of nanocomposite scaffolds (Figure 7). The amount of silver release was directly affected by the initial amount of silver doped inside the scaffolds. The linear release profile of silver-sulfadiazine was observed as ~38.8% after 60 min and ~90.6% after 260 min, and then the release of silver-sulfadiazine was reduced, whereas the release of silver-sulfadiazine from NCS-1 was observed to be ~11.3% after 60 min and ~47.5% after 260 min, and later it became continuous. It is also observed that the release of silver-sulfadiazine becomes reduce release to continuous release after 260 min from NCS-4 to NCS-1. The different silver-sulfadiazine release profile may be due to different physicochemical behavior of the nanocomposite scaffolds. Since the polymeric matrix of the nanocomposite scaffolds is similar to the extracellular matrix, and it interacted with the silver sulfadiazine drug through hydrogen-bonding or via weak van der Waals forces of attraction [49]. The quick-release of silver-sulfadiazine may be due to increased swelling. It was found that, initially, the release of silver is fast and later it became slow. The more release of silver ion facilitates in combating severe disease-causing pathogens at the initial stage after bone grafts.

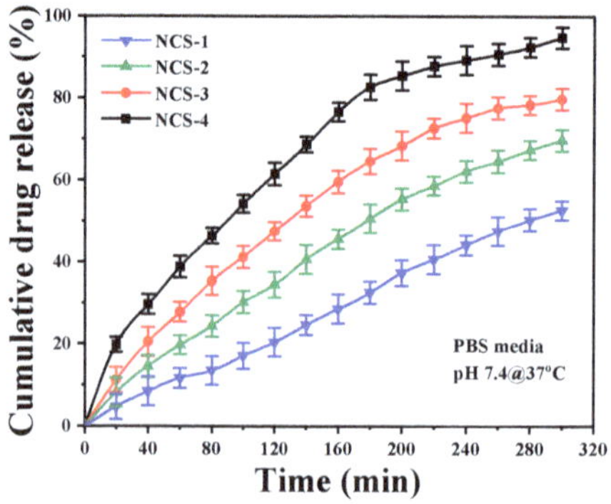

Figure 7. The drug release profile of the nanocomposite scaffold (NCS-4) under the controlled environment and physiological conditions.

4.8. In-Vitro Activities

4.8.1. Cell Viability and Optical Density

In-vitro cell viability assay and optical density (Figure 8A,B) of nanocomposite scaffolds were studied against *MC3T3-E1* cell lines. Different concentrations of nanocomposite scaffolds were prepared carefully (0.500, 1.000, 1.500, and 2.000 µm/mL) and viability was recorded at different intervals (24, 48, and 72 h) after incubation under standard in-vitro conditions [50]. Maximum cell viability and cell proliferation were recorded at 2 µg/mL for NCS-4 as it contains more GO contents. GO imparts important physicochemical properties that facilitated cellular compatibility. However, increasing extract concentration (2 µg/mL) was found to be more favorable for cell viability and optical density. The developed nanocomposite scaffolds displayed their support for cell viability and increased cell proliferation without causing any toxicity towards pre-osteoblast cells (Figure 8).

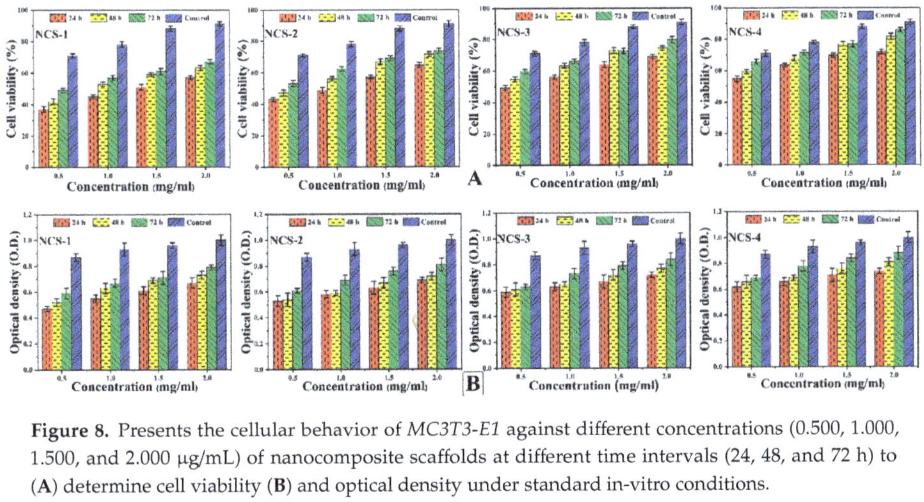

Figure 8. Presents the cellular behavior of *MC3T3-E1* against different concentrations (0.500, 1.000, 1.500, and 2.000 µg/mL) of nanocomposite scaffolds at different time intervals (24, 48, and 72 h) to (**A**) determine cell viability (**B**) and optical density under standard in-vitro conditions.

4.8.2. Cell Morphology

The cellular morphology studies of *MC3T3-E1* have been performed against nanocomposite scaffolds to determine their behavior towards cell growth and morphology (Figure 9). Different factors can influence cell differentiation by increasing biocompatibility over the surface of the substrate [51]. GO contains oxygen-based functional groups (–COOH, $-OSO_3$, –H, and –OH) and these functional groups are essential for H-bonding. An increasing amount of GO enhances surface properties and functionalities that encourage cell adherence, differentiation, and growth [52,53]. The surface morphology of NCS-4 is much rougher, better interconnected, and more porous than all nanocomposite scaffolds. The NCS-4, therefore, demonstrated more and clearer cell spreading over time. The cell adhesion can be regulated by integrin binding over a modified surface [54]. Literature reports also support increasing functionalities modulate fibronectin adsorption and integrin-binding in the order, OH > COOH > NH_2 > CH_3 [55]. The substantial change in absorbance was observed at different time intervals (24, 48, and 72 h). NCS-3 and NCS-4 present excellent cylindrical cell morphology as compared to control due to hydrophilic nature. Hydrophilic behavior facilitates communication of substrate; hosts bone DNA that promotes cell adherence and proliferation to encourage bone regeneration. Moreover, after 48 h, the cell spreading is quicker in cylindrical morphology, and, after 72 h, no considerable difference was found among all samples especially NCS-3 and NCS-4. This was due to a high GO content that enhanced their physicochemical characteristics than in the rest of the samples. It can be explained by the hydrophilic surface characteristics, which is the necessary behavior of biomaterials in tissue integration. NCS-4 is more hydrophilic that supported cell adherence through H-bonding that facilitated cell proliferation (Figure 5). It may be possible that increasing incubation time may increase the cell proliferation to the grown new bone that can repair and regenerate the fractured bone [56,57].

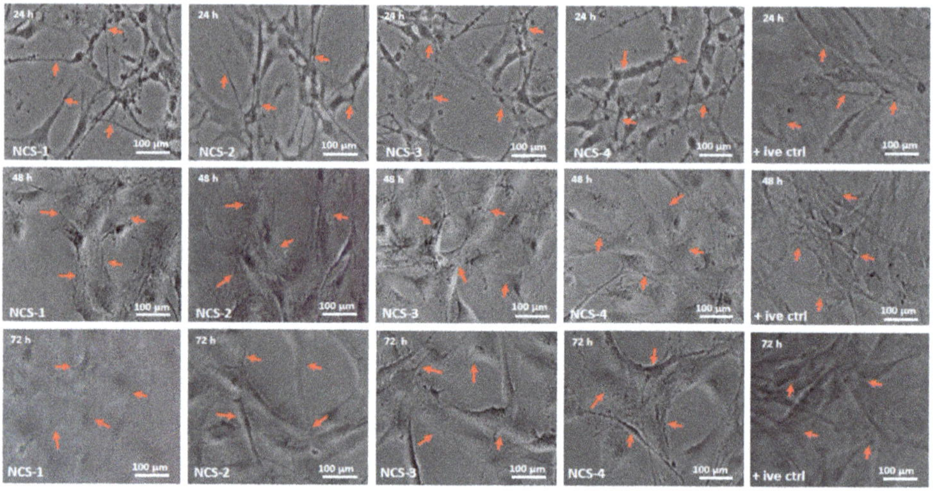

Figure 9. The cell morphology of *MC3T3-E1* against all nanocomposite scaffolds and +ive control under different times (24, 48, and 72 h). The red arrows are presenting the proper adherence and cylindrical shapes of the cells in all scaffold samples (NCS-1, NCS-2, NCS-3, and NCS-4) under standard in-vitro conditions.

4.8.3. SEM Analysis of Cell Culture and Attachment

The Figure 10 presents the morphology of *MC3T3-E1* cell adhered to the surface of all samples of nanocomposite scaffolds after 72 h of cell culture. It was found that cell adherence was increased as the amount of GO increased from NCS-1 to NCS-4. The culture cells spreading was found over the surface in cylindrical or polygonal form and also found in the pores of the scaffolds. Increasing cell culture time covered the surface and penetrated pores after culturing of 72 h, as observed in all scaffolds (presented by red arrows). The increasing cell adherence can be explained by increasing hydrophilic character that facilitated adherence of cells over the surface through hydrogen bonding [58]. It also can be explained by rough morphology obtained due to porosity and owed to GO-sheets that have several oxygen-based functional groups, readily available for H-bonding [59]. Thus, increasing time may increase the gene expression for cell adherence and proliferation for reconstruction and growth of new bone tissue.

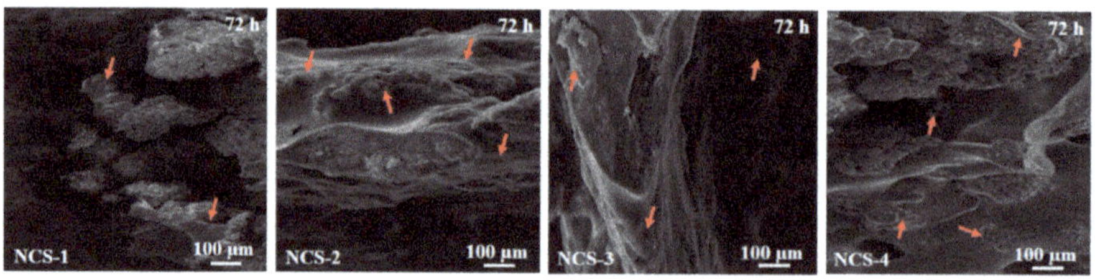

Figure 10. Presents the SEM analysis of cell culture and adherence for all nanocomposite scaffolds against *MC3T3-E1* after 72 h and red arrows are indicating the culture and adherence of osteoblast cells.

5. Conclusions

The novel porous foam-like nanocomposite scaffolds were fabricated through freeze drying from nanocomposite materials and coated with silver-sulphadiazine to deal with antibacterial activities. These nanocomposite materials revealed interesting bone-scaffold regeneration properties porous morphologies, hydrophilicity, and controlled biodegradation with worthy swelling behavior. The physicochemical properties of developed materials make them suitable to treat and repair fractured bone by enhanced osteo-regeneration and found to be biocompatible as depicted by in-vitro studies. The increasing amount of GO has improved physicochemical and biomechanical properties. The scaffolds sample NSC-4 with the maximum amount of GO exhibited a considerable cell adherence with cylinder interconnected cell morphology. The release of the antibacterial drug was found to be sustainable to address bacterial action after implantation. These nanocomposite scaffolds exhibit different mechanical properties that can be applied to repair fractured bone with different mechanical behavior. Hence, it can be concluded that these nanocomposite scaffolds (NSC-4) could serve as a potential biomaterial to repair fractured bone with load-bearing application in bone tissue engineering.

Author Contributions: Conceptualization, M.U.A.K. and S.I.A.R.; Data curation, M.U.A.K. and S.I.A.R.; Formal analysis, M.U.A.K. and S.I.A.R.; Funding acquisition, W.S.A.-A., S.N. and, S.I.A.R.; Investigation, M.U.A.K. and S.I.A.R. and M.R.A.K.; Methodology, M.U.A.K.; Project administration, M.U.A.K. and S.I.A.R.; Resources, S.I.A.R. and M.R.A.K.; Software, M.U.A.K.; Supervision, W.S.-A.A., M.R.A.K. and S.I.A.R.; Validation: M.U.A.K., S.I.A.R., and M.R.A.K.; Visualization, M.U.A.K. and M.R.A.K.; Writing—original draft, M.U.A.K.; Writing—review and editing, M.U.A.K., S.N. and S.I.A.R. All authors have read and agreed to the published version of the manuscript.

Funding: This research received no external funding.

Acknowledgments: The authors extend their appreciation to the Deputyship for Research and Innovation, Ministry of Education in Saudi Arabia, for funding this research work through the project number IFT20169.

Conflicts of Interest: The authors declare no conflict of interest.

References

1. Bose, S.; Roy, M.; Bandyopadhyay, A. Recent advances in bone tissue engineering scaffolds. *Trends Biotechnol.* **2012**, *30*, 546–554. [CrossRef] [PubMed]
2. De Witte, T.-M.; Fratila-Apachitei, L.E.; Zadpoor, A.A.; Peppas, N.A. Bone tissue engineering via growth factor delivery: From scaffolds to complex matrices. *Regen. Biomater.* **2018**, *5*, 197–211. [CrossRef] [PubMed]
3. Haj, J.; Haj Khalil, T.; Falah, M.; Zussman, E.; Srouji, S. An ECM-mimicking, mesenchymal stem cell-embedded hybrid scaffold for bone regeneration. *BioMed. Res. Int.* **2017**, *2017*, 1–12. [CrossRef] [PubMed]
4. Xu, Y.; Xia, D.; Han, J.; Yuan, S.; Lin, H.; Zhao, C. Design and fabrication of porous chitosan scaffolds with tunable structures and mechanical properties. *Carbohydr. Polym.* **2017**, *177*, 210–216. [CrossRef] [PubMed]
5. Sanz, M.; Dahlin, C.; Apatzidou, D.; Artzi, Z.; Bozic, D.; Calciolari, E.; De Bruyn, H.; Dommisch, H.; Donos, N.; Eickholz, P. Biomaterials and regenerative technologies used in bone regeneration in the craniomaxillofacial region: Consensus report of group 2 of the 15th European Workshop on Periodontology on Bone Regeneration. *J. Clin. Periodontol.* **2019**, *46*, 82–91. [CrossRef]
6. Aslam Khan, M.U.; Mehboob, H.; Abd Razak, S.I.; Yahya, M.Y.; Mohd Yusof, A.H.; Ramlee, M.H.; Sahaya Anand, T.J.; Hassan, R.; Aziz, A.; Amin, R. Development of polymeric nanocomposite (xyloglucan-co-methacrylic acid/hydroxyapatite/sio$_2$) scaffold for bone tissue engineering applications—in-vitro antibacterial, cytotoxicity and cell culture evaluation. *Polymers* **2020**, *12*, 1238. [CrossRef]
7. Khan, M.U.A.; Raza, M.A.; Mehboob, H.; Kadir, M.R.A.; Abd Razak, S.I.; Shah, S.A.; Iqbal, M.Z.; Amin, R. Development and in vitro evaluation of κ-carrageenan based polymeric hybrid nanocomposite scaffolds for bone tissue engineering. *RSC Adv.* **2020**, *10*, 40529–40542. [CrossRef]
8. Khan, M.U.A.; Haider, S.; Shah, S.A.; Abd Razak, S.I.; Hassan, S.A.; Kadir, M.R.A.; Haider, A. Arabinoxylan-co-AA/HAp/TiO_2 nanocomposite scaffold a potential material for bone tissue engineering: An in vitro study. *Int. J. Biol. Macromol.* **2020**, *151*, 584–594. [CrossRef]

9. Samura, M.; Hosoyama, T.; Takeuchi, Y.; Ueno, K.; Morikage, N.; Hamano, K. Therapeutic strategies for cell-based neovascularization in critical limb ischemia. *J. Transl. Med.* **2017**, *15*, 49. [CrossRef]
10. Christy, P.N.; Basha, S.K.; Kumari, V.S.; Bashir, A.; Maaza, M.; Kaviyarasu, K.; Arasu, M.V.; Al-Dhabi, N.A.; Ignacimuthu, S. Biopolymeric nanocomposite scaffolds for bone tissue engineering applications—A review. *J. Drug Deliv. Sci. Technol.* **2020**, *55*, 101452. [CrossRef]
11. Shavandi, A.; Silva, T.H.; Bekhit, A.A.; Bekhit, A.E.-D.A. Keratin: Dissolution, extraction and biomedical application. *Biomater. Sci.* **2017**, *5*, 1699–1735. [CrossRef] [PubMed]
12. Aravamudhan, A.; Ramos, D.M.; Nada, A.A.; Kumbar, S.G. Natural polymers: Polysaccharides and their derivatives for biomedical applications. In *Natural and Synthetic Biomedical Polymers*; Elsevier: Amsterdam, The Netherlands, 2014; pp. 67–89.
13. Khan, M.U.A.; Raza, M.A.; Razak, S.I.A.; Abdul Kadir, M.R.; Haider, A.; Shah, S.A.; Mohd Yusof, A.H.; Haider, S.; Shakir, I.; Aftab, S. Novel functional antimicrobial and biocompatible arabinoxylan/guar gum hydrogel for skin wound dressing applications. *J. Tissue Eng. Regen. Med.* **2020**, *14*, 1488–1501. [CrossRef] [PubMed]
14. Noreen, A.; Akram, J.; Rasul, I.; Mansha, A.; Yaqoob, N.; Iqbal, R.; Tabasum, S.; Zuber, M.; Zia, K.M. Pectins functionalized biomaterials; a new viable approach for biomedical applications: A review. *Int. J. Biol. Macromol.* **2017**, *101*, 254–272. [CrossRef] [PubMed]
15. Ninan, N.; Muthiah, M.; Park, I.-K.; Elain, A.; Thomas, S.; Grohens, Y. Pectin/carboxymethyl cellulose/microfibrillated cellulose composite scaffolds for tissue engineering. *Carbohydr. Polym.* **2013**, *98*, 877–885. [CrossRef] [PubMed]
16. Wang, Q.; Gu, Z.; Jamal, S.; Detamore, M.S.; Berkland, C. Hybrid hydroxyapatite nanoparticle colloidal gels are injectable fillers for bone tissue engineering. *Tissue Eng. Part A* **2013**, *19*, 2586–2593. [CrossRef]
17. Qi, C.; Deng, Y.; Xu, L.; Yang, C.; Zhu, Y.; Wang, G.; Wang, Z.; Wang, L. A sericin/graphene oxide composite scaffold as a biomimetic extracellular matrix for structural and functional repair of calvarial bone. *Theranostics* **2020**, *10*, 741. [CrossRef]
18. Ma, R.; Wang, Y.; Qi, H.; Shi, C.; Wei, G.; Xiao, L.; Huang, Z.; Liu, S.; Yu, H.; Teng, C. Nanocomposite sponges of sodium alginate/graphene oxide/polyvinyl alcohol as potential wound dressing: In vitro and in vivo evaluation. *Compos. Part B Eng.* **2019**, *167*, 396–405. [CrossRef]
19. Escudero, M.; Llorente, I.; Pérez-Maceda, B.; San José-Pinilla, S.; Sánchez-López, L.; Lozano, R.; Aguado-Henche, S.; de Arriba, C.C.; Alobera-Gracia, M.; García-Alonso, M. Electrochemically reduced graphene oxide on CoCr biomedical alloy: Characterization, macrophage biocompatibility and hemocompatibility in rats with graphene and graphene oxide. *Mater. Sci. Eng. C* **2020**, *109*, 110522. [CrossRef]
20. Narayanan, K.B.; Kim, H.D.; Han, S.S. Biocompatibility and hemocompatibility of hydrothermally derived reduced graphene oxide using soluble starch as a reducing agent. *Colloids Surf. B Biointerfaces* **2020**, *185*, 110579. [CrossRef]
21. Saghir, S.; Iqbal, M.S.; Hussain, M.A.; Koschella, A.; Heinze, T. Structure characterization and carboxymethylation of arabinoxylan isolated from Ispaghula (Plantago ovata) seed husk. *Carbohydr. Polym.* **2008**, *74*, 309–317. [CrossRef]
22. Standard, A. *D638: Standard Test Method for Tensile Properties of Plastics*; ASTM International: West Conshohocken, PA, USA, 2010.
23. Repetto, G.; Del Peso, A.; Zurita, J.L. Neutral red uptake assay for the estimation of cell viability/cytotoxicity. *Nat. Protoc.* **2008**, *3*, 1125. [CrossRef] [PubMed]
24. Paluszkiewicz, C.; Stodolak, E.; Hasik, M.; Blazewicz, M. FT-IR study of montmorillonite–chitosan nanocomposite materials. *Spectrochim. Acta Part A Mol. Biomol. Spectrosc.* **2011**, *79*, 784–788. [CrossRef] [PubMed]
25. Hutchins, K.M. Functional materials based on molecules with hydrogen-bonding ability: Applications to drug co-crystals and polymer complexes. *R. Soc. Open Sci.* **2018**, *5*, 180564. [CrossRef] [PubMed]
26. Andiappan, M.; Sundaramoorthy, S.; Panda, N.; Meiyazhaban, G.; Winfred, S.B.; Venkataraman, G.; Krishna, P. Electrospun eri silk fibroin scaffold coated with hydroxyapatite for bone tissue engineering applications. *Prog. Biomater.* **2013**, *2*, 6. [CrossRef] [PubMed]

27. Wang, J.; Wang, H.; Wang, Y.; Li, J.; Su, Z.; Wei, G. Alternate layer-by-layer assembly of graphene oxide nanosheets and fibrinogen nanofibers on a silicon substrate for a biomimetic three-dimensional hydroxyapatite scaffold. *J. Mater. Chem. B* **2014**, *2*, 7360–7368. [CrossRef] [PubMed]
28. Khan, M.U.A.; Al-Thebaiti, M.A.; Hashmi, M.U.; Aftab, S.; Abd Razak, S.I.; Abu Hassan, S.; Abdul Kadir, M.R.; Amin, R. Synthesis of silver-coated bioactive nanocomposite scaffolds based on grafted beta-glucan/hydroxyapatite via freeze-drying method: Anti-microbial and biocompatibility evaluation for bone tissue engineering. *Materials* **2020**, *13*, 971. [CrossRef]
29. Caló, E.; Khutoryanskiy, V.V. Biomedical applications of hydrogels: A review of patents and commercial products. *Eur. Polym. J.* **2015**, *65*, 252–267. [CrossRef]
30. Karahaliloglu, Z.; Kilicay, E.; Denkbas, E.B. Antibacterial chitosan/silk sericin 3D porous scaffolds as a wound dressing material. *Artif. Cells Nanomed. Biotechnol.* **2017**, *45*, 1172–1185. [CrossRef]
31. Riahi, N.; Liberelle, B.; Henry, O.; De Crescenzo, G. Impact of RGD amount in dextran-based hydrogels for cell delivery. *Carbohydr. Polym.* **2017**, *161*, 219–227. [CrossRef]
32. Murphy, C.M.; Haugh, M.G.; O'brien, F.J. The effect of mean pore size on cell attachment, proliferation and migration in collagen–glycosaminoglycan scaffolds for bone tissue engineering. *Biomaterials* **2010**, *31*, 461–466. [CrossRef]
33. Mandal, B.B.; Kundu, S.C. Cell proliferation and migration in silk fibroin 3D scaffolds. *Biomaterials* **2009**, *30*, 2956–2965. [CrossRef] [PubMed]
34. Lopez-Heredia, M.A.; Sohier, J.; Gaillard, C.; Quillard, S.; Dorget, M.; Layrolle, P. Rapid prototyped porous titanium coated with calcium phosphate as a scaffold for bone tissue engineering. *Biomaterials* **2008**, *29*, 2608–2615. [CrossRef] [PubMed]
35. Van Wachem, P.; Hogt, A.; Beugeling, T.; Feijen, J.; Bantjes, A.; Detmers, J.; Van Aken, W. Adhesion of cultured human endothelial cells onto methacrylate polymers with varying surface wettability and charge. *Biomaterials* **1987**, *8*, 323–328. [CrossRef]
36. Bauer, S.; Park, J.; von der Mark, K.; Schmuki, P. Improved attachment of mesenchymal stem cells on super-hydrophobic TiO_2 nanotubes. *Acta Biomater.* **2008**, *4*, 1576–1582. [CrossRef] [PubMed]
37. Neira, I.S.; Kolen'ko, Y.V.; Kommareddy, K.P.; Manjubala, I.; Yoshimura, M.; Guitián, F. Reinforcing of a calcium phosphate cement with hydroxyapatite crystals of various morphologies. *ACS Appl. Mater. Interfaces* **2010**, *2*, 3276–3284. [CrossRef]
38. Kim, D.K.; Kim, J.I.; Hwang, T.I.; Sim, B.R.; Khang, G. Bioengineered osteoinductive broussonetia Kazinoki/silk fibroin composite scaffolds for bone tissue regeneration. *ACS Appl. Mater. Interfaces* **2017**, *9*, 1384–1394. [CrossRef] [PubMed]
39. Wang, S.; Yang, Y.; Zhao, Z.; Wang, X.; Mikos, A.G.; Qiu, Z.; Song, T.; Sun, X.; Zhao, L.; Zhang, C. Mineralized collagen-based composite bone materials for cranial bone regeneration in developing sheep. *ACS Biomater. Sci. Eng.* **2017**, *3*, 1092–1099. [CrossRef]
40. Tserki, V.; Matzinos, P.; Kokkou, S.; Panayiotou, C. Novel biodegradable composites based on treated lignocellulosic waste flour as filler. Part I. Surface chemical modification and characterization of waste flour. *Compos. Part A Appl. Sci. Manuf.* **2005**, *36*, 965–974. [CrossRef]
41. Lei, B.; Shin, K.-H.; Noh, D.-Y.; Jo, I.-H.; Koh, Y.-H.; Choi, W.-Y.; Kim, H.-E. Nanofibrous gelatin–silica hybrid scaffolds mimicking the native extracellular matrix (ECM) using thermally induced phase separation. *J. Mater. Chem.* **2012**, *22*, 14133–14140. [CrossRef]
42. Porter, J.R.; Ruckh, T.T.; Popat, K.C. Bone tissue engineering: A review in bone biomimetics and drug delivery strategies. *Biotechnol. Prog.* **2009**, *25*, 1539–1560. [CrossRef]
43. Sivashankari, P.; Moorthi, A.; Abudhahir, K.M.; Prabaharan, M. Preparation and characterization of three-dimensional scaffolds based on hydroxypropyl chitosan-graft-graphene oxide. *Int. J. Biol. Macromol.* **2018**, *110*, 522–530. [CrossRef] [PubMed]
44. Shuai, C.; Li, Y.; Yang, W.; Yu, L.; Yang, Y.; Peng, S.; Feng, P. Graphene oxide induces ester bonds hydrolysis of poly-L-lactic acid scaffold to accelerate degradation. *Int. J. Bioprinting* **2020**, *6*, 249. [CrossRef]
45. Agrawal, C.; McKinney, J.; Lanctot, D.; Athanasiou, K. Effects of fluid flow on the in vitro degradation kinetics of biodegradable scaffolds for tissue engineering. *Biomaterials* **2000**, *21*, 2443–2452. [CrossRef]
46. Kouhi, M.; Prabhakaran, M.P.; Shamanian, M.; Fathi, M.; Morshed, M.; Ramakrishna, S. Electrospun PHBV nanofibers containing HA and bredigite nanoparticles: Fabrication, characterization and evaluation of mechanical properties and bioactivity. *Compos. Sci. Technol.* **2015**, *121*, 115–122. [CrossRef]

47. Hoo, S.P.; Loh, Q.L.; Yue, Z.; Fu, J.; Tan, T.T.; Choong, C.; Chan, P.P. Preparation of a soft and interconnected macroporous hydroxypropyl cellulose methacrylate scaffold for adipose tissue engineering. *J. Mater. Chem. B* **2013**, *1*, 3107–3117. [CrossRef]
48. Wu, S.; Duan, B.; Lu, A.; Wang, Y.; Ye, Q.; Zhang, L. Biocompatible chitin/carbon nanotubes composite hydrogels as neuronal growth substrates. *Carbohydr. Polym.* **2017**, *174*, 830–840. [CrossRef]
49. Bibby, D.C.; Davies, N.M.; Tucker, I.G. Mechanisms by which cyclodextrins modify drug release from polymeric drug delivery systems. *Int. J. Pharm.* **2000**, *197*, 1–11. [CrossRef]
50. Allen, R.J.; Waclaw, B. Bacterial growth: A statistical physicist's guide. *Rep. Prog. Phys.* **2018**, *82*, 016601. [CrossRef]
51. Bet, M.; Goissis, G.; Vargas, S.; Selistre-de-Araujo, H. Cell adhesion and cytotoxicity studies over polyanionic collagen surfaces with variable negative charge and wettability. *Biomaterials* **2003**, *24*, 131–137. [CrossRef]
52. Dadsetan, M.; Pumberger, M.; Casper, M.E.; Shogren, K.; Giuliani, M.; Ruesink, T.; Hefferan, T.E.; Currier, B.L.; Yaszemski, M.J. The effects of fixed electrical charge on chondrocyte behavior. *Acta Biomater.* **2011**, *7*, 2080–2090. [CrossRef]
53. Makohliso, S.A.; Valentini, R.F.; Aebischer, P. Magnitude and polarity of a fluoroethylene propylene electret substrate charge influences neurite outgrowth in vitro. *J. Biomed. Mater. Res.* **1993**, *27*, 1075–1085. [CrossRef] [PubMed]
54. Tang, L.; Thevenot, P.; Hu, W. Surface chemistry influences implant biocompatibility. *Curr. Top. Med. Chem.* **2008**, *8*, 270–280. [CrossRef] [PubMed]
55. Keselowsky, B.G.; Collard, D.M.; García, A.J. Integrin binding specificity regulates biomaterial surface chemistry effects on cell differentiation. *Proc. Natl. Acad. Sci. USA* **2005**, *102*, 5953–5957. [CrossRef] [PubMed]
56. Teymouri, S.; Calejo, M.T.; Hiltunen, M.; Sorkio, A.; Juuti-Uusitalo, K.; Skottman, H.; Kellomäki, M. Collagen-immobilized polyimide membranes for retinal pigment epithelial cell adherence and proliferation. *Cogent Chem.* **2017**, *3*, 1292593. [CrossRef]
57. Fan, Y.; Cui, F.; Chen, L.; Zhai, Y.; Xu, Q. Improvement of neural cell adherence to silicon surface by hydroxyl ion implantation. *Surf. Coat. Technol.* **2000**, *131*, 355–359. [CrossRef]
58. Lumetti, S.; Manfredi, E.; Ferraris, S.; Spriano, S.; Passeri, G.; Ghiacci, G.; Macaluso, G.; Galli, C. The response of osteoblastic MC3T3-E1 cells to micro-and nano-textured, hydrophilic and bioactive titanium surfaces. *J. Mater. Sci. Mater. Med.* **2016**, *27*, 68. [CrossRef]
59. Wang, P.; Yu, T.; Lv, Q.; Li, S.; Ma, X.; Yang, G.; Xu, D.; Liu, X.; Wang, G.; Chen, Z. Fabrication of hydroxyapatite/hydrophilic graphene composites and their modulation to cell behavior toward bone reconstruction engineering. *Colloids Surf. B Biointerfaces* **2019**, *173*, 512–520. [CrossRef]

coatings

Article

Characterization of Hydroxyapatite/Chitosan Composite Coating Obtained from Crab Shells on Low-Modulus Ti–25Nb–8Sn Alloy through Hydrothermal Treatment

Hsueh-Chuan Hsu [1], Shih-Ching Wu [1], Chien-Yu Lin [2] and Wen-Fu Ho [2,*]

1 Department of Dental Technology and Materials Science, Central Taiwan University of Science and Technology, Taichung 40601, Taiwan
2 Department of Chemical and Materials Engineering, National University of Kaohsiung, Kaohsiung 81148, Taiwan
* Correspondence: fujii@nuk.edu.tw

Abstract: In this study, hydroxyapatite/chitosan (HA/CS) composite coatings were prepared by hydrothermal treatment on the surface of low-modulus Ti–25Nb–8Sn alloy to improve the surface bioactivity of the alloy. HA, the main mineral composition of the human skeleton, has excellent bioactivity and is often used as a surface coating on biometal implants. CS, a natural polymer with good antibacterial, hydrophilic and non-toxic characteristics, is often used as dermal regeneration templates, hemostatic agents and drug delivery systems. In this experiment, a natural crab shell was used as a raw material to prepare the HA/CS composite coating by alkali treatment and hydrothermal reaction at various temperatures. The microstructure, morphology and phase composition of the coating surfaces were analyzed by XRD, SEM, and FTIR, and the sample coated with HA/CS was soaked in simulated body fluid (SBF) to evaluate its bioactivity. The experimental results showed that the HA/CS composite coatings through hydrothermal treatment at various temperatures can be successfully fabricated on the surface of the Ti alloy. HA on the coating surface exhibited mainly spherical particles and contained A- and B-type carbonate. When the hydrothermal temperature was up to 200 °C, the spherical particles were approximately 20–40 nm. An ultrasonic vibration test was used to evaluate the adhesion of the coatings, showing that the CS exhibited significantly improved adhesion capacity to the substrate. After being soaked in SBF for 7 days, apatite was deposited on the entire surfaces of the HA/CS coatings, indicating that the coating possesses excellent bioactivity.

Keywords: bioactivity; chitosan; composite coating; hydrothermal treatment; hydroxyapatite; titanium alloy

Citation: Hsu, H.-C.; Wu, S.-C.; Lin, C.-Y.; Ho, W.-F. Characterization of Hydroxyapatite/Chitosan Composite Coating Obtained from Crab Shells on Low-Modulus Ti–25Nb–8Sn Alloy through Hydrothermal Treatment. *Coatings* **2023**, *13*, 228. https://doi.org/10.3390/coatings13020228

Academic Editor: Liviu Duta

Received: 30 December 2022
Revised: 15 January 2023
Accepted: 17 January 2023
Published: 18 January 2023

1. Introduction

Because titanium (Ti) alloys have good biocompatibility, appropriate mechanical properties, and excellent corrosion resistance, they are widely used as a biomedical implants [1–4]. Ti has excellent biocompatibility because it readily reacts with oxygen in the air to form a dense oxide layer (TiO_2) on the surface. Additionally, the oxide layer can retard the release of metal ions, reducing the harm to humans [2]. Although Ti alloys exhibit many excellent properties, they are inherently bioinert and do not bond well to bone. In order to ameliorate this problem, bioactive coatings, such as hydroxyapatite (HA), bioglass, and titania, are usually made on Ti surfaces. HA has been widely used as a coating layer [5], and it is the main mineral component of the human skeleton, with excellent biocompatibility, biological activity, osseoconductivity and non-toxicity [6,7]. The adhesion of the coating to the substrate is an important issue that affects the long-term stability of the implant. Due to the mismatch of the thermal expansion coefficient between ceramic coating and metal substrate, residual stress was induced, which causes the coating to peel off easily [8–11].

Some literature indicated that the interfacial bonding can be improved by adding polymers to the coatings, such as chitosan (CS), collagen, gelatin and polylactide [12–15].

CS is a natural polysaccharide polymer, similar in structure to glucosamine and collagen, which can be easily derived from chitin by N-deacetylation with hot alkali. CS is currently used as dermal regeneration templates, hemostatic agents and drug delivery systems [16–18]. In addition, CS is widely used in biomedical materials because it is a non-toxic biopolymer and shows highly hydrophilic and good antibacterial properties. The antibacterial activity of CS comes from the electrostatic interaction. The amino group of CS is positively charged and interacts with the negatively charged microbial cell membrane, resulting in the leakage of bacterial intracellular components, thus making CS have good antibacterial properties [19–21].

Recently, HA/CS composites are often used as scaffolds, dressings, and coatings for implants. Jugowiec et al. [11] prepared nano-HA/CS by electrophoretic deposition on the surface of Ti–13Nb–13Zr alloy. The experimental results showed that the composite coating had better adhesion with the Ti alloy substrate and can improve the corrosion resistance of the Ti alloy. Pang et al. [19] fabricated HA/CS composite coatings on 316L stainless steel substrates using cathodic electrophoretic deposition. The CS with inherent binding properties plays a vital role in the good bonding between the coatings and the substrates, and the prepared composite coatings provided corrosion protection for the stainless steel substrates. Tang et al. [22] prepared carbonated HA/CS coatings on Ti–6Al–4V substrates using electrophoresis deposition. The carbonated HA/CS coatings exhibited moderately hydrophilic surfaces due to the presence of the CS. They also showed that the composite coating exhibited better cell morphology, adhesion, spreading, and proliferation compared to the carbonated HA coating.

Pure Ti and Ti–6Al–4V alloys have been widely used as biomedical materials, but they exhibit a low wear resistance, a high elastic modulus, and a low shear strength [23,24]. Therefore, many new Ti alloys have been developed in the past few decades to overcome these problems. A new Ti–25Nb–8Sn alloy developed by our research group exhibited much lower elastic modulus (52 GPa) than those of Ti–6Al–4V ELI alloy (110 GPa), Co–Cr–Mo alloy (230 GPa) and stainless steel (205 GPa) [25]. When the elastic modulus of the metal implant is much higher than that of the surrounding bone tissue, a stress shielding effect is generated, resulting in the absorption of the adjacent bone tissue and the loosening of the implant [26]. The main purpose of this study is to modify the surface of Ti–25Nb–8Sn alloy by preparing a bioactive coating. In this work, crab shells were used as raw materials to prepare the composite coating of HA/CS on the Ti–25Nb–8Sn alloy surface by hydrothermal method. A general approach is to prepare the raw materials of the coating by adding CS and HA separately, but the present experiment simultaneously synthesized HA and CS directly from the crab shell. Additionally, the HA and CS are obtained from crab shells in this work, which is different from the use of chemical agents in most literature. The hypothesis of this study is that CS can act as a bonding layer between the substrate and the composite coating. This study investigated the effects of hydrothermal temperatures on the features of the HA/CS coatings, including their morphology, microstructure, and phase composition. Additionally, the adhesion between the coating and the substrate was assessed, and in vitro bioactivity assay was evaluated. The bioactivity referred to in this study means that the material is capable of forming hydroxyl apatite minerals on its surface in artificially simulated body fluid (SBF) [27].

2. Materials and Methods

The Ti–25Nb–8Sn alloy used in this experiment was made of Ti sheet (99.7% in purity, thickness 0.5 mm), niobium wire (99.95% in purity, diameter 0.5 mm) and tin wire (99.95% in purity, diameter 1 mm) by using arc melting and a vacuum pressurized casting system under argon atmosphere. All the metals were purchased from Ultimate Materials Technology Co., Ltd., Hsinchu, Taiwan. High-purity argon (99.999%) was used as a protective atmosphere and adjusted to the appropriate pressure. The oxide layers on the ingots were removed using an electric dental handpiece (UM500SBT, NSK, Tokyo, Japan), and then these specimens were washed in C_2H_5OH by ultrasonic cleaning for 10 min. Melting was performed five times in a copper crucible to achieve chemical homogeneity. After the sixth melting step, the molten alloy was quickly poured into a graphite mold placed in the casting chamber. The detailed casting procedure was described previously [28].

Ti–25Nb–8Sn specimens were ground with 100-, 400-, and 600-grit sandpaper. After ultrasonic cleaning for 10 min in C_2H_5OH, the samples were subjected to alkali treatment by placing them in a 5 M NaOH solution at 60 °C for 24 h in a constant temperature water bath. The specimens were then washed twice with distilled water for 5 min each time.

Discarded mud crab shells (genus Scylla) were collected from local restaurants in Taiwan; 2 g of crab shell powder was added to 36.5 wt.% HCl aqueous solution, and then H_3PO_4 was added. Followed by deacetylation at 105 °C for 60 min with NaOH solution, the synthesized product was repeatedly cleaned with deionized water and then filtered by suction. The final product was dried in an oven at 45 °C for 24 h to obtain the HA/CS powder. Afterward, the above-mentioned alkali-treated Ti alloy samples were put into a Teflon jar with HA/CS solution containing 1.5 g of HA/CS powder, 50 mL of deionized water, and 10 mL of CH_3COOH. The lid of the Teflon jar was locked, followed by placing it into an autoclave for a hydrothermal reaction. The hydrothermal temperatures were set to 100, 125, 150, 175, or 200 °C, and the holding time was 24 h. Finally, the HA/CS-coated Ti alloy was washed with deionized water and placed in a 45 °C oven for 24 h.

To assess the effect of CS on the adhesion of the coating to the substrate, a qualitative measure was performed using an ultrasonic vibration test [29]. In this experiment, the HA/CS-coated sample prepared by hydrothermal treatment at 100 °C for 24 h was adopted. An HA-coated sample prepared by the same hydrothermal condition was used as a control. The samples were tested in an ultrasonic oscillator (DC200H, DELTA, New Taipei City, Taiwan) at a power of 200 W for 10 min. The surface morphology was then observed using a field emission scanning electron microscope (FE-SEM; S-4800, Hitachi, Tokyo, Japan) to understand the cohesiveness of the coating. Before and after the ultrasonic vibration test, the weights of the test samples were, respectively, measured and their weight loss was calculated.

In this study, specimens were immersed in SBF to assess their bioactivity. The samples were soaked in SBF solution at 37 °C in pH 7.4 for 1 and 7 days. After their immersion in SBF for various periods, the samples were removed and washed with deionized water, and then dried. The SBF solution has similar ionic strength values to human plasma and is based on the formulation proposed by Kokubo and Takadama [30].

The phase analyses of the surface coatings were analyzed by a high-resolution X-ray diffractometer (D8 Discover, Bruker, Berlin, Germany), using Cu K_α radiation (λ = 1.5405 Å) and operating at 40 kV and 40 mA. The surface morphologies of the samples were observed using a field emission scanning electron microscope (FE-SEM; S-4800, Hitachi, Tokyo, Japan) at a voltage of 10 or 15 kV. The chemical compositions of the coatings were examined using an energy dispersive spectrometer (EDS) at a voltage of 15 kV. The functional groups of the coatings were analyzed using a Fourier transform infrared spectrometer (FTIR; Cary 630, Agilent, Santa Clara, CA, USA).

3. Results

3.1. Morphology of HA/CS Composite Coatings

The FE-SEM images of the coating surfaces of Ti–25Nb–8Sn alloy prepared by alkali treatment and hydrothermal reaction under various hydrothermal temperatures for 24 h are shown in Figure 1. The low-magnification images of Figure 1a–e showed that the surfaces of all the HA/CS coatings were flat and free of cracks. When the hydrothermal temperatures were 125 and 175 °C, some areas with larger agglomerated HA particles can be observed. From the high-magnification SEM photos of Figure 1f–j, it was clearly observed that the entire surfaces of the composite coatings were covered by spherical HA particles. Below 150 °C, the size of HA particles was approximately 5–10 nm; however, it was increased to 20–40 nm when the reaction temperature rose to 200 °C.

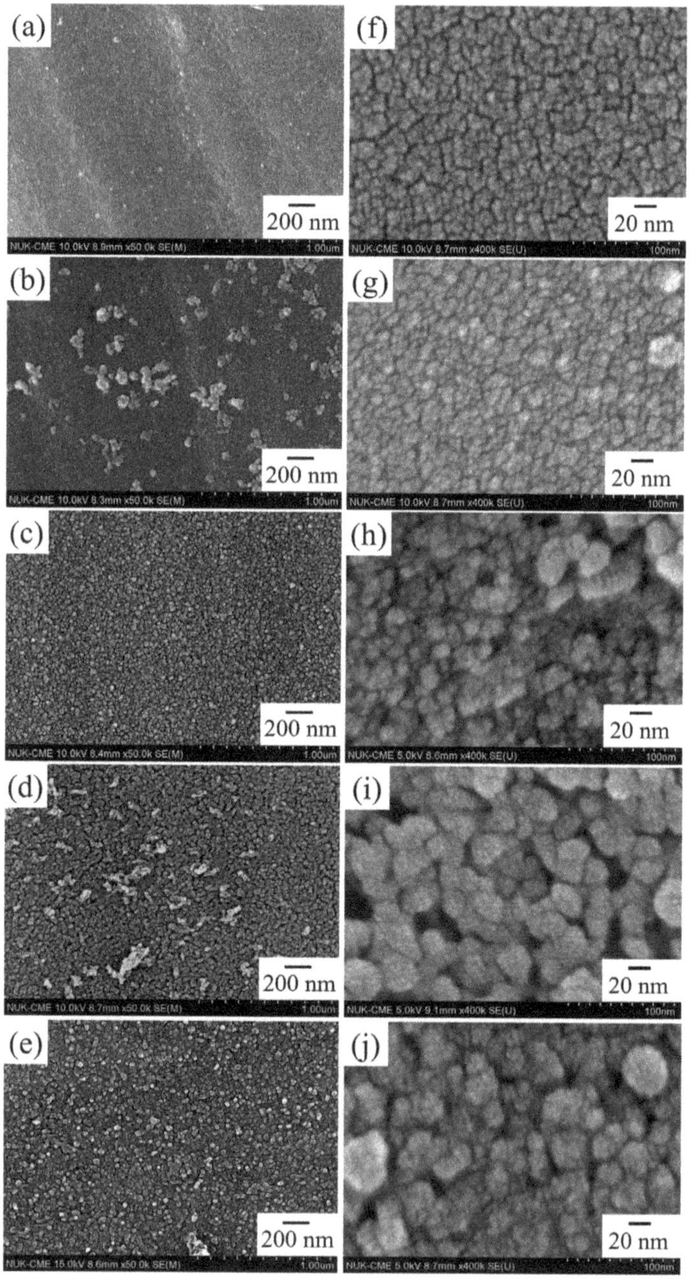

Figure 1. FE-SEM images of the coating surfaces of Ti–25Nb–8Sn alloy prepared by alkali treatment and hydrothermal reaction under various hydrothermal temperatures for 24 h. (**a**,**f**) 100 °C; (**b**,**g**) 125 °C; (**c**,**h**) 150 °C; (**d**,**i**) 175 °C; (**e**,**j**) 200 °C.

Figure 2 is the FE-SEM images of a HA/CS composite coating prepared under the hydrothermal temperature of 200 °C after scratching with a diamond blade. Since the surface coatings of the samples prepared at various hydrothermal temperatures had similar results, post-scratch photos only for the sample taken at 200 °C were provided. It is observed from the FE-SEM image at low magnification (Figure 2a) that the CS was well attached to the surface of the alloy substrate, and the HA layer was located above the CS coating. The high-magnification SEM image of the scratched coating surface (Figure 2b) showed that the CS was draped over the surface of the Ti substrate, which was attributed to the excellent binding and film-forming property of the CS [31].

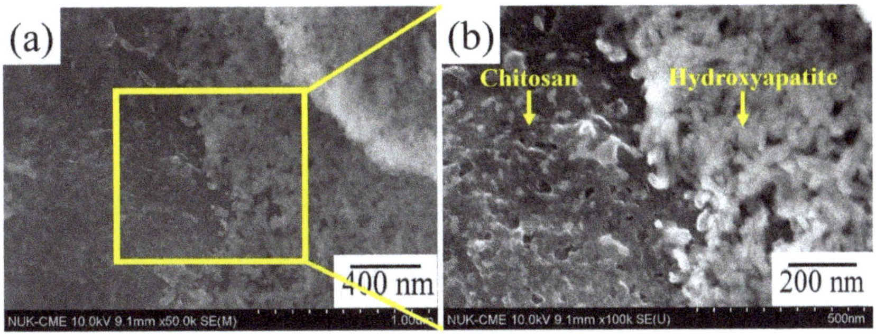

Figure 2. FE-SEM images of a HA/CS composite coating on Ti–25Nb–8Sn alloy prepared under the hydrothermal temperature of 200 °C after scratching with a diamond blade. (**a**) Low magnification image; (**b**) high magnification image.

3.2. Structure of HA/CS Composite Coatings

Figure 3 shows the XRD patterns of HA/CS composite coatings on Ti–25Nb–8Sn alloy after alkali treatment and hydrothermal reaction at different temperatures. The diffraction peaks at 28.1°, 28.94°, 31.8°, 32.92° and 34.77° corresponded to the HA phase. At the hydrothermal temperatures of 175 and 200 °C, the diffraction peaks of sodium titanate and anatase phases became more obvious, mainly formed from alkali treatment [32]. The diffraction peaks of CS were weaker and broader, mainly due to the fact that the CS belongs to an amorphous-crystalline polymer [13,22,33]. The diffraction peaks of HA were also much weaker and broader, which may be related to the thinner HA coating and the nanometer size of the crystals [9,12,13]. Additionally, Xue et al. [34] indicated that the broadening of the diffraction peaks of the HA may be due to the substitution of carbonate ions. In the current study, the HA contained carbonate ions and the result will be discussed later in the FTIR section.

Figure 4 shows the FTIR spectra of HA/CS coatings on Ti–25Nb–8Sn alloy after alkali treatment and hydrothermal reaction at different temperatures. The results showed that the band at 1653 cm^{-1} was for amide I (C=O), the bands at 1552, 1589, 1600, and 1658 cm^{-1} were assigned to amide II (N–H), and the band at 1293 cm^{-1} corresponded to amide III (–CN). The C–O group appeared at 1031 cm^{-1}. All of these bands were characteristic peaks of CS. The main functional groups of the HA are PO_4^{3-} at 665, 974, 1014, 1034 cm^{-1} and OH^- at 1700–1900 and 3600–3900 cm^{-1}. The band located at 1406, 1426 cm^{-1} was attributed to B-type carbonate-containing HA (replacement of PO_4^{3-} groups by CO_3^{2-} groups). On the other hand, A-type carbonate-containing HA (replacement of OH^- groups by CO_3^{2-} groups) was characterized by peaks at 1465 and 1500 cm^{-1} [35].

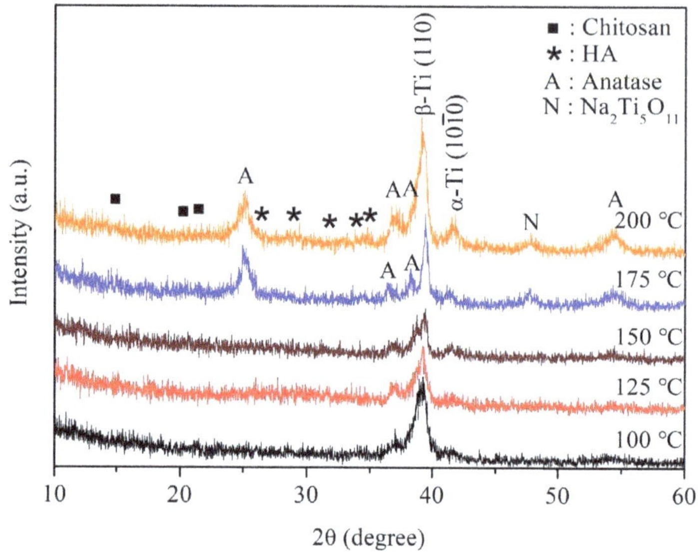

Figure 3. XRD patterns of HA/CS composite coatings on Ti–25Nb–8Sn alloy after alkali treatment and hydrothermal reaction at various temperatures.

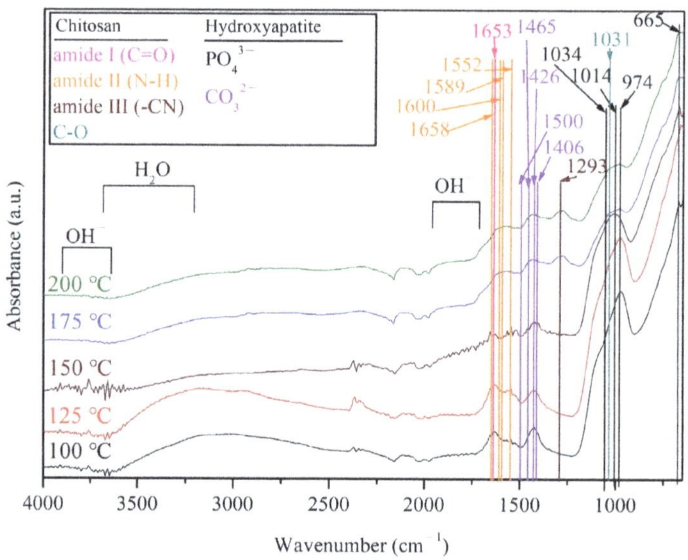

Figure 4. FTIR spectra of HA/CS composite coatings on Ti–25Nb–8Sn alloy after alkali treatment and hydrothermal reaction at various temperatures.

3.3. Adhesion of HA/CS Composite Coatings

Figure 5 is SEM images of the coating surfaces before and after ultrasonic cleaning for 10 min, and both the pure HA coating and the HA/CS composite coating prepared by hydrothermal treatment at 100 °C for 24 h were examined here. The results showed that the HA coating composed of single-phase HA (Figure 5b) was significantly peeled off when compared to the HA/CS composite coating (Figure 5d). Figure 6 shows the weight reductions in both the HA coating and the HA/CS composite coating after ultrasonic vibration for 10 min. The weight loss of the HA coating was about 20%, while that of the HA/CS composite coating was only approximately 0.05%. Additionally, at 125 and 175 °C, the samples had relatively smaller weight loss, which may be related to the larger agglomerated HA particles on the coatings under these temperatures.

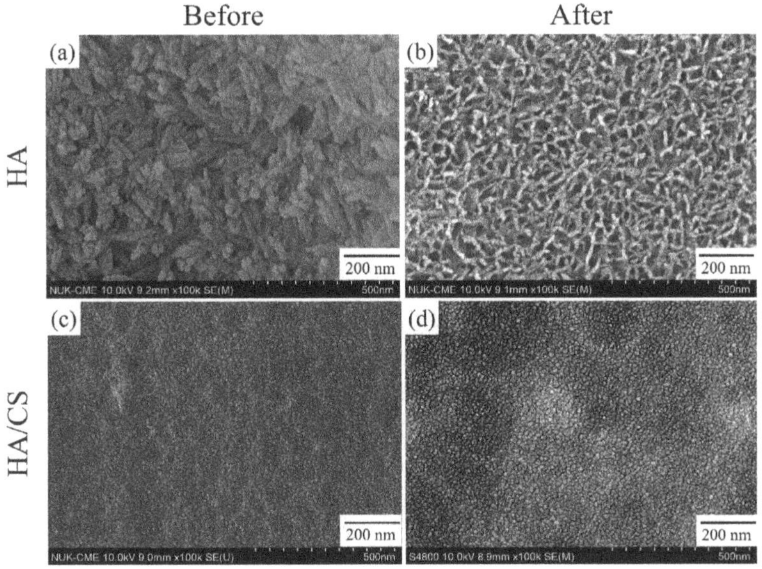

Figure 5. SEM images of coating surfaces before and after ultrasonic cleaning for 10 min, and both pure HA coating and HA/CS composite coating prepared by hydrothermal treatment at 100 °C for 24 h. (**a**) HA coating surface before ultrasonic cleaning; (**b**) HA coating surface after ultrasonic cleaning; (**c**) HA/CS coating surface before ultrasonic cleaning; (**d**) HA/CS coating surface after ultrasonic cleaning.

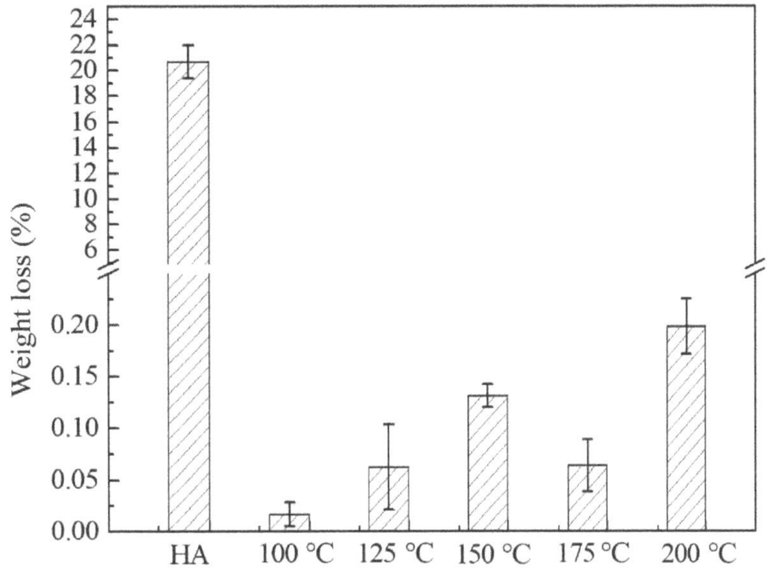

Figure 6. Weight reductions of both pure HA coating and HA/CS composite coating after ultrasonic cleaning for 10 min.

3.4. Characterization of HA/CS Composite Coating

In this study, Kokubo's solution (SBF solution) proposed by Kokubo and Takadama [30] was used for in vitro bioactivity assay. Figure 7 is the FE-SEM photos of the HA/CS coating surfaces, which were subjected to alkali treatment and hydrothermal reaction at different temperatures, and immersed in SBF for 1 day and 7 days. After soaking in SBF for 1 day, all the specimens had spherical apatite deposits on the surfaces, although the original coatings were still observed on the surfaces of the samples prepared by hydrothermal treatment. As the immersion time was prolonged to 7 days, all the specimen surfaces were covered with a large number of apatite precipitates. Thus, the result indicates that the HA/CS composite coatings hydrothermally synthesized in the present study have good bioactivity.

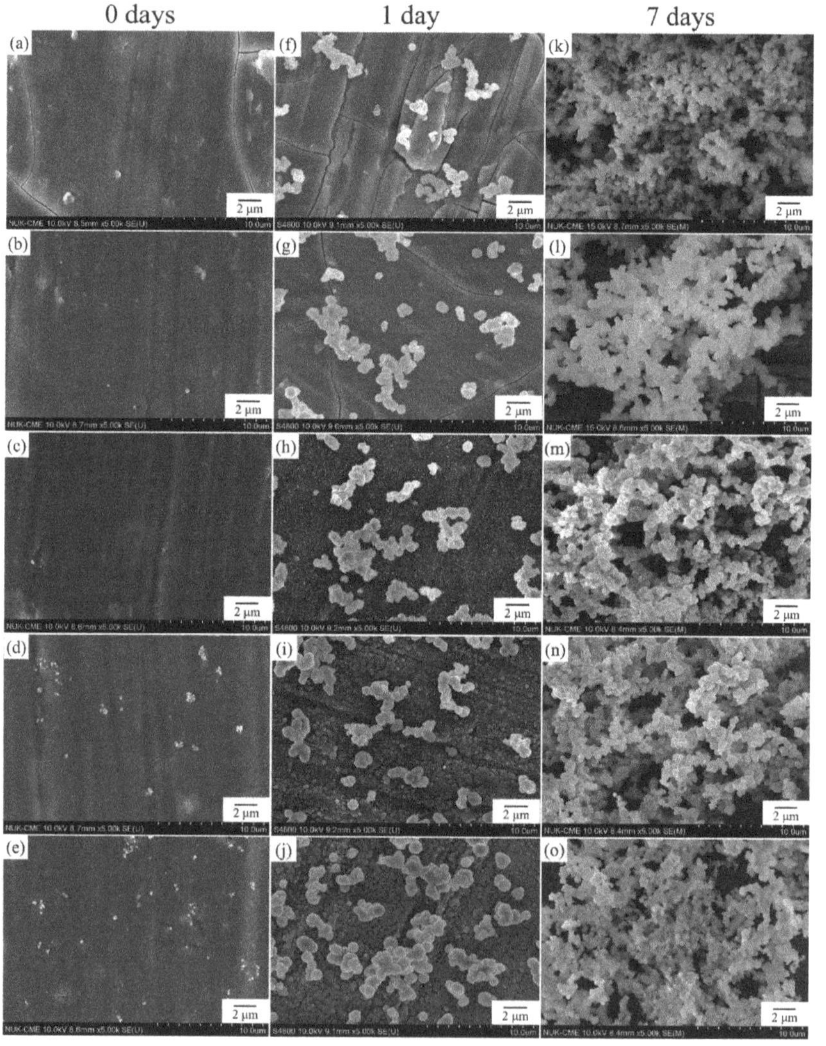

Figure 7. FE-SEM photos of HA/CS coating surfaces subjected to alkali treatment and hydrothermal reaction at various temperatures after immersion in SBF for 1 day and 7 days. (**a**,**f**,**k**) 100 °C; (**b**,**g**,**l**) 125 °C; (**c**,**h**,**m**) 150 °C; (**d**,**i**,**n**) 175 °C; (**e**,**j**,**o**) 200 °C.

Figure 8 is FTIR spectra of HA/CS coatings on the surfaces of Ti–25Nb–8Sn alloy after alkali treatment and hydrothermal reaction at various temperatures, followed by soaking in SBF for 7 days. The major stretching vibration of carbonate groups (CO_3^{2-}) appeared at 1418, 1462, and 1535 cm^{-1}. The samples treated at hydrothermal temperatures of 175 and 200 °C showed a significantly higher intensity of CO_3^{2-} bands located in the range of 2000–2300 cm^{-1}, resulting from more apatite deposition after soaking in SBF.

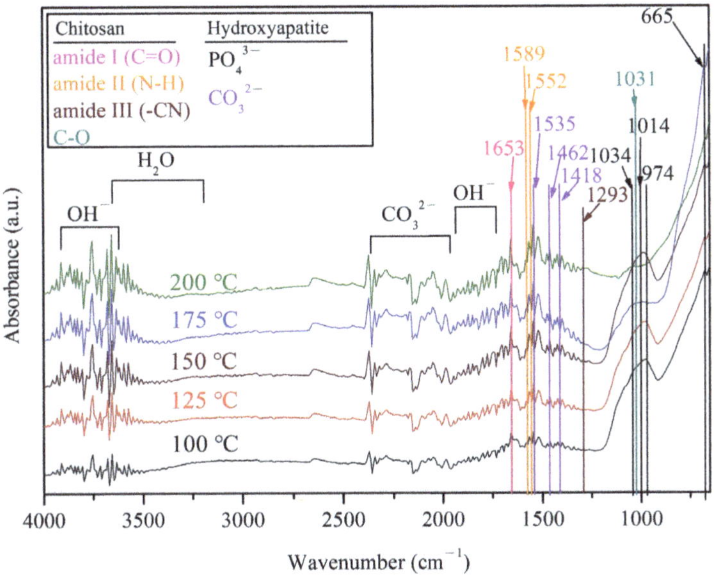

Figure 8. FTIR spectra of HA/CS composite coatings on Ti–25Nb–8Sn alloy after alkali treatment and hydrothermal reaction at various temperatures after soaking in SBF for 7 days.

4. Discussion

As shown in Figure 2, a large number of rod-shaped HA particles dispersed uniformly in the CS matrix. The rod-like HA particles had a diameter of 30–50 nm and a length of 40–60 nm. Pang et al. [19] prepared HA/CS composite coatings on 316L stainless steel substrates by electrophoretic deposition, which had similar morphologies of HA particles and CS layer to that in the present work. When adding 8.0 g/L HA into 0.5 g/L CS solution, the morphology of the HA nanoparticles was observed to be agglomerated with rod-like crystals.

In this study, CS becomes a cationic polyelectrolyte after dissolving CS in CH_3COOH (HAc) through the following equation [36,37]:

$$CS - NH_2 + HAc \rightarrow CS - NH_3^+ + Ac^- \quad (1)$$

Abundant hydroxyl ion (OH^-) groups were formed on the surface of NaOH-treated Ti–25Nb–8Sn through the Na^+ and H_3O^+ ion exchange [38]. The surface OH^- groups induced a neutralization of the positively charged CS macromolecules and the formation of an insoluble CS deposit on the Ti substrate:

$$CS - NH_3^+ + OH^- \rightarrow CS - NH_2 + H_2O \quad (2)$$

Additionally, the HA nanoparticles were positively charged due to the CS absorbed by the HA nanoparticles [19]. Because the velocity of migration of positively charged HA/CS was slower than that of the positively charged CS, the CS layer was precipitated preferentially on the surface of the Ti substrate. It is worth noting that the CS coating has good adhesion to stainless steel, Ti and other substrates [19,31,36].

In this work, AB-type carbonate-containing HA was prepared in the HA/CS coatings (Figure 4). The incorporation of carbonate into HA can cause a change in the crystal structure and result in a decrease in crystallinity and an increase in solubility, which led to an increase in local concentrations of calcium and phosphorus ions that are necessary for new bone formation. Therefore, carbonate-substituted HA has better biological activity [39].

The CS exhibited significantly improved adhesion capacity to the substrate (Figures 5 and 6). Moskalewicz et al. [16] produced the HA/CS composite coating on Ti–6Al–7Nb alloy substrate by electrophoretic deposition. They asserted that the good adhesion of this coating to the substrate is mainly due to the CS matrix, which has a structure similar to that of glycosaminoglycan and can favor the bonding between the coating and the Ti substrate.

Tiyyagura et al. [40] examined the surfaces of CS coating on the porous magnesium monoliths after soaking in SBF. They found that the dominant presence of CO_3^{2-} bands (2000–2300 cm^{-1}) was due to the obvious apatite deposition, which is in agreement with the results in the present study (Figure 8). In addition, the specimens prepared by hydrothermal temperatures of 175 and 200 °C showed a significant low-intensity PO_4^{3-} band at 1000–1100 cm^{-1} (Figure 8), possibly with thicker carbonate layers (B-type carbonate) or amorphous apatite [40]. Therefore, in the present study the samples prepared by hydrothermal reaction at 175 and 200 °C exhibited faster apatite formation capability, i.e., superior bioactivity.

In this study, alkali treatment and hydrothermal reaction were used to modify the surface of Ti alloy with bioactivity, which is characterized by simplicity and industrial applicability. Nevertheless, the thickness control of the coating needs to be further optimized through hydrothermal parameters. In this experiment, crab shells were used to directly synthesize HA and CS, which is different from the method of adding CS and HA, respectively, adopted in most literature [19,22,41]. However, the main limitation of this technique may lie in the degradability of biomass materials, especially CS. Therefore, the adhesion between the coating and the substrate may deteriorate, causing the implant to fail.

Biomaterial surface properties can regulate cellular responses and affect implant osseointegration [42]. It is recommended that further studies include surface energy, topography, wettability, cellular testing, and animal testing.

5. Conclusions

The main purpose of this study is to prepare bioactive coating on the surface of Ti–25Nb–8Sn alloy with low elastic modulus by using crab shells as a raw material. FE-SEM images showed that larger agglomeration of HA particles on the coating surfaces was clearly observed at hydrothermal temperatures of 125 and 175 °C. When the temperature was below 150 °C, the size of the spherical HA particles was about 5–10 nm. As the temperature rose up to 200 °C, the size of the particles became 20–40 nm. HR-XRD analyses showed that sodium titanate, anatase, CS and HA were found on the surface of the Ti alloy. FTIR analyses presented that the main functional groups of CS were amide I (C=O), amide II (N–H) and amide III (–CN). The functional groups of HA were mainly PO_4^{3-}, OH^-, and CO_3^{2-}. Ultrasonic vibration test indicated that the peeling off of the single-phase HA coating was observed obviously. By contrast, the HA/CS composite coating exhibited better adhesion. The SBF immersion test demonstrated that the HA/CS composite coating had good bioactivity. Additionally, the samples prepared by hydrothermal reactions at 175 and 200 °C exhibited faster apatite formation capability.

Author Contributions: Conceptualization, H.-C.H., C.-Y.L. and W.-F.H.; methodology, H.-C.H., S.-C.W., C.-Y.L. and W.-F.H.; validation, H.-C.H. and W.-F.H.; formal analysis, C.-Y.L. and W.-F.H.; investigation, C.-Y.L. and W.-F.H.; resources, H.-C.H., S.-C.W. and W.-F.H.; data curation, H.-C.H., S.-C.W. and C.-Y.L.; writing and original draft preparation, C.-Y.L. and W.-F.H.; supervision and writing—review and editing, H.-C.H. and W.-F.H. All authors have read and agreed to the published version of the manuscript.

Funding: This research was funded partially by National University of Kaohsiung.

Institutional Review Board Statement: Not applicable.

Informed Consent Statement: Not applicable.

Data Availability Statement: Not applicable.

Acknowledgments: The authors acknowledge the partial financial support of National University of Kaohsiung.

Conflicts of Interest: The authors declare no conflict of interest.

References

1. Arias-González, F.; Rodríguez-Contreras, A.; Punset, M.; Manero, J.M.; Barro, Ó.; Fernández-Arias, M.; Lusquiños, F.; Gil, J.; Pou, J. Laser-Deposited Beta Type Ti-42Nb Alloy with Anisotropic Mechanical Properties for Pioneering Biomedical Implants with a Very Low Elastic Modulus. *Materials* **2022**, *15*, 7172. [CrossRef] [PubMed]
2. Bocchetta, P.; Chen, L.-Y.; Tardelli, J.D.C.; Reis, A.C.d.; Almeraya-Calderón, F.; Leo, P. Passive Layers and Corrosion Resistance of Biomedical Ti-6Al-4V and β-Ti Alloys. *Coatings* **2021**, *11*, 487. [CrossRef]
3. Asri, R.I.M.; Harun, W.S.W.; Samykano, M.; Lah, N.A.C.; Ghani, S.A.C.; Tarlochan, F.; Raza, M.R. Corrosion and surface modification on biocompatible metals: A review. *Mater. Sci. Eng. C* **2017**, *77*, 1261–1274. [CrossRef] [PubMed]
4. Sarraf, M.; Ghomi, E.R.; Alipour, S.; Ramakrishna, S.; Sukiman, N.L. A state-of-the-art review of the fabrication and characteristics of titanium and its alloys for biomedical applications. *Bio-Des. Manuf.* **2022**, *5*, 371–395. [CrossRef] [PubMed]
5. Schwartz, A.; Kossenko, A.; Zinigrad, M.; Gofer, Y.; Borodianskiy, K.; Sobolev, A. Hydroxyapatite Coating on Ti-6Al-7Nb Alloy by Plasma Electrolytic Oxidation in Salt-Based Electrolyte. *Materials* **2022**, *15*, 7374. [CrossRef]
6. Almulhim, K.S.; Syed, M.R.; Alqahtani, N.; Alamoudi, M.; Khan, M.; Ahmed, S.Z.; Khan, A.S. Bioactive Inorganic Materials for Dental Applications: A Narrative Review. *Materials* **2022**, *15*, 6864. [CrossRef]
7. Harun, W.S.W.; Asri, R.I.M.; Alias, J.; Zulkifli, F.H.; Kadirgama, K.; Ghani, S.A.C.; Shariffuddin, J.H.M. A comprehensive review of hydroxyapatite-based coatings adhesion on metallic biomaterials. *Ceram. Int.* **2018**, *44*, 1250–1268. [CrossRef]
8. Jagadeeshanayaka, N.; Awasthi, S.; Jambagi, S.C.; Srivastava, C. Bioactive surface modifications through thermally sprayed hydroxyapatite composite coatings: A review of selective reinforcements. *Biomater. Sci.* **2022**, *10*, 2484–2523. [CrossRef]
9. He, Y.; Zhang, Y.; Zhang, J.; Jiang, Y.; Zhou, R. Fabrication and characterization of Ti-13Nb-13Zr alloy with radial porous Ti-HA coatings for bone implants. *Mater. Lett.* **2017**, *209*, 543–546. [CrossRef]
10. Rajesh, K.; Ghosh, S.; Islam, A.; Rangaswamy, M.K.; Haldar, S.; Roy, P.; Keshri, A.K.; Lahiri, D. Multilayered porous hydroxyapatite coating on Ti6Al4V implant with enhanced drug delivery and antimicrobial properties. *J. Drug Deliv. Sci. Technol.* **2022**, *70*, 103155. [CrossRef]
11. Jugowiec, D.; Lukaszczyk, A.; Cieniek, L.; Kowalski, K.; Rumian, L.; Pietryga, K.; Kot, M.; Pamula, E.; Moskalewicz, T. Influence of the electrophoretic deposition route on the microstructure and properties of nano-hydroxyapatite/chitosan coatings on the Ti-13Nb-13Zr alloy. *Surf. Coat. Technol.* **2017**, *324*, 64–79. [CrossRef]
12. Atak, B.H.; Buyuk, B.; Huysal, M.; Isik, S.; Senel, M.; Metzger, W.; Cetin, G. Preparation and characterization of amine functional nano-hydroxyapatite/chitosan bionanocomposite for bone tissue engineering applications. *Carbohydr. Polym.* **2017**, *164*, 200–213. [CrossRef] [PubMed]
13. Hahn, B.D.; Park, D.S.; Choi, J.J.; Ryu, J.H.; Yoon, W.H.; Choi, J.H.; Kim, H.E.; Kim, S.G. Aerosol deposition of hydroxyapatite-chitosan composite coatings on biodegradable magnesium alloy. *Surf. Coat. Technol.* **2011**, *205*, 3112–3118. [CrossRef]
14. Kang, S.; Haider, A.; Gupta, K.C.; Kim, H.; Kang, I. Chemical Bonding of Biomolecules to the Surface of Nano-Hydroxyapatite to Enhance Its Bioactivity. *Coatings* **2022**, *12*, 999. [CrossRef]
15. Sharma, S.; Gupta, V.; Mudgal, D. Current trends, applications, and challenges of coatings on additive manufacturing based biopolymers: A state of art review. *Polym. Compos.* **2022**, *43*, 6749–6781. [CrossRef]
16. Moskalewicz, T.; Kot, M.; Seuss, S.; Kedzierska, A.; Czyrska-Filemonowicz, A.; Boccaccini, A.R. Electrophoretic deposition and characterization of HA/Chitosan nanocomposite coatings on Ti6Al7Nb alloy. *Met. Mater. Int.* **2015**, *21*, 96–103. [CrossRef]
17. Lu, X.; Li, X.; Yu, J.; Ding, B. Nanofibrous hemostatic materials: Structural design, fabrication methods, and hemostatic mechanisms. *Acta Biomater.* **2022**, *154*, 49–62. [CrossRef]
18. Bashir, S.M.; Ahmed Rather, G.; Patrício, A.; Haq, Z.; Sheikh, A.A.; Shah, M.Z.u.H.; Singh, H.; Khan, A.A.; Imtiyaz, S.; Ahmad, S.B.; et al. Chitosan Nanoparticles: A Versatile Platform for Biomedical Applications. *Materials* **2022**, *15*, 6521. [CrossRef]
19. Pang, X.; Zhitomirsky, I. Electrophoretic deposition of composite hydroxyapatite-chitosan coatings. *Mater. Charact.* **2007**, *58*, 339–348. [CrossRef]
20. Atay, H.Y.; Celik, E. Investigations of antibacterial activity of chitosan in the polymeric composite coatings. *Prog. Org. Coat.* **2017**, *102*, 194–200. [CrossRef]

21. Li, J.; Fu, J.; Tian, X.; Hua, T.; Poon, T.; Koo, M.; Chan, W. Characteristics of chitosan fiber and their effects towards improvement of antibacterial activity. *Carbohydr. Polym.* **2022**, *280*, 119031. [CrossRef] [PubMed]
22. Tang, S.; Tian, B.; Guo, Y.J.; Zhu, Z.A.; Guo, Y.P. Chitosan/carbonated hydroxyapatite composite coatings: Fabrication, structure and biocompatibility. *Surf. Coat. Technol.* **2014**, *251*, 210–216. [CrossRef]
23. Bălţatu, M.S.; Vizureanu, P.; Bălan, T.; Lohan, M.; Ţugui, C.A. Preliminary Tests for Ti-Mo-Zr-Ta Alloys as Potential Biomaterials. *IOP Conf. Ser. Mater. Sci. Eng.* **2018**, *374*, 012023. [CrossRef]
24. Baltatu, I.; Sandu, A.V.; Vlad, M.D.; Spataru, M.C.; Vizureanu, P.; Baltatu, M.S. Mechanical Characterization and In Vitro Assay of Biocompatible Titanium Alloys. *Micromachines* **2022**, *13*, 430. [CrossRef] [PubMed]
25. Hsu, H.C.; Wu, S.C.; Hsu, S.K.; Syu, J.Y.; Ho, W.F. The structure and mechanical properties of as-cast Ti-25Nb-xSn alloys for biomedical applications. *Mater. Sci. Eng. A* **2013**, *568*, 1–7. [CrossRef]
26. Al Zoubi, N.F.; Tarlochan, F.; Mehboob, H.; Jarrar, F. Design of Titanium Alloy Femoral Stem Cellular Structure for Stress Shielding and Stem Stability: Computational Analysis. *Appl. Sci.* **2022**, *12*, 1548. [CrossRef]
27. Darvell, B.W. Bioactivity—Symphony or Cacophony? A Personal View of a Tangled Field. *Prosthesis* **2021**, *3*, 75–84. [CrossRef]
28. Wong, K.-K.; Hsu, H.-C.; Wu, S.-C.; Hung, T.-L.; Ho, W.-F. Structure, Properties, and Corrosion Behavior of Ti-Rich TiZrNbTa Medium-Entropy Alloys with $\beta+\alpha''+\alpha'$ for Biomedical Application. *Materials* **2022**, *15*, 7953. [CrossRef]
29. Yu, H.N.; Hsu, H.C.; Wu, S.C.; Hsu, C.W.; Hsu, S.K.; Ho, W.F. Characterization of Nano-Scale Hydroxyapatite Coating Synthesized from Eggshells Through Hydrothermal Reaction on Commercially Pure Titanium. *Coatings* **2020**, *10*, 112. [CrossRef]
30. Kokubo, T.; Takadama, H. How useful is SBF in predicting in vivo bone bioactivity? *Biomaterials* **2006**, *27*, 2907–2915. [CrossRef]
31. Sun, F.; Pang, X.; Zhitomirsky, I. Electrophoretic deposition of composite hydroxyapatite-chitosan-heparin coatings. *J. Mater. Process. Technol.* **2009**, *209*, 1597–1606. [CrossRef]
32. Kim, H.M.; Miyaji, F.; Kokubo, T.; Nakamura, T. Preparation of bioactive Ti and its alloys via simple chemical surface treatment. *J. Biomed. Mater. Res. A* **1996**, *32*, 409–417. [CrossRef]
33. Sadiq, N.M.; Aziz, S.B.; Kadir, M.F.Z. Development of Flexible Plasticized Ion Conducting Polymer Blend Electrolytes Based on Polyvinyl Alcohol (PVA): Chitosan (CS) with High Ion Transport Parameters Close to Gel Based Electrolytes. *Gels* **2022**, *8*, 153. [CrossRef]
34. Xue, C.B.; Chen, Y.Z.; Huang, Y.Z.; Zhu, P.Z. Hydrothermal synthesis and biocompatibility study of highly crystalline carbonated hydroxyapatite nanorods. *Nanoscale Res. Lett.* **2015**, *10*, 316. [CrossRef]
35. Jokanovic, V.; Jokanovic, B.; Markovic, D.; Zivojinovic, V.; Pasalic, S.; Izvonar, D.; Plavsic, M. Kinetics and sintering mechanisms of hydro-thermally obtained hydroxyapatite. *Mater. Chem. Phys.* **2008**, *111*, 180–185. [CrossRef]
36. Pang, X.; Zhitomirsky, I. Electrodeposition of composite hydroxyapatite-chitosan films. *Mater. Chem. Phys.* **2005**, *94*, 245–251. [CrossRef]
37. Fu, X.; Chang, X.; Ding, Z.; Xu, H.; Kong, H.; Chen, F.; Wang, R.; Shan, Y.; Ding, S. Fabrication and Characterization of Eco-Friendly Polyelectrolyte Bilayer Films Based on Chitosan and Different Types of Edible Citrus Pectin. *Foods* **2022**, *11*, 3536. [CrossRef]
38. Ho, W.F.; Lai, C.H.; Hsu, H.C.; Wu, S.C. Surface modification of a low-modulus Ti-7.5Mo alloy treated with aqueous NaOH. *Surf. Coat. Technol.* **2009**, *203*, 3142–3150. [CrossRef]
39. Safarzadeh, M.; Chee, C.F.; Ramesh, S. Effect of carbonate content on the in vitro bioactivity of carbonated hydroxyapatite. *Ceram. Int.* **2022**, *48*, 18174–18179. [CrossRef]
40. Tiyyagura, H.R.; Rudolf, R.; Gorgieva, S.; Fuchs-Godec, R.; Boyapati, V.R.; Mantravadi, K.M.; Kokol, V. The chitosan coating and processing effect on the physiological corrosion behaviour of porous magnesium monoliths. *Prog. Org. Coat.* **2016**, *99*, 147–156. [CrossRef]
41. Song, L.; Gan, L.; Xiao, Y.F.; Wu, Y.; Wu, F.; Gu, Z.W. Antibacterial hydroxyapatite/chitosan complex coatings with superior osteoblastic cell response. *Mater. Lett.* **2011**, *65*, 974–977. [CrossRef]
42. Rupp, F.; Liang, L.; Geis-Gerstorfer, J.; Scheideler, L.; Hüttig, F. Surface characteristics of dental implants: A review. *Dent. Mater.* **2018**, *34*, 40–57. [CrossRef] [PubMed]

coatings

MDPI

Article

Development of Apatite Nuclei Precipitated Carbon Nanotube-Polyether Ether Ketone Composite with Biological and Electrical Properties

Chihiro Ishizaki, Takeshi Yabutsuka * and Shigeomi Takai

Department of Fundamental Energy Science, Graduate School of Energy Science, Kyoto University, Kyoto 606-8501, Japan

* Correspondence: yabutsuka@energy.kyoto-u.ac.jp; Tel.: +81-75-753-9129

Received: 26 January 2020; Accepted: 21 February 2020; Published: 24 February 2020

Abstract: We aimed to impart apatite-forming ability to carbon nanotube (CNT)-polyether ether ketone (PEEK) composite (CNT-PEEK). Since CNT possesses electrical conductivity, CNT-PEEK can be expected to useful not only for implant materials but also biosensing devices. First of all, in this study, CNT-PEEK was treated with sulfuric acid to form fine pores on its surface. Then, the hydrophilicity of the substrate was improved by oxygen plasma treatment. After that, the substrate was promptly immersed in simulated body fluid (SBF) which was adjusted at pH 8.40, 25.0 °C (alkaline SBF) and held in an incubator set at 70.0 °C for 1 day to deposit fine particles of amorphous calcium phosphate, which we refer to as 'apatite nuclei'. When thus-treated CNT-PEEK was immersed in SBF, its surface was spontaneously covered with hydroxyapatite within 1 day by apatite nuclei deposited in the fine pores and high apatite-forming ability was successfully demonstrated. The CNT-PEEK also showed conductivity even after the above treatment and showed smaller impedance than that of the untreated CNT-PEEK substrate.

Keywords: carbon nanotube-polyether ether ketone composite; apatite nuclei; apatite-forming ability; electrical conductivity

1. Introduction

Polyether ether ketone (PEEK) is one of the most attractive engineering plastics with great material properties such as heat resistance, chemical resistance and radiation resistance and has attracted much attention as an artificial bone material with high mechanical toughness, low elastic modulus and light weight in the orthopedic or dental fields [1]. Generally, organic polymers can be easily combined with other functional materials with electrical, magnetical or biological properties because of their excellent processability. In the case of PEEK, several types of functional composites have been widely developed recently. Among them, carbon nanotube (CNT)-PEEK composite (CNT-PEEK) possesses electrical conductivity derived from CNTs in addition to the mechanical characteristics of PEEK. Hence, CNT-PEEK can be expected to be useful not only as an implant material but also in biosensing devices by utilizing its electrical conductivity. If the electrical properties of CNT-PEEK and biological properties of the bioactive materials such as calcium phosphates can be combined, development of not only artificial bone but also high-performance small devices using cells as sensor elements such as biosensors and tip devices may be achieved.

For impartation of biological functionality to CNT-PEEK while maintaining its electrical conductivity, the introduction of functional materials by a surface modification process is more desirable than by an elementary mixing process. Several types of surface modification techniques for impartation of biological activity to PEEK such as TiO_2 coating by sol-gel process [2,3], sulfuric acid treatment [4] and sulfuric acid and subsequent calcium chloride treatment [5] have been proposed in recent years. Although the TiO_2 coating attained sufficient apatite-forming ability in simulated body fluid (SBF) [6–9] with inorganic ion concentrations nearly equal to those of human blood plasma, this method required a sandblasting process to fix the film on the surface of the substrate. Hence, it is considered that fine structures near the surface should be controlled carefully. In addition, sulfuric acid treatment and sulfuric acid and calcium chloride treatment required more than 2 weeks to show apatite-forming ability in SBF.

When the temperature and pH of SBF are raised, on the other hand, fine particles of amorphous calcium phosphate, which we refer as 'apatite nuclei', are precipitated [10,11]. In previous studies, we formed apatite nuclei on pure PEEK, carbon fiber-reinforced PEEK and glass fiber-reinforced PEEK after sulfuric acid treatment and subsequent oxygen plasma treatment and found that the surfaces of the PEEK were spontaneously covered with apatite within 1 day [12–14]. In addition, the bone-bonding ability was better than that of a TiO_2 sol-gel coating [14].

Utilizing the above attractive biological function of apatite nuclei after surface modification of CNT-PEEK, we tried to fabricate bioactive CNT-PEEK device for orthopedic and biosensing applications by combination of electrical properties of CNT-PEEK and the biological properties of apatite nuclei. In this study, we aimed to establish a surface modification process for CNT-PEEK using apatite nuclei and to clarify the material properties from a viewpoint of apatite-forming ability and conductivity.

2. Materials and Methods

2.1. Outline of Experimental Procedure

First, CNT-PEEK was treated with sulfuric acid to form fine pores on its surface. Then, the hydrophilicity of the substrate was improved by oxygen plasma treatment. After that, the substrate was immersed in SBF which was adjusted higher pH than conventional SBF and subsequently the solution was heated to precipitate apatite nuclei. The apatite-forming ability of the samples was evaluated in SBF and adhesive strength of the formed apatite layer was evaluated. In order to evaluate conductivity of the substrates, in addition, impedance measurements were carried out. The details of the experimental procedure are described below.

2.2. Materials Fabrication

2.2.1. Substrate and Pre-Treatment

CNT-PEEK discs (TECAPEEK ELS nano black, Ensinger, Nufringen, Germany) with 10 mm in diameter and 2 mm in thickness which include 10% CNT in a PEEK matrix were used as samples. Both surfaces of the samples were polished by using #400 and #1200 SiC abrasive paper in order. Then the polished samples were washed ultrasonically in acetone, ethanol and pure water for 10 min in order and then air-dried. This sample is denoted as 'Sample N' hereafter.

2.2.2. Sulfuric Acid Treatment

The samples were immersed in concentrated sulfuric acid (Hayashi Pure Chemical, Osaka, Japan) at room temperature for 4 s in total and then washed with pure water. Then, the samples were washed ultrasonically in pure water for 10 min and then air-dried. This sample is denoted as 'Sample S' hereafter.

2.2.3. Oxygen Plasma Treatment

The surfaces of the Sample S were irradiated oxygen plasma at 200 W for 4 min by a glow-discharge equipment (Model BP-1, Samco Inc., Kyoto, Japan) to improve the hydrophilicity. This sample is denoted as 'Sample SG' hereafter.

2.2.4. 'Alkaline SBF' Treatment

SBF was prepared by dissolving reagent grade NaCl (FUJIFILM Wako Pure Chemical, Osaka, Japan), $NaHCO_3$ (Hayashi Pure Chemical), KCl (Hayashi Pure Chemical), $K_2HPO_4 \cdot 3H_2O$ (Nacalai Tesque, Kyoto, Japan), $MgCl_2 \cdot 6H_2O$ (Hayashi Pure Chemical), 1 mol·dm^{-3} HCl (Hayashi Pure Chemical), $CaCl_2$ (Hayashi Pure Chemical), Na_2SO_4 (Hayashi Pure Chemical) and $(CH_2OH)_3CNH_2$ (Hayashi Pure Chemical) in a pure water in order and adjusted at pH 7.40, 36.5 °C by dissolving 1 mol·dm^{-3} HCl based on the method certified as ISO 23317 [9].

The pH of SBF was increased to 8.40 by dissolving $(CH_2OH)_3CNH_2$ at 25.0 °C. This solution is denoted as 'alkaline SBF' hereafter. Sample SG was immersed in the 'alkaline SBF' and held in an incubator held at 70.0 °C for 1 day. After taken out, the samples were washed in pure water and then air-dried. This sample is denoted as 'Sample SGA' hereafter.

2.2.5. Analyses

We observed and analyzed the surfaces of the samples in the above each step using field emission scanning electron microscopy (FE-SEM) (SU6600, Hitachi High-technologies, Tokyo, Japan), energy dispersive X-ray spectroscopy (EDX) (XFlash®5010, Bruker, Billerica, MA, USA), thin film X-ray diffraction (TF-XRD) (Hyper-Rint, Rigaku, Tokyo, Japan) using Cu-Kα radiation at tube voltage and current of 50 kV, 300 mA, X-ray photoelectron spectroscopy (XPS) (JPS-9010TRX, JEOL, Tokyo, Japan) using Mg-Kα radiation at an acceleration voltage of 10 kV. For checking the reproducibility, more than 10 samples were analyzed per each condition. In addition, water contact angle was measured using contact angle meter (CAX-150, Kyowa Interface Science, Saitama, Japan) by using three samples for each condition. In the water contact angles measurement, one-way analysis of variance (ANOVA) followed by Tukey's multiple comparison tests was carried out to calculate p values and to evaluate significant difference between Samples N, S, SG and SGA each other.

2.3. Evaluation of Materials Properties

2.3.1. Evaluation of Apatite-Forming Ability

The samples processed above treatments were immersed in SBF with physiological condition (pH 7.40, 36.5 °C) and taken out after 1, 4, 7 and 14 days. The samples were washed in pure water and then air-dried. The surfaces of samples were observed and analyzed using FE-SEM, EDX and TF-XRD. In the EDX analyses, quantitative analyses were carried out using the ZAF correction method to evaluate changes in elemental composition on the surface of the sample and formation or growth of apatite in the SBF test. Adhesive strength between the substrates and the apatite layer formed by the immersion in SBF for 14 days was investigated by a modified ASTM C-633 method [15–18]. A couple of stainless-steel jigs were fixed to both surfaces of the substrates by using Araldite® glue (NICHIBAN, Tokyo, Japan) and a tensile load was applied at 1 mm·min^{-1} of a crosshead speed until a fracture occurred between apatite layers and the substrates by using a universal testing machine (AGS-H Autograph, Shimadzu, Kyoto, Japan). In this tensile test, five samples were tested. In the adhesive strength measurement, one-way ANOVA followed by the Tukey's tests was carried out between bioactive CNT-PEEK in this study and bioactive pure PEEK [12,13], bioactive carbon fiber reinforced PEEK [13] or bioactive glass fiber reinforced PEEK [13] reported in our previous papers.

2.3.2. Impedance Measurement

We measured the impedance of Samples N, S, SG and SGA by using an impedance analyzer (Z Hightester 3531, HIOKI, Nagano, Japan). In the impedance measurement, the surfaces of the samples were coated with gold by sputtering method and thus treated samples were used as electrodes. During the measurements, frequency range was controlled from 100 Hz to 5 MHz and measurement temperature was controlled at 25 °C.

3. Results and Discussion

3.1. Material Analyses

3.1.1. Changes in Surface Morphology during the Fabrication Process

Figure 1a,b shows the SEM micrograph and the EDX profile of the surface of Sample N, which is the untreated CNT-PEEK. Although there were scratches due to the polishing process, the surface morphology was almost flat. By the EDX, peaks of C and O were observed. The peak of Au derived from the gold coating for the observation was also observed.

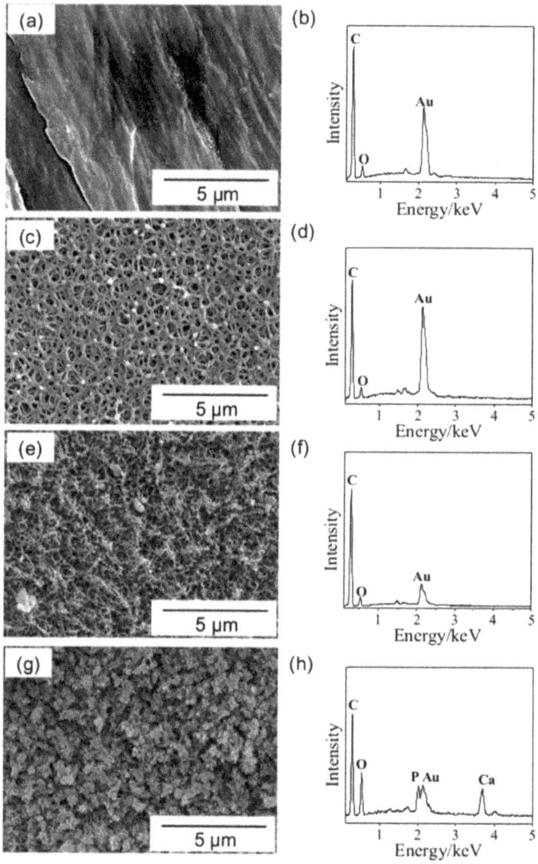

Figure 1. (**a**,**c**,**e**,**g**) SEM micrographs and (**b**,**d**,**f**,**h**) EDX profiles of the surfaces of (**a**,**b**) Sample N, (**c**,**d**) Sample S, (**e**,**f**) Sample SG, (**g**,**h**) Sample SGA.

Figure 1c,d shows a SEM micrograph and the EDX profile of the surface of Sample S, which is the sulfuric acid-treated CNT-PEEK. In comparison with the Sample N, morphological changes are clearly observed on the surface of Sample S. Fine reticulated pores with ca. 500 nm in diameter were formed on the entire surface. The shape of the reticulated pores formed in this study were similar to those formed on pure PEEK, carbon fiber reinforced PEEK and glass fiber reinforced PEEK in our previous studies [12–14]. Hence, it is considered that kind of materials contained in a PEEK matrix does not affect the pore formation. These results suggest that the reticulated pores were formed by the reaction of the PEEK. The EDX profile was almost similar to that of Sample N.

Figure 1e,f shows a SEM micrograph and the EDX profile of the surface of Sample SG, which is the sulfuric acid and subsequent oxygen plasma-treated CNT-PEEK. The reticulated pores formed by sulfuric acid treatment were broken after the oxygen plasma treatment. It is considered that the heat of the oxygen plasma caused such a morphological change. The EDX profile was almost similar to that of the Samples N and S.

Figure 1g,h shows a SEM micrograph and the EDX profile of the surface of Sample SGA, which is the sulfuric acid, oxygen plasma and subsequent alkaline SBF-treated CNT-PEEK. Precipitation of a large number of fine particles were observed on the entire surface of the substrate. Furthermore, peaks of P and Ca were newly detected in the EDX profile. Therefore, it is thought that the fine particles consisted of calcium phosphate. In other words, the apatite nuclei were formed on the surfaces of the substrate similar to the case of our previous studies [12–14]. As the pH of the solution rises, calcium phosphate formation proceeds due to chemical equilibrium ($10Ca^{2+} + 6PO_4^{3-} + OH^- = Ca_{10}(PO_4)_6OH_2$). In this study, furthermore, the experimental temperature (70.0 °C) was higher than the physiological temperature (36.5 °C). Hence, it is considered that the reaction of calcium phosphate formation was accelerated, and its nucleation was promoted.

3.1.2. Changes in Functional Groups during the Fabrication Process

Figure 2a shows the XPS narrow spectra around the binding energy of S2*p* of Samples N, S, SG and SGA. For Sample N, no peaks were observed. For Sample S, a peak derived from sulfo groups was confirmed.

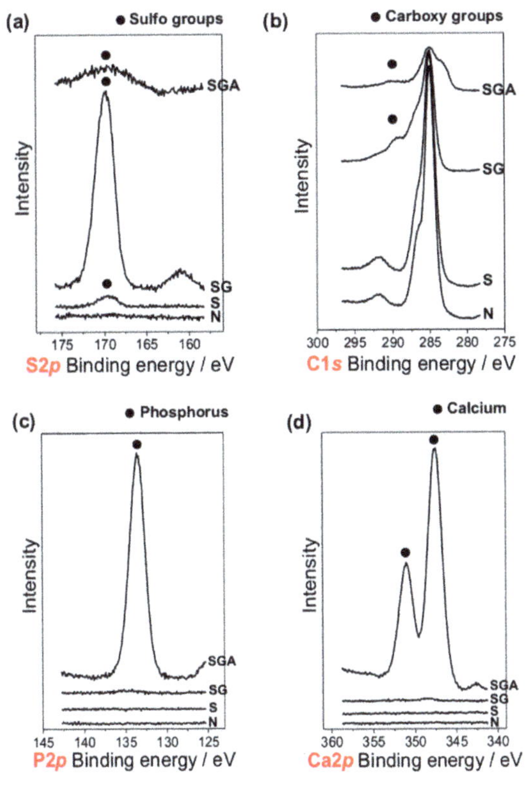

Figure 2. XPS narrow spectra around binding energy of (**a**) S2*p*, (**b**) C1*s*, (**c**) P2*p* and (**d**) Ca2*p* on the surface of Samples N, S, SG and SGA.

It is well known that aromatic compounds are sulfonated by the reaction with sulfuric acid [19]. Hence, it is considered that aromatic groups in PEEK reacted with sulfuric acid and sulfonation occurred. Interestingly, the sulfo group peak was significantly intensified in Sample SG. Such a result was also obtained for pure PEEK in our previous study [14]. It is speculated that such an increase of sulfo groups is related with oxidation reactions of sulfone-related components near the surface generated during the sulfuric acid treatment because abundant oxygen was supplied around the samples in the oxygen plasma treatment. For Sample SGA, the intensity of the sulfo groups peak decreased because the apatite nuclei were precipitated on the entire surface area of the substrate by the alkaline SBF treatment.

Figure 2b shows the XPS narrow spectra around the energy of C1*s* of Samples N, S, SG and SGA. For Samples SG, a slight peak derived from carboxy groups was detected. It is considered that the PEEK matrix was oxidized by the oxygen plasma. For Sample SGA, the intensity of the carboxy groups peak decreased because the apatite nuclei were precipitated on the entire surface area of the substrate by the alkaline SBF treatment.

Figure 2c,d shows the XPS narrow spectra around the binding energy of P2*p* and Ca2*p* of Samples N, S, SG and SGA. For Samples N, S and SG, no peaks were observed. For Sample SGA, peaks derived from phosphorus and calcium were clearly observed, indicating that apatite nuclei were precipitated on the surface of the substrate by the alkaline SBF treatment.

3.1.3. Hydrophilicity

Figure 3 shows the water contact angles of Samples N, S, SG and SGA. By applying one-way ANOVA followed by the Tukey's tests, Samples N (94.8° ± 9.7° for three samples) and S (92.7° ± 7.3° for three samples) did not show a statistically significant ($p > 0.05$) difference in water contact angle among each other. In addition, Samples SG (4.7° ± 0.3° for three samples) and SGA (94.8° ± 9.7° for three samples) also did not show a significant difference between each other ($p > 0.05$). In contrast, the values of Samples SG or SGA showed significantly lower contact angles than Samples N or S ($p < 0.01$). It was shown that the surfaces of Samples N and S, whose fabrication process did not include oxygen plasma treatment, were hydrophobic and those of Samples SG and SGA, whose fabrication process included oxygen plasma treatment, were super-hydrophilic. Such result was also obtained for pure PEEK in our previous study [14]. It is considered that the super hydrophilicity of Samples SG and SGA was caused by the introduction of carboxy groups and a significant increase of the sulfo groups during oxygen plasma treatment.

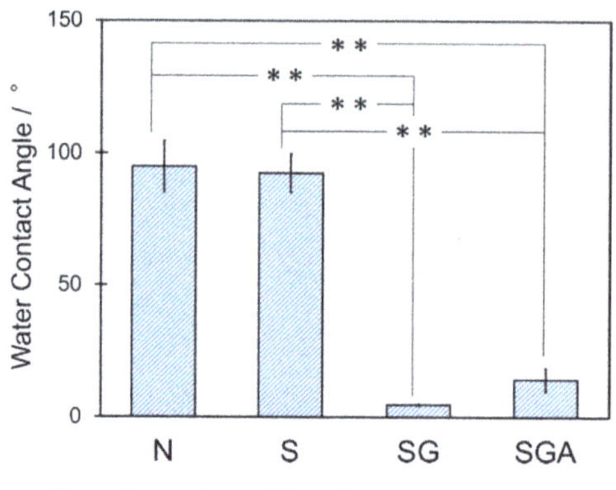

Figure 3. Water contact angle on the surface of Samples N, S, SG and SGA. The symbol '**' indicates $p < 0.01$ and the no symbol indicates $p > 0.05$ by one-way ANOVA followed by the Tukey's tests.

3.2. Evaluation of Material Properties

3.2.1. Apatite-Forming Ability

Figure 4 shows the TF-XRD profiles of Sample SGA before and after immersion in SBF for 1, 4 and 7 days, respectively. Before the immersion in SBF, apatite peaks were not detected. After immersing in SBF for 1 day, a slight apatite peak was detected around 26°. Moreover, the intensity and number of apatite peaks intensified after 4 and 7 days as the immersion time increased. This means that apatite nuclei induced apatite formation in SBF within 1 day and the induced apatite grew with increasing immersion time.

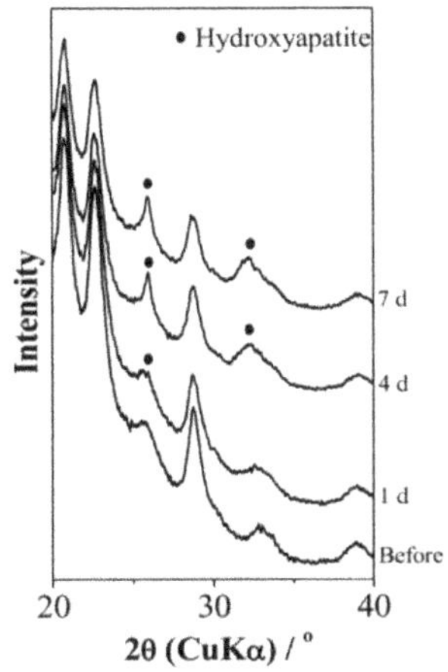

Figure 4. TF-XRD profiles of on the surface of Sample SGA before and after immersion in SBF for 1 day, 4 days and 7 days.

Figure 5 shows SEM micrographs and the EDX profiles of the surfaces of the Sample SGA after immersion in SBF for 1, 4 and 7 days. After the immersion in SBF, a film consisting of flake-like crystallites, characteristic of apatite formed in SBF, so-called 'bone-like apatite', covered the entire surfaces of the samples and strong peaks of P and Ca were detected. This result indicates that apatite formation was induced on the entire surface of the Sample SGA within 1 day in SBF. Table 1 shows the elemental composition of the surface of Sample SGA after immersion in SBF for each period calculated from the EDX profiles shown in Figure 5. It was verified that the atomic ratios of Ca, P and O, the constituents of apatite, were increased and that of C, the constituent of CNT-PEEK, decreased as the immersion period was extended. This means that once formed the apatite grew on the surface of the substrate in SBF. In addition, it is considered that other detected elements were derived from minute components inserted in crystal structures of apatite or contaminations attached on the surface of the samples because the atomic ratios were very minute in comparison with C, O, Ca and P and their changes could not show a clear increasing or decreasing tendency depending on the immersion period. Zhao et al. reported that sulfuric acid-treated PEEK showed apatite-forming ability in SBF within 28 days [4]. Miyazaki et al. reported that sulfuric acid and calcium chloride-treated PEEK showed apatite-forming ability in SBF within 14 days [5]. In this study, the induction period of apatite formation was within 1 day. Hence, it is confirmed that the apatite nuclei on the surface of CNT-PEEK largely contribute to the effective acceleration of apatite formation.

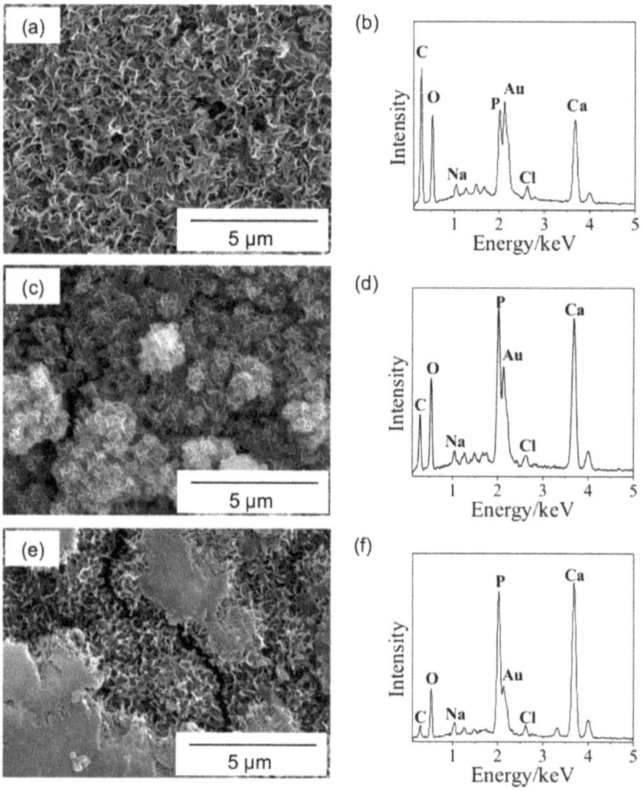

Figure 5. (**a**,**c**,**e**) SEM micrographs and (**b**,**d**,**f**) EDX profiles of the surface of Sample SGA after immersion in SBF for (**a**,**b**) 1 day, (**c**,**d**) 4 days and (**e**,**f**) 7 days.

Table 1. Elemental composition of the surface of Sample SGA after immersion in SBF for each period calculated from the EDX profiles shown in Figure 5 (except Au generated by sputtering).

Element	Elemental Composition/at%		
	1 Day	4 Days	7 Days
C	61.623	40.386	21.384
O	32.300	40.683	43.106
Ca	2.730	11.586	21.458
P	2.181	6.621	11.429
Na	0.662	0.452	1.213
Mg	0.172	0.017	0.177
Cl	0.164	0.231	0.517
K	0.001	0.001	0.715
Al	0.166	0.002	0.002
Si	0.001	0.020	0.002

The adhesive strength of the apatite layer formed by immersion in SBF for 14 days was 8.8 ± 5.0 MPa for five samples. In a previous study, we reported that the adhesive strength of the apatite layer in the case of pure PEEK, carbon fiber-reinforced PEEK and glass fiber reinforced PEEK were 6.6 ± 1.5 MPa for 12 samples, 7.7 ± 0.9 MPa for 11 samples and 6.7 ± 1.8 MPa for seven samples, respectively [12,13]. Although one-way ANOVA followed by the Tukey's tests was carried out between the CNT-PEEK investigated in this paper and the above other types of PEEK reported in our previous papers, significant differences were not observed among them ($p > 0.05$). Hence, it is considered that the CNT-PEEK statistically showed almost a similar level of adhesive strength in comparison with other types of PEEK composite. It is considered that such adhesive strength was attained by a mechanical interlocking effect between apatite inside the pores and the CNT-PEEK substrate, like in our previous study.

3.2.2. Impedance

Figure 6 shows the Nyquist plots of Samples N, S, SG and SGA. The impedance of Sample S was considerably larger than that of Sample N and the impedance was increased by the sulfuric acid treatment. It is considered that the number of CNTs near the surface of the substrate was decreased by the sulfuric acid treatment. The impedance of Sample SG was smaller than that of Sample S and the impedance was decreased by the oxygen plasma treatment. It is considered that CNTs existing inside the substrate were newly exposed by the changes in the surface morphology caused by the heat of the oxygen plasma. The impedance of Sample SGA was the smallest among these sample conditions. Interestingly, the impedance of Sample SGA was smaller than that of Sample N and it was suggested that the electrical conductivity was improved after the series of surface modifications. It is speculated that the apatite nuclei precipitated by the alkaline SBF treatment might possess an ionic conductivity slightly because of their extremely low crystallinity.

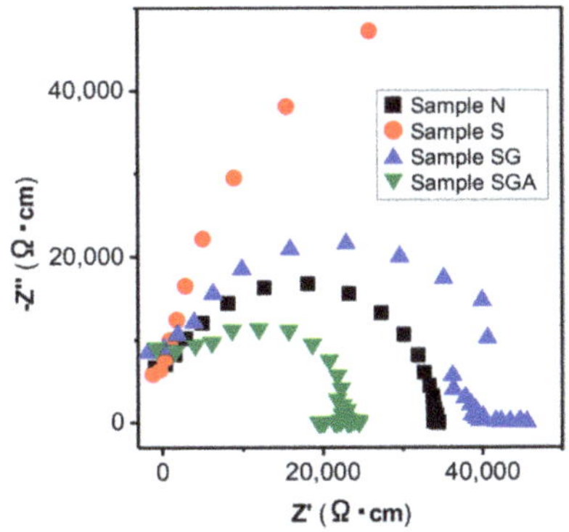

Figure 6. Nyquist plots of Samples N, S, SG and SGA.

We acknowledge several limitations in this study. First, we have not conducted animal tests yet. It is well known that SBF test can predict bone-bonding ability well [7], however, some researchers have pointed out several exceptions of the relationship between apatite-forming ability in SBF and bone-bonding ability in vivo [20]. For clinical applications, hence, it is essential to evaluate bone-bonding ability of this material by animal tests. About this point, however, we have obtained expected results in animal test when we applied similar fabrication methodology to pure PEEK [14]. Hence, we consider that also this material has a potential to exhibit similar osteoconductivity in vivo. Second, we have not conducted cell activity tests yet. Especially for biosensing applications, a comparative evaluation of cell responses on the surface of the substrate is essential. Third, we have not evaluated the relationship between amount of apatite nuclei on the substrate and the impedance of the substrate and have not optimized the conductivity required in biosensing applications yet. Finally, we carried out the impedance tests only in normal air circumstances in this study. Taking into consideration possible biosensing applications, evaluation in a humid or a cell environment is required. These points will be addressed in our future studies.

4. Conclusions

We successfully imparted apatite-forming ability to CNT-PEEK by the following three steps: First a sulfuric acid treatment to form fine pores, second an oxygen plasma treatment to improve the hydrophilicity and lastly an alkaline SBF treatment to form apatite nuclei. Thus-obtained samples were coated with hydroxyapatite within 1 day in SBF and showed high apatite-forming ability. The apatite film formed in SBF showed sufficient adhesive strength by a mechanical interlocking effect. Moreover, the treated CNT-PEEK showed a small impedance in comparison with the untreated CNT-PEEK and possessed electrical conductivity. This material is expected to be a new type of biomaterial with both biological and electrical properties, useful in the orthopedic and biosensing fields.

Author Contributions: Conceptualization, C.I. and T.Y.; Data curation, C.I.; Formal analysis, C.I. and T.Y.; Funding acquisition, T.Y.; Investigation, C.I. and T.Y.; Methodology, T.Y.; Project administration, T.Y.; Supervision, T.Y. and S.T.; Validation, C.I., T.Y. and S.T.; Writing—Original Draft, C.I.; Writing—Review and editing, T.Y. All authors have read and agreed to the published version of the manuscript.

Funding: This study was partly funded by JSPS KAKENHI (16K16401 and 19H02442), Murata Science Promotion Foundation Research Grant, and Kyoto University Research and Development Program "Ishizue".

Conflicts of Interest: The authors declare no conflict of interest.

References

1. Kurtz, S.M. An Overview of PEEK Biomaterials. In *PEEK Biomaterials Handbook*, 2nd ed.; Kurtz, S.M., Ed.; Elsevier: Amsterdam, The Netherlands, 2019; pp. 3–9.
2. Kizuki, T.; Matsushita, T.; Kokubo, T. Apatite-forming PEEK with TiO_2 surface layer coating. *J. Mater. Sci.-Mater. Med.* **2015**, *26*, 5359. [CrossRef]
3. Shimizu, T.; Fujibayashi, S.; Yamaguchi, S.; Yamamoto, K.; Otsuki, B.; Takemoto, M.; Tsukanaka, M.; Kizuki, T.; Matsuhita, T.; Kokubo, T.; et al. Bioactivity of sol-gel-derived TiO_2 coating on polyetheretherketone: In vitro and in vivo studies. *Acta Biomater.* **2016**, *35*, 305–317. [CrossRef] [PubMed]
4. Zhao, Y.; Wong, H.M.; Li, P.; Xu, Z.; Chong, E.Y.W.; Yan, C.H.; Yeung, K.W.K.; Chu, P.K. Cytocompatibility, osseointegration, and bioactivity of three-dimensional porous and nanostructured network on polyetheretherketone. *Biomaterials* **2013**, *34*, 9264–9277. [CrossRef] [PubMed]
5. Miyazaki, T.; Matsunami, C.; Shirosaki, Y. Bioactive carbon–PEEK composites prepared by chemical surface treatment. *Mater. Sci. Eng. C* **2017**, *70*, 71–75. [CrossRef] [PubMed]
6. Kokubo, T.; Kushitani, H.; Sakka, S.; Kitsugi, T.; Yamamuro, T. Solutions able to reproduce in vivo surface-structure changes in bioactive glass-ceramic A-W. *J. Biomed. Mater. Res.* **1990**, *24*, 721–734. [CrossRef] [PubMed]
7. Kokubo, T.; Takadama, H. How useful is SBF in predicting in vivo bone bioactivity? *Biomaterials* **2006**, *27*, 2907–2915. [CrossRef] [PubMed]
8. Takadama, H.; Kokubo, T. In vitro evaluation of bone bioactivity. In *Bioceramics and Their Clinical Applications*; Kokubo, T., Ed.; Woodhead Publishing: Cambridge, UK, 2008; pp. 165–182.
9. *Implants for Surgery—In Vitro Evaluation for Apatite-Forming Ability of Implant Materials*; ISO 23317; International Organization for Standardization: Geneva, Switzerland, 2014.
10. Yao, T.; Hibino, M.; Yamaguchi, S.; Okada, H. Method for Stabilizing Calcium Phosphate Fine Particles, Process for Production of Calcium Phosphate Fine Particles by Utilizing the Method, and Use Thereof. U.S. Patent 8178066, 15 May 2012. Japanese Patent 5261712, 10 May 2013.
11. Yao, T.; Yabutsuka, T. Material Having Pores on Surface, and Method for Manufacturing Same. Japanese Patent 6071895, 13 January 2017.
12. Yabutsuka, T.; Fukushima, K.; Hiruta, T.; Takai, S.; Yao, T. Effect of pores formation process and oxygen plasma treatment to hydroxyapatite formation on bioactive PEEK prepared by incorporation of precursor of apatite. *Mater. Sci. Eng. C* **2017**, *81*, 349–358. [CrossRef] [PubMed]
13. Yabutsuka, T.; Fukushima, K.; Hiruta, T.; Takai, S.; Yao, T. Fabrication of bioactive fiber-reinforced PEEK and MXD6 by incorporation of precursor of apatite. *J. Biomed. Mater. Res. B Appl. Biomater.* **2018**, *106*, 2254–2265. [CrossRef] [PubMed]

14. Masamoto, K.; Fujibayashi, S.; Yabutsuka, T.; Hiruta, T.; Otsuki, B.; Okuzu, Y.; Goto, K.; Shimizu, T.; Shimizu, Y.; Ishizaki, C.; et al. In vivo and in vitro bioactivity of a "precursor of apatite" treatment on polyetheretherketone. *Acta Biomater.* **2019**, *91*, 48–59. [CrossRef] [PubMed]
15. Lacefield, W.R. Hydroxyapatite coatings. In *An Introduction to Bioceramics*, 2nd ed.; Imperial College Press: London, UK, 2013; pp. 331–347.
16. Kim, H.-M.; Miyaji, F.; Kokubo, T.; Nakamura, T. Bonding strength of bonelike apatite layer to Ti metal substrate. *J. Biomed. Mater. Res.* **1997**, *38*, 121–127. [CrossRef]
17. Miyazaki, T.; Kim, H.-M.; Kokubo, T.; Ohtsuki, C.; Kato, H.; Nakamura, T. Enhancement of bonding strength by graded structure at interface between apatite layer and bioactive tantalum metal. *J. Mater. Sci. Mater. Med.* **2002**, *13*, 651–655. [CrossRef]
18. Juhasz, J.A.; Best, S.M.; Kawashita, M.; Miyata, N.; Kokubo, T.; Nakamura, T.; Bonfield, W. Bonding strength of the apatite layer formed on glass-ceramic apatite-wollastonite-polyethylene composites. *J. Biomed. Mater. Res. A* **2003**, *67*, 952–959. [CrossRef]
19. Solomons, T.W.G.; Fryhle, C.B. Reactions of Aromatic Compounds. In *Organic Chemistry*, 7th ed.; Wiley: New York, NY, USA, 2000; pp. 661–713.
20. Kawashita, M.; Hayashi, J.; Li, Z.; Miyazaki, T.; Hashimoto, M.; Hihara, H.; Kanetaka, H. Adsorption characteristics of bovine serum albumin onto alumina with a specific crystalline structure. *J. Mater. Sci. Mater. Med.* **2014**, *25*, 453–459. [CrossRef] [PubMed]

coatings

MDPI

Article

Characterization of Nano-Scale Hydroxyapatite Coating Synthesized from Eggshells Through Hydrothermal Reaction on Commercially Pure Titanium

Hsing-Ning Yu [1], Hsueh-Chuan Hsu [2], Shih-Ching Wu [2], Cheng-Wei Hsu [3], Shih-Kuang Hsu [2] and Wen-Fu Ho [4,*]

1 Department of Surgery, Division of Orthopaedics, Zuoying Branch of Kaohsiung Armed Forces General Hospital, Kaohsiung 81342, Taiwan; david4229@mail.ngh.com.tw

2 Department of Dental Technology and Materials Science, Central Taiwan University of Science and Technology, Taichung 40601, Taiwan; hchsu@ctust.edu.tw (H.-C.H.); scwu@ctust.edu.tw (S.-C.W.); sksheu@ctust.edu.tw (S.-K.H.)

3 Department of Materials Science and Engineering, Da-Yeh University, Changhua 51591, Taiwan; a10019990@gmail.com

4 Department of Chemical and Materials Engineering, National University of Kaohsiung, Kaohsiung 81148, Taiwan

* Correspondence: fujii@nuk.edu.tw; Tel.: +886-7-591-9276

Received: 12 January 2020; Accepted: 28 January 2020; Published: 28 January 2020

Abstract: Commercially pure titanium (c.p. Ti) is often used in biomedical implants, but its surface cannot usually combine with the living bone. A coating of hydroxyapatite (HA) on the surface of titanium implants provides excellent mechanical properties and has good biological activity and biocompatibility. For optimal osteocompatibility, the structure, size, and composition of HA crystals should be closer to those of biological apatite. Our results show that the surface of c.p. Ti was entirely covered by rod-like HA nanoparticles after alkali treatment and subsequent hydrothermal treatment at 150 °C for 48 h. Nano-sized apatite aggregates began to nucleate on HA-coated c.p. Ti surfaces after immersion in simulated body fluid (SBF) for 6 h, while no obvious precipitation was found on the uncoated sample. Higher apatite-forming ability (bioactivity) could be acquired by the samples after HA coating. The HA coating featured bone-like nanostructure, high crystallinity, and carbonate substitution. It can be expected that HA coatings synthesized from eggshells on c.p. Ti through a hydrothermal reaction could be used in dental implant applications in the future.

Keywords: commercially pure titanium (c.p. Ti); hydroxyapatite; eggshell; hydrothermal reaction; nanoparticle; simulated body fluid (SBF)

1. Introduction

Commercially pure titanium (c.p. Ti) has been used in dentistry, mainly due to its resistance to corrosion, superior biocompatibility, and favorable mechanical properties [1]. However, Ti is regarded as a bioinert metal, which cannot form a chemical bond with bone, and this biological inactivity often results in fibrous tissue surrounding the implanted device [2]. In order to improve both osseointegration rates and longevity of Ti implants, hydroxyapatite (HA) coating could promote the formation of real bonds with the surrounding bone tissue. As previously reported, HA-coated Ti has shown better long-term clinical survival rates than uncoated Ti due to its chemical similarity with natural bone tissue and its high biocompatibility [3,4]. In order to shorten the surgical healing time of dental implants, a rapid and reliable bonding with the bone is highly desirable.

Among various surface-coating methods, the plasma spray technique is widely used for coating implants with bioactive ceramics, especially HA coatings. However, there have been problems associated with the poor clinical performances of conventional plasma-sprayed coatings [5]. Therefore, different methods such as electrophoretic deposition, thermal spray deposition, sol–gel coating, and biomimetic coating have been developed to coat Ti implants [6–10]. In particular, the biomimetic apatite coating method is one of the most promising coating technologies, since it can prepare the HA coatings on implant surfaces at a relatively low temperature [6,11]. A bone-like HA coating is formed on a Ti surface by immersing the implant in a Ca-, P-containing solution [11].

In recent years, natural materials such as animal bones, corals, eggshells, and oyster shells have been used to synthesize HA. Wu et al. [12] reported the successful fabrication of HA from oyster shell powder by solid-state reaction and subsequently heat treatment at 900–1100 °C. Recently, our research group [13] proposed an environmentally beneficial and cost-effective method of preparing HA or tri-calcium phosphate (TCP) bioceramics from eggshell biowaste. The eggshell mainly consisted of a major component, i.e., calcium carbonate (94%), besides calcium phosphate (1%), magnesium carbonate (1%), and organic matter (4%) [14]. It is worth mentioning that the crystal structure and composition of HA prepared from natural sources like eggshells are similar to those of human bones because eggshells notably contain several trace elements (Na, Mg, Sr, etc.). Thus, eggshell could be a valuable and promising raw material [15]. In spite of these advantages, there are still few studies concerning ceramic coatings made from natural sources and biowastes. Qaid et al. prepared HA coatings on Ti–6Al–4V substrates using micro-arc oxidation (MAO) in electrolytes at various concentrations of eggshells-derived HA [16]. Roudan et al. used an electrophoretic deposition method to deposit HA from eggshells on a Ti substrate and studied the thermal stability of the resulting coating [17]. Also, eggshells-derived HA was prepared on low-modulus Ti–5Nb–5Mo alloy by hydrothermal heating processing [18].

In this study, we hypothesized that eggshell-derived HA coating, prepared on c.p. Ti surface after alkali treatment, could promote apatite formation and bioactivity. The samples were immersed in simulated body fluid (SBF) for a period of time to evaluate their bioactivity. In order to test this hypothesis, the surface of c.p. Ti was coated with HA by hydrothermal treatment followed by heat treatment in air, using eggshell biowaste as the source of Ca. The hydrothermal reaction could enhance the crystallinity of the product at relatively low temperatures without significantly increase of grain size. The characteristics of the HA coatings after hydrothermal treatment for different periods of time from 0 to 48 h were then analyzed.

2. Materials and Methods

2.1. Research Material

C.p. Ti (Grade II) was cut to make plates with the dimension of 5.0 × 5.0 × 0.7 mm. The metal surfaces were sanded to the final level using 600-grit paper and then ultrasonically cleaned twice with ethanol and deionized water for 5 min. An alkali treatment, which increased hydrophilicity, was performed by soaking the specimens in 50 mL solution of 5 M NaOH at 60 °C for 24 h. Finally, the samples were ultrasonically cleaned twice in deionized water for 5 min each time and dried in air at 45 °C for 24 h.

The eggshell powders (2 g) were put into 15 mL of deionized water and completely dissolved after adding 5 mL of hydrochloric acid. The solution was then stirred at 200 rpm for 40 min. A total of 0.85 mL of H_3PO_4 (85%) was added dropwise while the mixture was continually stirred for 40 min. The molar Ca/P ratio was set to 1.67. A NH_4OH solution was added to maintain the pH of the solution at 10, which resulted in a white HA suspension. Details of this method can be found in some previous works [12,18]. The pretreated c.p. Ti samples were placed into the HA suspension and sealed in Teflon-lined stainless-steel autoclaves. The hydrothermal treatments were performed at 150 °C for 12, 24, and 48 h. The autoclave was allowed to naturally cool to room temperature. After the respective periods of reaction, the hydrothermally treated samples were removed, rinsed with deionized water, and dried at 45 °C in air. The sample without a hydrothermally treated HA coating was used as a control. Finally, all the samples were heat-treated at 500 °C for 2 h to increase the crystallinity of the HA coating.

2.2. Research Methodology

In this experiment, the samples before and after the heat treatment were placed in deionized water and subjected to ultrasonic vibration at 200 W for 3 min to determine the effect of heat treatment on the adhesion between the coating and the substrate [19]. After ultrasonic cleaning, the surface morphology of the coating was observed using field-emission scanning electron microscopy (FE-SEM; JSM-6700F, JEOL, Tokyo, Japan). After the hydrothermal reactions, the samples were immediately immersed in 30 mL of SBF to examine the bioactivity of the HA-coated c.p. Ti. For comparison purposes, the same immersion test was performed on alkali-treated c.p. Ti without hydrothermal treatment. The bioactivity of all samples was evaluated by soaking the samples in SBF for 6, 12, and 24 h, maintained in a water bath at 37 °C. Reagent-grade chemicals, NaCl, $NaHCO_3$, KCl, $K_2HPO_4{\cdot}3H_2O$, $MgCl_2{\cdot}6H_2O$, $CaCl_2$, and Na_2SO_4, were sequentially dissolved in deionized water to prepare SBF, using the method described by Kokubo and Takadama [20]. Table 1 lists the ion concentrations of SBF used in this experiment and that of human blood plasma [20]. The SBF was renewed every 2 days to ensure its fixed ion concentration. At each soaking time point, the samples were taken out, washed with deionized water, and then air-dried. Before and after the various treatments and the soaking in SBF, the c.p. Ti surfaces were analyzed using FE-SEM (JSM-6700F, JEOL, Tokyo, Japan) and X-ray diffractometry (XRD; D8-Discover, BRUKER, Karlsruhe, Germany). The coated surfaces of the specimens were examined using Fourier-transform infrared (FT-IR; FTS-40, Bio-Rad, Cambridge, MA, USA) reflection spectroscopy. After 24 h of immersion, the chemical surfaces were examined using energy-dispersive X-ray spectroscopy (EDS) attached to an SEM. Changes in weight over the time periods (6, 12, and 24 h) were recorded and calculated according to the dry weight (Ws) of the samples and the original dry weights (Wo) according to the equation:

$$\text{Weight gain (\%)} = (\text{Ws}/\text{Wo} - 1) \times 100\% \quad (1)$$

Table 1. Ion concentrations (mM) of simulated body fluid (SBF) used in this experiment and human blood plasma [20].

	Na^+	K^+	Mg^{2+}	Ca^{2+}	Cl^-	HPO_4^{2-}	SO_4^{2-}	HCO_3^-
Blood plasma	142.0	5.0	1.5	2.5	103.0	1.0	0.5	27.0
SBF	142.0	5.0	1.5	2.5	147.8	1.0	0.5	4.2

3. Results

3.1. Characterization of Surfaces under Various Treatment Conditions

Figure 1 shows FE-SEM images of the c.p. Ti surfaces subjected to hydrothermal treatment for 12, 24, and 48 h. A nanoscale fine network structure, composed of many feather-like, elongated features placed perpendicularly to the surface, formed on the c.p. Ti surface after the initial NaOH treatment, as shown in Figure 1a. When c.p. Ti was hydrothermally treated at 150 °C, the treatment time had an obvious effect on the formation of nano-HA. Few particles were formed on the c.p. Ti surface after 12 h (Figure 1b). More rod-like particles were formed and completely covered the surface after treatment time reached 24 and 48 h (Figure 1c,d). As the time of the hydrothermal reaction increased, the amount and size of the rod-shaped HA particles also increased significantly.

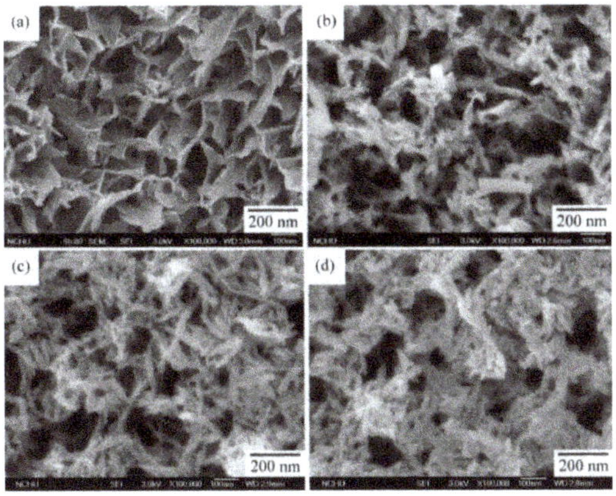

Figure 1. Field-emission scanning electron microscopy (FE-SEM) photographs of the surfaces of commercially pure titanium (c.p. Ti) subjected to hydrothermal treatment at 150 °C for (**a**) 0, (**b**) 12, (**c**) 24, and (**d**) 48 h.

Figure 2 shows the SEM images of the non-heat-treated surfaces of the HA-coated c.p. Ti samples and the samples that were heat-treated at 500 °C for 2 h in air. After heat treatment, the HA particles still retained a rod-like nanostructure, while the size of the nanoparticles slightly increased to around 90 nm in length and 23 nm in width. Figure 3 shows the XRD patterns of the non-heat-treated and heat-treated HA-coated c.p. Ti samples. For the non-heat-treated samples, besides the HA and Ti peaks, only broader peaks could be identified for the surface after NaOH treatment. After heat treatment at 500 °C for 2 h, peaks corresponding to the anatase phase of TiO_2 were detected, as shown in Figure 3b.

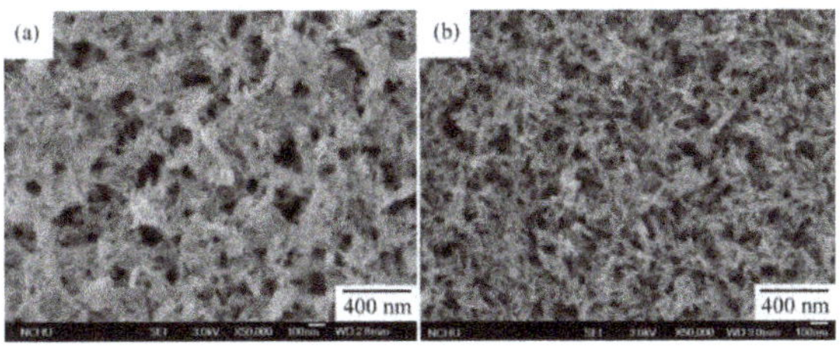

Figure 2. FE-SEM photographs of HA-coated c.p. Ti samples after hydrothermal treatment at 150 °C for 48 h: non-heat-treated surface (a) and heat-treated surface at 500 °C for 2 h (**b**).

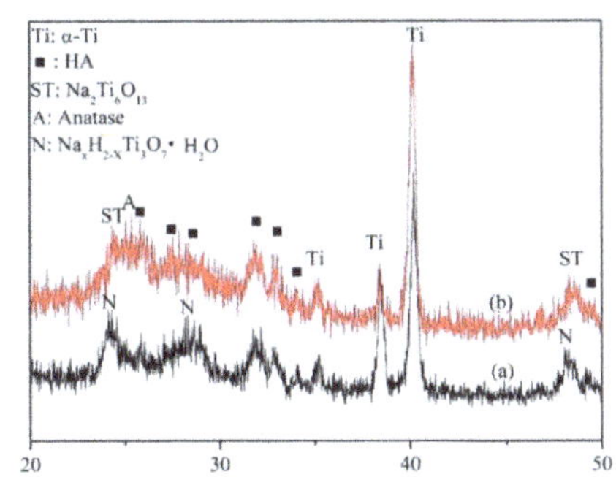

Figure 3. X-ray diffractometry (XRD) patterns of HA-coated c.p. Ti samples after hydrothermal treatment at 150 °C for 48 h: non-heat-treated surface (**a**) and heat-treated surface at 500 °C for 2 h (**b**).

The FT-IR spectra of the non-heat-treated and heat-treated HA-coated c.p. Ti samples are shown in Figure 4. The non-heat-treated and heat-treated samples had similar FT-IR spectra, indicating that the same functional groups were observed in the HA coatings. Bands assigned to OH^-, PO_4^{3-}, CO_3^{2-}, and H_2O were present. The FT-IR bands also identified the functional groups that are characteristic of carbonate-incorporated HA.

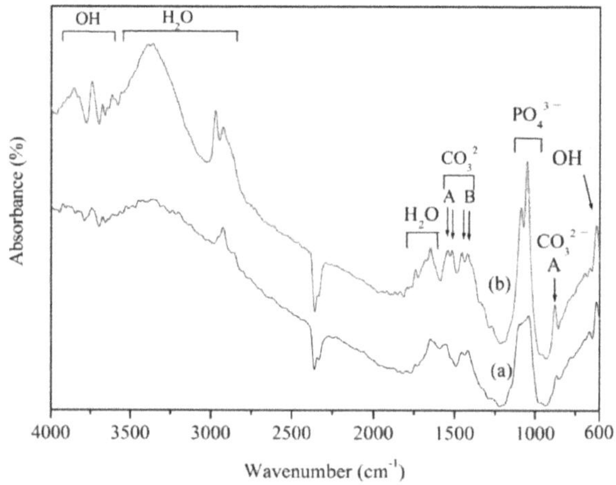

Figure 4. Fourier-transform infrared (FT-IR) spectra of HA-coated c.p. Ti samples after hydrothermal treatment at 150 °C for 48 h: non-heat-treated surface (**a**) and heat-treated surface at 500 °C for 2 h (**b**).

In order to examine the adhesion between the c.p. Ti substrate and the nano-HA layer, both the non-heat-treated and the heat-treated samples were ultrasonically cleaned at 200 W for 3 min in deionized water. The experimental results are shown in Figure 5. The coating of the heat-treated sample appeared to be strongly adhesive to the c.p. Ti substrate. The loosely bound HA coating on the non-heat-treated sample was almost completely removed during ultrasonic cleaning, exposing the original alkali-treated surface, while the nano-HA crystals remained intact on the coated surface of the heat-treated samples. Accordingly, the adhesive strength of the HA coating on the surface of the heat-treated samples was relatively high due to heat treatment.

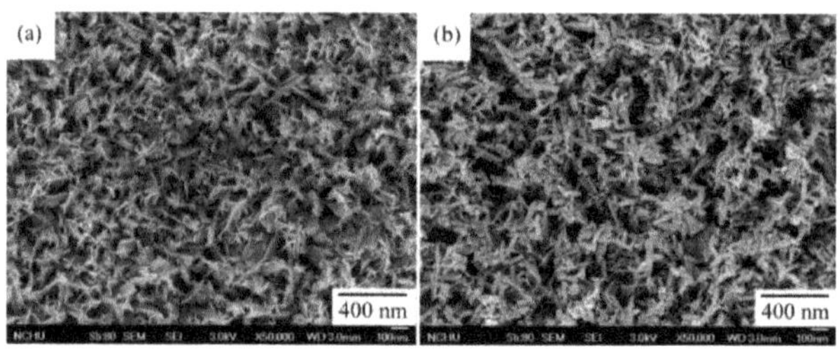

Figure 5. FE-SEM images showing the effect of ultrasonic cleaning on the surface morphology of HA-coated c.p. Ti samples after hydrothermal treatment at 150 °C for 48 h: non-heat-treated surface (**a**) and heat-treated surface at 500 °C for 2 h (**b**).

3.2. Apatite-inducing Ability of HA-Coated c.p. Ti

Figure 6 shows FE-SEM micrographs of the surfaces of both NaOH-treated c.p. Ti and HA-coated c.p. Ti that were subjected to heat treatment at 500 °C for 2 h after subsequent soaking in SBF for 6, 12, and 24 h. The bioactivity of the HA-coated c.p. Ti was evaluated by examining apatite's formation on the surface of the specimens after soaking in SBF. NaOH-treated c.p. Ti without an HA coating was tested for comparison. Nano-sized particles in the form of aggregates were first observed on the HA-coated Ti surface after 6 h of immersion in SBF, while no obvious precipitation was found on the metal without an HA coating. In contrast, precipitates were observed on the NaOH-treated sample after immersion in SBF for 12 h. In the same immersion period, larger numbers of nano-spherulites were deposited on the HA-coated Ti. Dense, compact, and larger sized round-shaped apatite clusters were formed on the HA-coated Ti samples after only 24 h of immersion. Conversely, rod-like apatite nanoparticles in a loosely compact film covered the NaOH-treated Ti samples. This study found that the morphology of apatite deposits was closely related to the surface modification conditions and the underlying substrate material and topography.

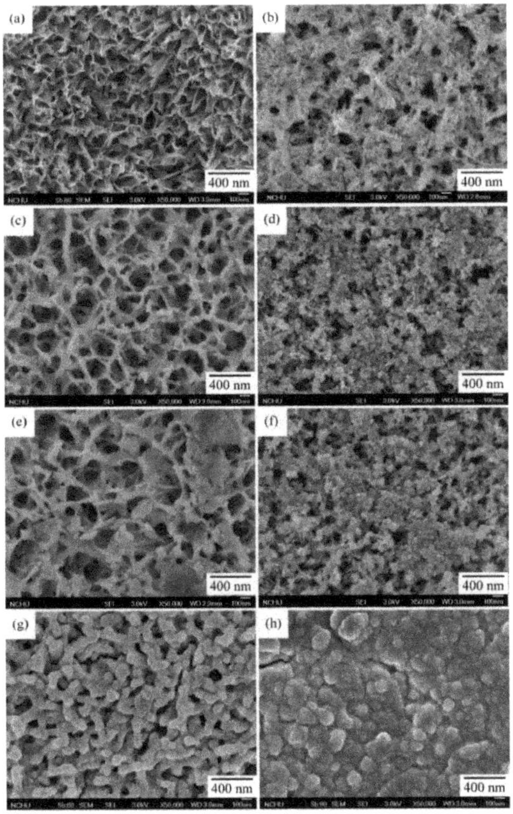

Figure 6. FE-SEM photographs of the surfaces of NaOH-treated c.p. Ti (**a**, **c**, **e**, **g**) and HA-coated c.p. Ti samples (**b**, **d**, **f**, **h**) that were subjected to heat treatment at 500 °C for 2 h after subsequent soaking in SBF for 6, 12 and 24 h, respectively.

Figure 7 shows the results of EDS analysis of both NaOH-treated c.p. Ti and HA-coated c.p. Ti that were heat-treated at 500°C for 2 h, after subsequent soaking in SBF for 24 h. EDS analysis identified the mineral phase of the depositions on the surfaces of the two groups of samples as mainly composed of calcium and phosphorous, with some magnesium uptaken from the SBF. The results for the HA-coated c.p. Ti before soaking in SBF, used as a control, are shown in Figure 7c. It is clear that the intensities of the Ca and P apatite peaks were greater for the HA-coated sample, when compared with those of the control.

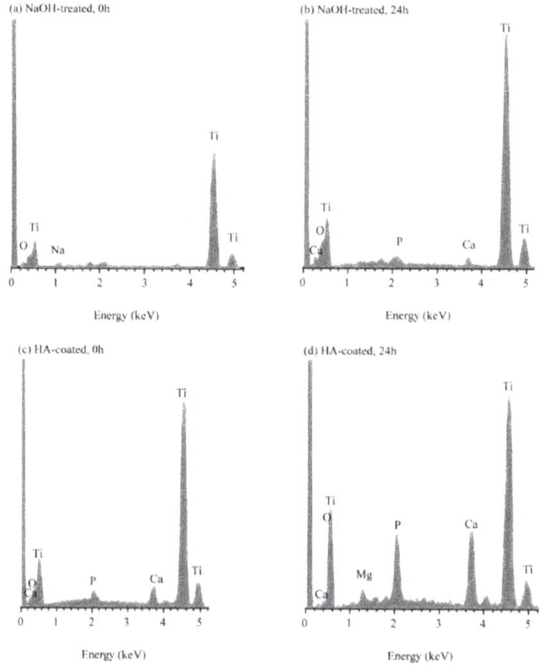

Figure 7. Energy-dispersive X-ray spectroscopy (EDS) of the surfaces of NaOH-treated c.p. Ti (**a**,**b**) and HA-coated c.p. Ti (**c**,**d**) subjected to heat treatment at 500°C for 2 h before and after subsequent soaking in SBF for 24 h.

Figure 8 shows the weight change of the apatite deposition on the surfaces of both NaOH-treated c.p. Ti and HA-coated c.p. Ti that were subjected to heat treatment at 500 °C for 2 h after subsequent soaking in SBF for 6, 12, and 24 h. The samples were dried at 45 °C for 24 h prior to weighing. The weight gain of the apatite deposition increased significantly with the immersion time for both groups. The HA-coated substrate showed an increase in the deposition rate compared with its non-HA-coated counterpart. It was noted that the non-HA-coated substrate showed little or no increase in weight up to 6 h of immersion in SBF. The result confirmed that the samples with HA coating had a higher apatite-forming ability, which suggests better bioactivity.

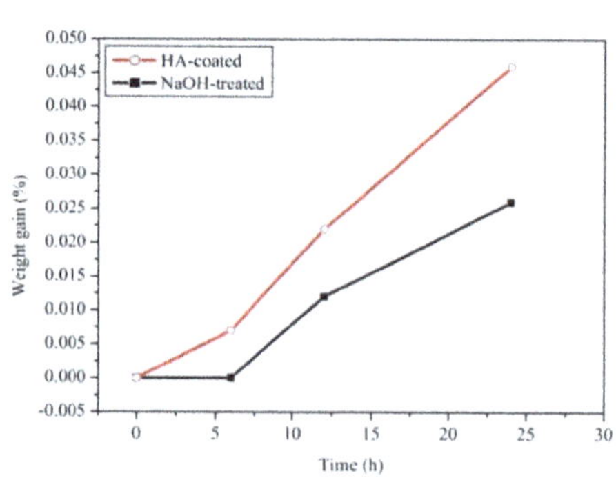

Figure 8. Weight gain as a function of soaking time for the surfaces of NaOH-treated c.p. Ti and HA-coated c.p. Ti that were subjected to heat treatment at 500°C for 2 h after subsequent soaking for different periods in SBF.

4. Discussion

This study found that the amount and size of rod-shaped HA particles on the c.p. Ti surface increased with the treatment time, from 12 h to 48 h (Figure 1). A possible explanation for this result is that, when the hydrothermal reaction time is extended, a large number of hydroxyl groups are promoted on the c.p. Ti surface, which is the key step to initiate the growth of HA on c.p. Ti [21]. The ability to form nano-HA on Ti surface could be promoted by increasing the temperature and time of the hydrothermal treatment. Hu et al. [21] and Nakagawa et al. [22] reported that, when Ti was hydrothermally treated in an HA suspension solution, negatively charged basic OH and acidic OH groups were formed on Ti surface. The abundant Ti–OH groups were negatively charged and could combine with the positively charged ions of Ca^{2+} in the suspension. As the Ca^{2+} ions accumulated on the surface, the surface became positively charged and reacted with the negatively charged phosphate ions, thus nucleating HA.

The presence of a rough surface with many micropores on the nano-HA coated samples may effectively enhance the fixation of an implant to bone [23]. After the hydrothermal reaction, the aggregated particles had a tiny rod-like nanostructure with an average length of less than 100 nm. Liu and Webster indicated that nano-HA coatings exhibit enhanced osteoblast cell functions due to the large effective surface area and surface nano-roughness [24]. Therefore, the nano-HA coating prepared by the hydrothermal method is very promising as an application prospect for implant materials. In this study, the c.p. Ti sample hydrothermally treated at 150 °C for 48 h with a wholly HA-coated surface was analyzed for its coating characteristics and apatite-forming ability in SBF.

It has been asserted [25,26] that the surface of Ti forms sodium hydrogen titanate ($Na_xH_{2-x}Ti_3O_7{\cdot}H_2O$) after NaOH treatment, which is subsequently transformed into sodium titanate ($Na_2Ti_6O_{13}$) and anatase through heat treatment. Also, after heat treatment at 500 °C for 2 h, peaks corresponding to the anatase phase of TiO_2 were detected, as shown in Figure 3. Wang et al. [27] investigated c.p. Ti treated with a $H_2O_2/TaCl_5$ solution at 80 °C for various times and found that anatase was the primary phase detected in the surface layers after heating between 300 °C and 600 °C, suggesting that anatase is necessary to trigger early apatite formation in SBF.

In Figure 4, carbonate substituted the OH^- (A-type) and phosphate (B-type) groups, which resulted in the transformation of HA into an AB-type carbonated structure. This carbonated HA is similar to bone [28]. The most notable fact is that the FT-IR spectra of the samples reported in Figure 4 resemble those found in previous research [12], in which nano-sized HA powders were prepared with eggshells through a hydrothermal reaction at 150 °C. For both non-heat-treated and heat-treated c.p. Ti samples, the functional group OH^- was also found on the surfaces: this is a proven positive factor for bone-like apatite formation [29]. The result indicated that the OH^- group formed on the heat-treated c.p. Ti surface seemed to be more abundant, which might be attributed to the formation of the anatase phase after heat treatment at 500 °C. Previous studies have shown that only weakly absorbed OH^- disappears at lower heat-treatment temperatures, while OH^- that is strongly bounded with Ti remains on the surface at temperatures up to 600 °C and more [30,31]. Moreover, it has been observed that only strongly bounded OH^- groups are responsible for the bioactivity of Ti surfaces [32,33].

The results of Figure 5 indicated that the adhesive strength of the HA coating was improved after heat treatment. Lu et al. [34] also reported that the adhesive strength of HA/Ag coatings on a Ti substrate was significantly raised after heat treatment. Additionally, a high degree of crystallinity (86%) in the HA coating was observed after heat treatment at 500 °C. It is generally believed that crystallinity is the dominant factor for the dissolution of an HA coating [35]. HA coatings with low crystallinity have been shown to be more easily degraded in body fluids, and Hu et al. [36] have also confirmed that more cells are absorbed and proliferate on well-crystallized HA coatings.

In a study by Nishigawa et al., the effect of silica-coated zirconia surfaces on bond strength was also examined by using the ultrasonic cleaning test [37]. In the present study, the adhesion between the c.p. Ti substrate and the HA coating was examined in the non-heat-treated and heat-treated specimens using an ultrasonic washing process in deionized water. Due to the limitations of this test method, we could not obtain quantitative data to determine if the coating had sufficient adhesion to survive implantation. In our future research, a micro-scratch tester will be adopted to quantitatively evaluate the adhesion strength of the coating on the substrate.

Due to the chemical and structural similarities to bone and dental minerals, HA exhibits excellent biocompatibility and possesses the ability to stimulate the formation of new bones, which makes it a potential material of choice for coating metals for orthopedic, dental, and other biomedical applications [21]. When the structure, size, and composition of HA crystals are closer to those of biological apatite, optimal osteocompatibility of HA can be achieved [38]. In this study, the features of the HA coatings such as nano-structure, high crystallinity, and carbonate substitution led to coatings exhibiting bone-like structures. Additionally, HA prepared from eggshell powders contains several important trace elements, such as Na, Mg, and Sr, which play a key role in bone metabolism and cell proliferation [12,39–41].

5. Conclusions

In this study, nano-HA coatings on c.p. Ti were successfully developed through a hydrothermal reaction and heat treatment using eggshell biowaste as a source of Ca. When c.p. Ti was hydrothermally treated at 150 °C, the reaction time had an effect on the formation of nano-HA particles. The amount and size of the prepared HA particles increased with the treatment time. The c.p. Ti surface was entirely covered by HA nanoparticles after 48 h, and the aggregated HA particles exhibited a tiny rod-like nanostructure, with an average particle length of less than 100 nm. After subsequent heat treatment, the morphology of the HA particles was retained, while their size slightly increased to around 90 nm in length and 23 nm in width. The FT-IR bands identified functional groups that were characteristic of AB-type carbonate-incorporated HA. The HA coating achieved a high degree of crystallinity (86%) after heat treatment at 500 °C for 2 h and strongly adhered to the c.p. Ti substrate. Dense, compact, and larger sized round-shaped apatite granules accumulated on the HA-coated Ti samples after only 24 h of immersion, while rod-like apatite nanoparticles with a loosely compact film covered the NaOH-treated Ti samples. A higher apatite-forming ability could be acquired by the samples coated with HA, suggesting better bioactivity.

Author Contributions: Conceptualization, W.-F.H., H.-C.H., and S.-C.W.; methodology, W.-F.H., H.-C.H., and S.-C.W.; software, C.-W.H.; validation, W.-F.H. and H.-N.Y.; formal analysis, H.-C.H.; investigation, W.-F.H. and C.-W.H.; resources, W.-F.H., H.-N.Y., and H.-C.H.; data curation, S.-K.H.; writing—original draft preparation, W.-F.H.; writing—review and editing, H.-N.Y.; visualization, S.-K.H.; supervision, H.-N.Y.; project administration, W.-F.H. and H.-N.Y.; funding acquisition, W.-F.H. and H.-N.Y. All authors have read and agreed to the published version of the manuscript.

Funding: Partial financial support for this research was provided by the Zuoying Branch of Kaohsiung Armed Forces General Hospital and National University of Kaohsiung, grant number ZBH 108-09.

Conflicts of Interest: The authors declare no conflicts of interest.

References

1. Yeo, I.-S.L. Modifications of dental implant surfaces at the microand nano-level for enhanced osseointegration. *Materials* **2020**, *13*, 89. [CrossRef] [PubMed]
2. Katta, P.P.K.; Nalliyan, R. Corrosion resistance with self-healing behavior and biocompatibility of Ce incorporated niobium oxide coated 316L SS for orthopedic applications. *Surf. Coat. Technol.* **2019**, *375*, 715–726. [CrossRef]
3. Qi, J.; Yang, Y.; Zhou, M.; Chen, Z.; Chen, K. Effect of transition layer on the performance of hydroxyapatite/titanium nitride coating developed on Ti–6Al–4V alloy by magnetron sputtering. *Ceram. Int.* **2019**, *45*, 4863–4869. [CrossRef]
4. Kar, A.; Raja, K.S.; Misra, M. Electrodeposition of hydroxyapatite onto nanotubular TiO_2 for implant applications. *Surf. Coat. Technol.* **2006**, *201*, 3723–3731. [CrossRef]
5. Odhiambo, J.G.; Li, W.G.; Zhao, Y.T.; Li, C.L. Porosity and its significance in plasma-sprayed coatings. *Coatings* **2019**, *9*, 460. [CrossRef]

6. Barrère, F.; van der Valk, C.M.; Meijer, G.; Dalmeijer, R.A.J.; de Groot, K.; Layrolle, P. Osteointegration of biomimetic apatite coating applied onto dense and porous metal implants in femurs of goats. *J. Biomed. Mater. Res. B* **2003**, *67*, 655–665. [CrossRef]
7. Aydın, İ.; Bahçepınar, A.İ.; Kırman, M.; Çipiloğlu, M.A. HA coating on Ti6Al7Nb alloy using an electrophoretic deposition method and surface properties examination of the resulting coatings. *Coatings* **2019**, *9*, 402. [CrossRef]
8. Vilardell, A.M.; Cinca, N.; Garcia-Giralt, N.; Dosta, S.; Cano, I.G.; Nogués, X.; Guilemany, J.M. In-vitro comparison of hydroxyapatite coatings obtained by cold spray and conventional thermal spray technologies. *Mater. Sci. Eng. C* **2020**, *107*, 110306. [CrossRef]
9. Boyd, A.R.; Duffy, H.; McCann, R.; Meenan, B.J. Sputter deposition of calcium phosphate/titanium dioxide hybrid thin films. *Mater. Sci. Eng. C* **2008**, *28*, 228–236. [CrossRef]
10. Azari, R.; Rezaie, H.R.; Khavandi, A. Investigation of functionally graded HA-TiO_2 coating on Ti–6Al–4V substrate fabricated by sol-gel method. *Ceram. Int.* **2019**, *45*, 17545–17555. [CrossRef]
11. Qu, H.; Wei, M. The effect of temperature and initial pH on biomimetic apatite coating. *J. Biomed. Mater. Res. B* **2008**, *87*, 204–212. [CrossRef] [PubMed]
12. Wu, S.C.; Tsou, H.K.; Hsu, H.C.; Hsu, S.K.; Liou, S.P.; Ho, W.F. A hydrothermal synthesis of eggshell and fruit waste extract to produce nanosized hydroxyapatite. *Ceram. Int.* **2013**, *39*, 8183–8188. [CrossRef]
13. Wu, S.C.; Hsu, H.C.; Wu, Y.N.; Ho, W.F. Hydroxyapatite synthesized from oyster shell powders by ball milling and heat treatment. *Mater. Character.* **2011**, *62*, 1180–1187. [CrossRef]
14. Balasubramanian, V.; Daniel, T.; Henry, J.; Sivakumar, G.; Mohanraj, K. Electrochemical performances of activated carbon prepared using eggshell waste. *S. N. Appl. Sci.* **2020**, *2*, 127. [CrossRef]
15. Boutinguiza, M.; Pou, J.; Comesaña, R.; Lusquiños, F.; de Carlos, A.; León, B. Biological hydroxyapatite obtained from fish bones. *Mater. Sci. Eng. C* **2012**, *32*, 478–486. [CrossRef]
16. Qaid, T.H.; Ramesh, S.; Yusof, F.; Basirun, W.J.; Ching, Y.C.; Chandran, H.; Krishnasamy, S. Micro-arc oxidation of bioceramic coatings containing eggshell-derived hydroxyapatite on titanium substrate. *Ceram. Int.* **2019**, *45*, 18371–18381. [CrossRef]
17. Roudan, M.A.; Ramesh, S.; Wong, Y.H.; Chandran, H.; Krishnasamy, S.; Teng, W.D.; Bang, L.T. Sintering behavior and characteristic of bio-based hydroxyapatite coating deposited on titanium. *J. Ceram. Proc. Res.* **2017**, *18*, 640–645.
18. Hsu, H.C.; Wu, S.C.; Hsu, S.K.; Hsu, C.W.; Ho, W.F. Bone-like nano-hydroxyapatite coating on low-modulus Ti–5Nb–5Mo alloy using hydrothermal and post-heat treatments. *Thin Solid Films* **2019**, *687*, 137463. [CrossRef]
19. Wu, X.; Weng, D.; Zhao, S.; Chen, W. The deposition of γ-Al_2O_3 layers on ceramic and metallic supports for the preparation of structured catalysts. *Surf. Coat. Technol.* **2005**, *190*, 434–439. [CrossRef]
20. Kokubo, T.; Takadama, H. How useful is SBF in predicting in vivo bone bioactivity? *Biomaterials* **2006**, *27*, 2907–2915. [CrossRef]
21. Hu, X.; Shen, H.; Cheng, Y.; Xiong, X.; Wang, S.; Fang, J.; Wei, S. One-step modification of nano-hydroxyapatite coating on titanium surface by hydrothermal method. *Surf. Coat. Technol.* **2010**, *205*, 2000–2006. [CrossRef]
22. Nakagawa, M.; Zhang, L.; Udoh, K.; Matsuya, S.; Ishikawa, K. Effects of hydrothermal treatment with $CaCl_2$ solution on surface property and cell response of titanium implants. *J. Mater. Sci. Mater. Med.* **2005**, *16*, 985–991. [CrossRef] [PubMed]
23. Habibovic, P.; Yuan, H. 3D microenvironment as essential element for osteoinduction by biomaterials. *Biomaterials* **2005**, *26*, 3565–3575. [CrossRef] [PubMed]
24. Liu, H.; Webster, T.J. Nanomedicine for implants: A review of studies and necessary experimental tools. *Biomaterials* **2007**, *28*, 354–369. [CrossRef] [PubMed]
25. Hsu, H.C.; Wu, S.C.; Hsu, S.K.; Chuang, S.H.; Ho, W.F. Surface modification of commercially pure Ti treated with aqueous NaOH treatment and ethyl alcohol aging. *J. Med. Biol. Eng.* **2013**, *33*, 331–336. [CrossRef]
26. Yamaguchi, S.; Takadama, H.; Matsushita, T.; Nakamura, T.; Kokubo, T. Preparation of bioactive Ti–15Zr–4Nb–4Ta alloy from HCl and heat treatments after an NaOH treatment. *J. Biomed. Mater. Res. A* **2011**, *97*, 135–144. [CrossRef]
27. Wang, X.-X.; Hayakawa, S.; Tsuru, K.; Osaka, A. Improvement of the bioactivity of H_2O_2/$TaCl_5$-treated titanium after subsequent heat treatment. *J. Biomed. Mater. Res.* **2000**, *52*, 171–176. [CrossRef]
28. Kunze, J.; Müller, L.; Macak, J.M.; Greil, P.; Schmuki, P.; Müller, F.A. Time-dependent growth of biomimetic apatite on anodic TiO_2 nanotubes. *Electrochim. Acta* **2008**, *53*, 6995–7003. [CrossRef]

29. Li, P.; Ohtsuki, C.; Kokubo, T.; Nankanishi, K.; Soga, N. The role of hydrated silica, titania, and alumina in inducing apatite on implants. *J. Biomed. Mater. Res.* **1994**, *28*, 7–15. [CrossRef]
30. Busca, G.; Sausey, H.; Saur, O.; Lavalley, J.C.; Lorenzelli, V. FT-IR characterization of the surface acidity of different titanium dioxide anatase preparations. *Appl. Catal.* **1985**, *14*, 245–260. [CrossRef]
31. Martra, G. Lewis acid and base sites at the surface of microcrystalline TiO_2 anatase: Relationships between surface morphology and chemical behaviour. *Appl. Catal. A* **2000**, *200*, 275–285. [CrossRef]
32. Lu, X.; Wang, Y.; Yang, X.; Zhang, Q.; Zhao, Z.; Wenig, L.T.; Leng, Y. Spectroscopic analysis of titanium surface functional groups under various surface modification and their behaviours in vitro and in vivo. *J. Biomed. Mater. Res. A* **2008**, *84*, 523–534. [CrossRef] [PubMed]
33. Lu, X.; Zhang, H.P.; Leng, Y.; Fang, L.; Qu, S.; Feng, B.; Weng, J.; Huang, N. The effects of hydroxyl groups on Ca adsorption on rutile surface: A first-principles study. *J. Mater. Sci. Mater. Med.* **2010**, *21*, 1–10. [CrossRef] [PubMed]
34. Lu, X.; Zhang, B.; Wang, Y.; Zhou, X.; Weng, J.; Qu, S.; Feng, B.; Watari, F.; Ding, Y.; Leng, Y. Nano-Ag-loaded hydroxyapatite coatings on titanium surfaces by electrochemical deposition. *J. R. Soc. Interface* **2011**, *8*, 529–539. [CrossRef]
35. Sun, L.; Berndt, C.C.; Khor, K.A.; Cheang, H.N.; Gross, K.A. Surface characteristics and dissolution behavior of plasma-sprayed hydroxyapatite coating. *J. Biomed. Mater. Res.* **2002**, *62*, 228–236. [CrossRef]
36. Hu, Q.; Tan, Z.; Liu, Y.; Tao, J.; Cai, Y.; Zhang, M.; Pan, H.; Xu, X.; Tang, R. Effect of crystallinity of calcium phosphate nanoparticles on adhesion, proliferation, and differentiation of bone marrow mesenchymal stem cells. *J. Mater. Chem.* **2007**, *17*, 4690–4698. [CrossRef]
37. Nishigawa, G.; Maruo, Y.; Irie, M.; Oka, M.; Yoshihara, K.; Minagi, S.; Nagaoka, N.; Yoshida, Y.; Suzuki, K. Ultrasonic cleaning of silica-coated zirconia influences bond strength between zirconia and resin luting material. *Dent. Mater. J.* **2008**, *27*, 842–848. [CrossRef]
38. Barbosa, M.C.; Messmer, N.R.; Brazil, T.R.; Marciano, F.R.; Lobo, A.O. The effect of ultrasonic irradiation on the crystallinity of nano-hydroxyapatite produced via the wet chemical method. *Mater. Sci. Eng. C* **2013**, *33*, 2620–2625. [CrossRef]
39. Ho, W.F.; Hsu, H.C.; Hsu, S.K.; Hung, C.W.; Wu, S.C. Calcium phosphate bioceramics synthesized from eggshell powders through a solid state reaction. *Ceram. Int.* **2013**, *39*, 6467–6473. [CrossRef]
40. Boanini, E.; Gazzano, M.; Bigi, A. Ionic substitutions in calcium phosphates synthesized at low temperature. *Acta Biomater.* **2010**, *6*, 1882–1894. [CrossRef]
41. Rude, R.K.; Gruber, H.E. Magnesium deficiency and osteoporosis: Animal and human observations. *J. Nutr. Biochem.* **2004**, *15*, 710–716. [CrossRef] [PubMed]

MDPI
St. Alban-Anlage 66
4052 Basel
Switzerland
www.mdpi.com

Coatings Editorial Office
E-mail: coatings@mdpi.com
www.mdpi.com/journal/coatings

www.ingramcontent.com/pod-product-compliance
Lightning Source LLC
Chambersburg PA
CBHW041950260926
51850CB00019B/2283
9783725804573